Diss. ETH No. 18920

Design and Optimization of Distributed Multiuser Cooperative Wireless Networks

A dissertation submitted to the

SWISS FEDERAL INSTITUTE OF TECHNOLOGY (ETH)

ZURICH

for the degree of

Doctor of Sciences

presented by

CELAL EŞLİ

MSc. Boğaziçi University

born April 21, 1981

citizen of Turkey

accepted on the recommendation of

Prof. Dr. Armin Wittneben, examiner

Prof. Dr. Hakan Deliç, co-examiner

Assoc. Prof. Dr. Elza Erkip, co-examiner

2010

Reihe Series in Wireless Communications
herausgegeben von:
Prof. Dr. Armin Wittneben
Eidgenössische Technische Hochschule
Institut für Kommunikationstechnik
Sternwartstr. 7
CH-8092 Zürich

E-Mail: wittneben@nari.ee.ethz.ch
Url: http://www.nari.ee.ethz.ch/

Bibliografische Information der Deutschen Nationalbibliothek

Die Deutsche Nationalbibliothek verzeichnet diese Publikation in der Deutschen Nationalbibliografie; detaillierte bibliografische Daten sind im Internet über http://dnb.d-nb.de abrufbar.

ISBN 978-3-8325-2484-5
ISSN 1611-2970

Logos Verlag Berlin GmbH
Comeniushof, Gubener Str. 47,
10243 Berlin
Tel.: +49 030 42 85 10 90
Fax: +49 030 42 85 10 92
INTERNET: http://www.logos-verlag.de

To my dear parents İlyas and Sevgi
and my angels Zekiye and Selen...

Abstract

Employing multiple antennas at the transmitter and receiver sides has been identified as the key enabler for high spectral efficiency in point-to-point communication, since it facilitates multiplexing of several data streams in space rather than in time/frequency. In this work, we are interested in achieving spatial multiplexing (SM) through efficient cooperative relaying schemes, even if both the transmit and receive antennas are distributed. To this end, we focus on wireless multiuser networks and aim at designing novel cooperative communication protocols, developing corresponding transmission and signal processing techniques, and optimizing the network performance.

This thesis is presented in two main parts. In the first part, we consider coherent multiuser *amplify-and-forward* (AF) relaying, where a set of source-destination (S-D) terminal pairs communicate concurrently over the same physical channel and a set of AF relay nodes assist the communication in a half-duplex scheme. It is known that multiuser interference can be cancelled with sufficient spatial degrees of freedom, i.e. relays. When additional relays are introduced to the network, we show that further multiple-input multiple-output (MIMO) gains can be achieved in a distributed manner through efficient (complex) relay gain optimizations. We choose two objective functions to maximize: sum rate and minimum link rate (max-min fairness). Moreover, it is shown that distributed diversity can be attained at the destinations through either relay selection or max-min type relay gain optimizations. Such gain allocation schemes require global knowledge of the channel coefficients between all participating nodes, which in practice may well diminish the spatial multiplexing gain. Addressing this issue, we introduce a new *distributed* gradient based gain allocation scheme, which substantially reduces this overhead. Key to this is the proof, that the gradient of the destination signal-to-interference-plus-noise ratio can be calculated in a distributed manner based on local channel state information (CSI) at the relays and limited feedback from the destinations.

The two *sine qua non* assumptions for the efficiency of coherent multiuser relaying are perfect CSI knowledge per relay and globally phase-synchronous relays. We take a practical look on these two assumptions and assume that there are imperfections with both. After modeling the corresponding data mismatches within given uncertainty sets, we follow a worst-case framework and design semidefinite programming (SDP) based robust counterparts for max-min beamforming.

Lastly in order to have a bounded CSI dissemination overhead independent from the number of relays, we propose to partition the relays into multiple independent clusters. We consider either homogeneous relay clusters where each cluster independently manages the multiuser interference, or heterogeneous clusters where we establish a hierarchy in between different clusters each with a specific gain (array or spatial multiplexing) to achieve. We obtain effective diversity gain through cluster and time specific phase rotations in the former case, and through relay selection for clusters in the latter.

We shift our focus to *decode-and-forward* (DF) relaying in the second part of the thesis, and address both multiuser one- and two-way MIMO relaying. First we consider two MIMO terminals exchanging information via a single MIMO relay node. We extend the two-way protocol to multiple antenna equipped nodes with a primary focus on network coding based signal combination at the relay. In order to support unbalanced relay-to-terminal rates, we propose a *modified* network coding based approach, where we apply *zero padding per symbol* on the information sequence to be transmitted to the weak terminal, prior to XOR addition, and further we provide *a priori* decoding information to the corresponding terminal. Afterwards, we assume transmit CSI at the terminals/relay and design the transmit covariances that characterize the capacity boundaries of two-way relaying. We design SDP based covariance optimization schemes, each of which maximizes sum or minimum of the two terminal-to-terminal information rates. Moreover, we consider *imperfect* transmit CSI for the special case of single antenna terminals. Therein, we derive the worst-case broadcast capacity region, and further propose robust counterparts for the two optimization problems designed for perfect transmit CSI.

Finally, we extend one- and two-way DF relaying to the case of simultaneous MIMO communication of multiple S-D pairs. The relay is equipped with sufficient multiple antennas to resolve the MIMO multiple access channel spatially in the uplink and also to manage the interference through the downlink. Through the broadcast phase of multiuser one-way relaying, we apply conventional zero-forcing beamforming based MIMO broadcasting techniques. However, for the two-way case, we propose a novel two-level terminal separation for broadcasting: *bit-level* and *spatial-level*. Consequently, we design a modular relay transmit covariance optimization scheme that aims at maximizing the sum rate of all terminal-to-terminal rates optimally over the two relaying phases.

Kurzfassung

Die Verwendung von mehreren Sende- und Empfangsantennen wurde in den letzten Jahren als wichtigster Ansatz zur Verbesserung der spektralen Effizienz ausgemacht, da sie ermöglicht mehrere Datenströme gleichzeitig und im gleichen Frequenzband mittels räumlicher Multiplexverfahren (Spatial Multiplex) zu übertragen. In dieser Arbeit wird untersucht, wie eine räumliche Mehrfachübertragung mit verteilten Sende- und Empfangsantennen durch die Anwendung von effizienten und kooperativen Relaying-Protokollen erreicht werden kann. Der Fokus wird dabei auf drahtlose Netzwerke mit mehreren Benutzern gelegt, und es wird das Ziel verfolgt, neuartige kooperative übertragungsprotokolle zu entwerfen, die dafür notwendigen übertragungs- und Signalverarbeitungsalgorithmen zu entwickeln und die Netzwerkperformance zu optimieren.

Diese Dissertation ist in zwei Teile gegliedert. Im ersten Teil wird kohärentes Amplify-and-Forward (AF) Relaying für mehrere Benutzer betrachtet: Die gleichzeitige Kommunikation über den gleichen physikalischen Kanal einer Menge von Quell- und Zielknoten wird mit einer Menge von AF Relays im Halbduplex-Betrieb unterstützt. Es ist bekannt, dass die Interferenzen durch die Mehrfachbenutzung vollständig beseitigt werden können, falls eine gewisse mindest Anzahl von Relays zur Verfügung steht. In dieser Dissertation wird gezeigt, dass die Optimierung der Verstärkungsfaktoren zusätzlicher Relays weitere Multiple-Input Multiple-Output (MIMO) Gewinne bringt. Zur Optimierung werden zwei Kostenfunktionen betrachtet: entweder die summierte Datenübertragungsrate aller Verbindungen oder die kleinste Datenübertragungsrate aller einzelnen Verbindungen (Max-Min Fairness). Darüber hinaus wird gezeigt, dass verteilte Diversität durch die Auswahl von Relays oder die Max-Min Optimierung der Verstärkungsfaktoren erzielt werden kann. Solche Berechnungsmethoden der Relay-Verstärkungsfaktoren benötigen die allgemeine Kenntnis aller Kanalkoeffizienten zwischen allen Knoten im Netzwerk. Diese Vorraussetzung reduziert die erreichbaren Spatial Multiplex Gewinne in der Praxis erheblich. Um dieses Problem zu lösen, stellen wir ein neuartiges, verteiltes, gradientenbasiertes Optimierungsverfahren vor. Der Schlüssel zu dieser Methode ist der Beweis, dass der Gradient des Signal-zu-Interferenz-und-Rausch-Verhältnisses eines Zielknotens mit Hilfe von ausschliesslich lokalem Kanalwissen sowie eingeschränktem Feedback der Zielknoten am Relay berechnet werden kann.

Die zwei notwendigen Annahmen, welche die Effizienz von kohärentem AF Relaying für mehrere Benutzer gewährleisten sind perfekte Kanalkenntis und phasensynchrone Relays. In der Praxis können beide Annahmen nicht erfüllt werden. In dieser Dissertation werden diese Imperfektionen zuerst mittels eines Unsicherheitsbereiches modelliert und danach wird eine robuste Variante des auf der Semidefiniten Programmierung (SDP) basierenden Max-Min Beamformings entwickelt.

Schliesslich schlagen wir vor, die Relays in mehrere unabhängige Gruppen aufzuteilen, um die erforderliche Verteilung von Kanalkoeffizienten im Netzwerk zu begrenzen und von der gesamten Anzahl von Relays unabhängig zu machen. Einerseits werden homogene Relay-Gruppen betrachtet, wobei jede Gruppe unabhängig von den anderen die Mehrbenutzerinterferenzen eliminiert. Anderseits werden auch heterogene Relay-Gruppen betrachtet, wobei jede Gruppe einen speziellen MIMO Gewinn (Array oder Spatial Multiplexing) erzielen soll. Darüber hinaus wird die Diversität, im ersten Fall durch gruppen- und zeitabhängige Phasenrotationen und im zweiten Fall durch die Auswahl von Relays, erhöht.

Im zweiten Teil dieser Dissertation wird Decode-and-Forward (DF) MIMO Relaying für mehrere Benutzer im One-Way und Two-Way Betrieb untersucht. Zunächst betrachten wir zwei MIMO Endknoten, die ihre Daten über ein einzelnes MIMO Relay austauschen. Es wird das Two-Way Protokoll auf Knoten mit mehreren Antennen erweitert, wobei das Hauptaugenmerk auf die auf Network Coding basierende Signalverknüpfung am Relay gelegt wird. Um ungleiche Datenraten vom Relay zu den Zielknoten zu unterstützen, wird eine modifizierte auf Network Coding basierende Methode vorgeschlagen, bei der vor der XOR Addition ein Zero-Padding pro Symbol auf die Informationssequenz angewendet wird, die zum schwachen Zielknoten übertragen wird. Dies wird zusätzlich als a priori Information am entsprechenden Zielknoten zur Decodierung genutzt. Anschliessend wird betrachtet, dass Sender und Relays über die Kenntnis der Kanalkoeffizienten verfügen und es werden die Sendekovarianzen entworfen, welche die Kapazitätsgrenzen des Two-Way Relayings kennzeichnen. Es werden auf SDP basierende Optimierungsschemata entworfen, die jeweils die Summenrate oder das

Minimum der zwei Punkt-zu-Punkt übertragungsraten maximieren. Ausserdem wird für ein Netzwerk mit Knoten mit nur einer Antenne der Fall betrachtet, dass die Kenntnis der Kanalkoeffizienten am Sender nicht perfekt ist. Hierzu wird die Worst-case Broadcast Capacity Region hergeleitet und weiter werden die robusten Gegenstücke zu den zwei Optimierungsproblemen für perfekte Kanalkenntnis entworfen.

Zum Abschluss wird One-Way und Two-Way DF Relaying erweitert für den Fall der simultanen MIMO Kommunikation von mehreren Quell- und Zielknoten. Dazu ist das Relay mit ausreichend Mehrfachantennen ausgestattet um den MIMO Multiple Access Kanal im Uplink räumlich aufzulösen und die Interferenz durch den Downlink zu handhaben. In dem Broadcast Zeitabschnitt des Mehrbenutzer-One-Way Relayings wenden wir herkömmliche Zero-Forcing Beamforming basierte MIMO Broadcasting Techniken an. Für den Two-Way Fall hingegen wird eine neue Knotentrennung auf zwei Ebenen vorgeschlagen: Auf Bit- und auf Raumebene. Infolgedessen wird ein modulares Relay Optimierungsschema der Sendekovarianz entworfen, das darauf ausgerichtet ist die Summenrate von allen Punkt-zu-Punkt Raten optimalerweise über die beiden Relaying Phasen zu maximieren.

Acknowledgments

First and foremost, I want to give special thanks to my supervisor Prof. Armin Wittneben for giving me the opportunity to work with him and his joyful research environment. During my PhD work, he has provided me not only numerous insightful advices and fruitful discussions, but also constant support and motivation whenever I needed it. I would like to thank Assoc. Prof. Elza Erkip and Prof. Hakan Deliç for taking part in my dissertation committee. Prof. Hakan Deliç deserves particular attention for encouraging me to do a PhD and keeping an eye on me ever since.

The Communication Technology Laboratory has been a very friendly and intellectual research environment for me as a graduate student, through which I got to know truly talented and competent colleagues. I am indebted to each former and current member of Wireless Communications Group for their friendship and collaboration in research. My special thanks go to my office-mates Christoph Steiner and Georgios Psaltopoulos for literally being good friends in and out of the office: it has been a pleasure sharing room-F110 (aka. my home) with you guys!

My out-of-office time during these last years would have been much duller without my friends and fellows. I want to thank to my Turkish community in Zurich - especially to Cemal, Serra, Şevket (hocam), and Frank (yeah master, I count you as a Turk) - for all memorable moments. Akin Şahin deserves my deepest gratitude for being a very close friend I can always rely on: I feel so lucky to get to know you my bro!

I would like to thank my comrade Ender Konukoğlu for accompanying me in all the good and bad times of my last seven years. I would also like to take this opportunity to thank my extended family members Aycan, Bahadır, Barbara, Burcu, Çağlar, Derya, Gizem, Hakan, Pınar, Seçil, Serhan, Sinan for all their support and friendship.

Finally, I want to sincerely thank to my dear parents İlyas and Sevgi, and my beautiful sisters Zekiye and Selen for their endless support through my academic career. They made this thesis possible and my life worthwhile. I love you all!

Contents

Chapter 1

Introduction

1.1. Motivation

The central theme of interest in this work revolves around efficient design of cooperative relaying protocols for wireless multiuser networks. To this end, we first need to get to know our playground and its current borderline. We start with a very brief overview of current trends in wireless communications, which is followed by more specific topics of cooperative and multiuser communications. Subsequently, we touch upon recent efforts to incorporate cooperation with multiuser setups.

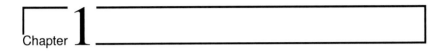

Trends in Wireless Communications It is in the nature of human-beings that each expects discrepant data services with the only common will that they want the connectivity to be ubiquitous, fast and low-cost. In the early stage of wireless standardization, an on-demand approach was taken such that independent and disjoint wireless systems have been designed for each specific need for wireless or wireline data. The widest spread wireline network is the public switched telephone network, which is now gradually being replaced by the lines of global data-coverage-enabler, the Internet. On the wireless side of communications, the cellular networks have been evolving for three decades through 1G, 2G and 3G, to provide very-long-distance wireless connectivity, which, in general, is dominated by voice-based traffic. In principle, wireless local area networks (WLAN) establish wireless links between several nodes in a limited-coverage area. However, the appreciation for WLAN has accelerated as it enables, though in a short-to-medium coverage distance, mobile and wireless connection to the broad wireline coverage of the Internet. Contrary to the previous two infrastructure based wireless systems, wireless ad-hoc networks are gaining a growing popularity due to their self-organizing structure. With the advantage of reduced deployment costs, wireless ad-hoc networks provide a seamless connection between wireless mobile nodes without a need for pre-installed infrastructure. Some typical examples are wireless personal area networks, e.g., Bluetooth, Zigbee, and wireless sensor networks for military applications, traffic control, parking, surveillance, etc. [3,32].

That is to say, we are surrounded by numerous wireless radio networks in our daily lives, each with different focus, range, spectrum, and available transmit power. However, it is a common conjecture that in order to fully utilize the available limited bandwidth for attaining maximal throughput, the incorporation of these heterogeneous communications media is unavoidable for upcoming wireless systems. To this end, design proposals for 4G and beyond should aim at collating these heterogeneous wireless systems within a hierarchical structure overlaying the Internet protocol.

Future wireless communication systems are expected to exhibit very large node densities, and thus require highly efficient spectrum utilization in order to keep the per-node-throughput at the prospected levels. Further, efficient management of transmit energy consumption is also supposed to play a major role for the envisioned very-high throughput and its corresponding coverage. In the current cellular networks, spectral efficiency is enhanced through frequency reuse in different cells and/or spread-spectrum techniques. However, it is anticipated that upcoming cellular systems will operate at higher frequencies, e.g., in many European countries 2.6 GHz spectrum has been already licensed for 4G, where the difficulty of sustainable coverage will be emphasized even more. An immediate solution to address the coverage issue can be either increasing the number of base stations (smaller cells) or adjusting the transmit power of the base station [93]. However, the former would boost the infrastructure deployment costs up, whereas the latter would constrain the inter-cell interference management.

In wireless ad-hoc networks, where the nodes are already limited in power (and therefore in coverage), *multihop* links emerge as vital for connectivity and acceptable throughput levels. That is, if the destination of a transmit message is not in the close proximity of the source node, the neighbouring nodes assist the communication by simply forwarding a function of source information. Moreover, as the transmission range of each intermediate node is shortened, the transmit power levels are reduced correspondingly. Such a power-efficient link establishment provides the opportunity to reuse the same physical channel, and hence, improves the spectral efficiency [37].

In the nineties, the use of multiple antennas at transmitter and receiver has been identified as the key enabler for high spectral efficiency in point-to-point communication, since it facilitates multiplexing of several data streams in space rather than in time or frequency. Moreover, improved *link reliability* and *array gain* can be efficiently achieved through proper signal combination schemes at the destination [24]. In order to benefit from these gains in local area networks, the IEEE 802.11n standard has been established, which integrates multiple antennas to the previous 802.11 standards [132]. However, the implementation of wireless multiple-input multiple-output (MIMO) systems are limited by the maximal feasible number of transmit and receive antennas at the respective nodes due to practical reasons such as current operation carrier frequencies of the networks and small size of mobile units.

Cooperative Communications After it has been first investigated in the late seventies [35, 107], the *relay channel* did not appeal to the communication society for about twenty years. Aiming ultimately at range-extended connectivity, the multihop ad-hoc networks have been inherently employing cooperation in communications for the mutual-benefit of all nodes within the network. However, in the more recent past, it has been realized that cooperation can enable to achieve MIMO gains in a distributed manner [70, 108]. Since then, several cooperative communication protocols have been proposed to supersede or improve MIMO systems with an additional advantage that the communication range is extended through idle nodes, namely *relays*. In [70–72, 108–110] *cooperative diversity* (also referred as *user cooperation*) has been introduced, which aims at incorporating the distributed antennas attached to different user nodes. That is, distributed spatial diversity is exploited by generating virtual antenna arrays through distributed transmission and signal processing. In such a multihop information transfer, the primary communication between a source-destination (S-D) terminal pair is aided either by the other user terminals or idle relay nodes without private benefit.

Among several proposed relaying schemes, decode-and-forward (DF) and amplify-and-forward (AF) relaying are the most extensively investigated forwarding schemes in the literature [23, 51, 71, 72, 102, 109, 110, 149]. A relay can be conceived as a smart repeater, where DF is a digital and AF an analog one. In spite of the main drawback of noise amplification at the relays, AF relays are commonly appreciated for their simplicity and transparency. While DF relaying avoids the aforementioned drawback in the expense of additional complexity of decoding and re-encoding processes, it suffers from the induced error propagation. Both forwarding schemes will be addressed throughout this thesis. Another classification for relaying protocols is between *half-duplex* and *full-*

duplex relays. The former indicates that the relay node can not transmit and receive at the same time/frequency, which is due to the hardware related constraint that the signal powers of the transmit and receive signals are difficult to isolate. Therefore, the impractical full-duplex relays have been rather considered in information theoretical derivations of the relay channel [35].

A more detailed discussion and literature survey about cooperative diversity and capacity theorems on the relay channel can be found in [56] and [103], respectively.

Multiuser Communications We classify multiuser communications in three classes: one-to-many, many-to-one, many-to-many. The first two are respectively referred as broadcasting and multiple accessing, where either a single source terminal transmits private information to several destination terminals, or several source terminals transmit private information to a single destination terminal [119]. There are historically known trivial methods to separate different users in time or frequency, e.g, time division multiple access (TDMA), code division multiple access (CDMA), frequency division multiple access (FDMA), orthogonal FDMA (OFDMA). However, as time and frequency dimensions are confined due to delay and spectrum limitations, respectively, the spatial separation of multiple users using the same time and frequency has recently gained an increasing attention, i.e., space division multiple access (SDMA). Within the SDMA context, the one-to-many downlink and the many-to-one uplink cases are referred respectively as the broadcast channel (BC) and the multiple access channel (MAC). The current research intensively focuses on the capacity region of both BC and MAC, and on practical precoding schemes to achieve these capacities [29, 36, 49, 115, 125, 142].

However, much less is known for the interference management of the third class of multiuser communications, where there are multiple S-D pairs communicating with each other at the same time and frequency slot. This case is referred as *multiuser interference channel* in the literature. In the meanwhile, there are methods proposed, e.g., [28], that enable spatial multiplexing (SM) of distributed multiple S-D pairs through interference alignment and coding over time/frequency/space. Although such an approach achieves the ultimate limit of spatial degrees of freedom, i.e., in the order of half of the number of S-D pairs, it requires coding over very many slots of the chosen dimension, which is practically infeasible. Moreover, in [118], a multiuser setup, where the sources are partitioned into multiple clusters, and every cluster is dedicated to transmit a common message to a unique destination, has been studied. Therein, the feasibility of distributed SM within this setup has been asymptotically proved for a special case of real-valued channels through 1-bit feedbacks.

Cooperative Relaying for Multiuser Communications The simplest way to benefit from relaying in communication of multiple S-D pairs is through scheduling. Assuming a certain set of relays to participate, the available time or frequency resources are allocated to the pairs whose associated channel conditions via these relays are good enough to transmit with a higher rate than the others. Such an adaptive scheduling achieves *multiuser diversity* [56, 67]. On the other hand, employment of fixed relays through the downlink transmission of a cellular network, i.e., one-to-many communications, is receiving an increasing interest due to its substantial coverage and link capacity improvement in regions with non-line-of-sight or shadowing [30, 64].

Recently, it has been shown that relaying protocols can also be employed in multiuser interference channels for providing efficient interference management. A particularly simple (from a coding perspective) method for realizing distributed SM is based on AF relaying. The approach has been discussed in [133] for single-antenna equipped nodes and requires distributed relay antennas to assist the communication between terminal nodes without private benefit. The method mimics classical MIMO transmit beamforming in a distributed fashion, where the role of the spatial precoder is played by the distributed relay antennas. These impose specific amplifications on their received signal, such that individual pairs of terminals can communicate over effective interference free channels, and thus can employ standard single-input-single-output encoding and decoding techniques. Such a coherent relaying based distributed SM scheme is promising for future implementation of heterogeneous networks, since it enables coexistence of a large set of low-cost AF relays with a set of rather more complex terminal nodes. The corresponding capacity scaling laws of a more general multiuser cooperative setup enhanced by multiple antennas have been derived in [23, 80].

There is an unavoidable spectral efficiency loss, i.e., one half, in the aforementioned conventional single or multiuser two-hop relaying protocols (referred commonly as *one-way relaying*), since we require two channel uses for the transmission from source to destination. It has been shown in [99] that, in fact, this loss can be recovered for a special case of multiuser communications, where two terminals are exchanging information via a single relay. Referred as *two-way relaying* due to the bidirectional information flow, such a relaying scheme consists of two phases. In the first phase, both terminals transmit simultaneously via a multiple access scenario to the relay node. In the second phase, the relay broadcasts a common message which is obtained by combining the received messages. Since the nodes know their own transmitted signal, they subtract the back-propagated self-interference prior to decoding. Two-way relaying has been adapted for both AF and DF protocols and extended to the multiuser AF relaying in [101]. Achievable rate regions, optimal time-sharing between the two phases and relay selection have been investigated in [88–90]. Recently, it has also been extended to the case of multiple antenna equipped nodes in [55, 137].

In order to serve multiple S-D pairs simultaneously through one- or two-way DF protocols, i.e., distributed spatial multiplexing, we need multiple antenna equipped relay(s) for the sake of efficient decoding of the received signals from the source terminals and spatial separation of the transmit signals to the destination terminals. Alternatively, as proposed in [31], CDMA based separation can be employed between the S-D pairs.

1.2. Contribution and Outline

In this thesis, we focus on wireless multiuser networks, and aim at designing novel cooperative communication protocols, developing corresponding transmission and signal processing techniques, and optimizing the network performance. While doing so, we take the perspective of answering the following questions:

1. How can cooperative relaying be employed to achieve distributed spatial multiplexing in different wireless multiuser network setups?

2. How can further distributed MIMO gains be achieved?

3. What is the optimal resource allocation for the chosen figure-of-merit? Can we design computationally tractable algorithms to obtain optimal solutions?

4. What are the impacts of system imperfections? Can we design robust solutions?

5. Can we exploit the trade-off between complexity and performance?

The thesis is presented in two main parts. In the first part, we consider a wireless ad-hoc setup and investigate coherent multiuser AF relaying with an extensive scope. We shift our focus to DF relaying in the second part, and address both multiuser one- and two-way MIMO relaying with applications to both multihop cellular and local area networks. The organization of the thesis and the contribution of each chapter are summarized as follows:

Chapter 2 Taking a unified perspective, Chapter 2 presents an extensive overview of fundamentals on coherent multiuser AF relaying, and further enriches the content with recent advances. Specifically, we consider a wireless ad-hoc network, where a set of single-antenna S-D terminal pairs communicate concurrently over the same physical channel, and a set of single-antenna AF relay nodes assist the communication in a half-duplex scheme. We first investigate the sufficient conditions for efficient multiuser interference cancellation, e.g., minimum number of required relays, channel state information (CSI) requirements per relay, relay sum transmit power constraint, and etc. Next, we derive the corresponding relay gain vector allocation through zero-forcing (ZF) beamforming for different cases of assumed CSI at the relays. Finally, we extend the ZF framework to a scenario where the communications within the primary cluster of S-D terminal pairs, is subject to out-of-cluster interference. Assuming multiple interference sources, we derive the conditions for interference-free multiuser communication within the primary S-D pairs cluster.

Chapter 3 When the distributed network is composed with more relays than the minimum required number of relays for efficient interference mitigation, we refer this scenario as an *excess number of relays*. In this chapter, we identify such a scenario as an efficient means to obtain additional degrees of freedom for further relay gain optimization. To this end, we first extend the ZF beamforming based gain allocation to the case of general beamforming without ZF constraintbi. Therein, we choose two objective functions to maximize: sum rate and minimum link rate. The corresponding maximization problems are solved either by gradient descent based algorithms or through efficient semidefinite programming (SDP) solutions. With these efficient designs of complex relay gain allocation, coherent relaying exploits the additional degrees of freedom to improve sum or outage rates and to provide fairness between different terminal pairs. Moreover, it is numerically shown that effective distributed diversity can be attained at the destinations through either relay selection or max-min type relay gain optimizations.

However, the *sine qua non* for the efficiency of coherent relaying and also for achieving the aforementioned gains, is the assumption of global CSI knowledge at each relay, which requires each relay to distribute its own local CSI to the others. Depending on the number of relays and channel variations, this requirement may introduce a major overhead, which in practice can well diminish the spatial multiplexing gain. Hence, in the second part of this chapter, we introduce a new *distributed* gradient based gain allocation scheme, which substantially minimizes this overhead. Key to this is the proof, that the gradient of the destination signal-to-interference-plus-noise ratio (SINR) can be calculated in a distributed manner based on local CSI at the relays and very limited feedback from the destinations. In order to minimize the number of iterations in the distributed gradient algorithm, we propose two distributed approaches to determine the optimal step-size for each iteration. They are based on the concept of measurement cycles, where the relays use the downlink channel to the destinations to evaluate matrix expressions. In order to further optimize the implementation performance of the algorithm, we focus on key issues like efficient starting vector computations and variants of step-size calculation. Finally, we discuss the convergence behaviour of the algorithm through computer simulations. We identify the sequential opening of the spatial channels as a key factor that impacts speed of convergence.

Chapter 4 In the previous chapters, we assume that the CSI estimates at the relays are perfect and the relays are phase-synchronous so that coherency of AF relaying is perfectly satisfied. However, it is highly possible to face imperfections while implementing the aforementioned designs in a practical system. Hence, in Chapter 4 we address the impact of these performance limiting data uncertainties. To this end, we first study the impact of random phase noise at the local oscillators (LO) of the relays, which is emphasized more in delay-and-forward schemes. Such a random impact destroys the coherence and causes the channel estimates at the relays,as well as the SINR at the destinations, to be uncertain. We propose a worst-case max-min robust relay gain allocation design based on SDP, where we assume that the phase uncertainties at the relays are bounded within a defined range determined according to the desired probability of assurance.

Next, we focus on the case of imperfect CSI knowledge at the relays, which may arise due to noisy channel estimation, quantization errors or simply being outdated. We model the channel estimates within spherical uncertainty sets. As the generalized problem of both uncertain source-to-relay and relay-to-destination CSI, is mathematically non-trivial to model, we rather focus on two simplified cases where either only source-to-relay or relay-to-destination links are assumed to be subject to uncertainty. Correspondingly, the robust counterparts of the max-min SINR problem for both cases are formulated as SDPs. In general for both phase and CSI imperfection cases, we show that the robust designs provide substantial worst-case performance gains over their non-robust counterparts.

Chapter 5 We have already noted that large number of relays provides additional gains to the network, but also reported that this benefit comes at the expense of local

CSI dissemination overhead. Addressing this, we propose the distributed calculation in Chapter 3. However, in this chapter, we take an alternative approach of partitioning the relays into multiple independent clusters with the essential goal of achieving similar gains of network-wide global CSI requiring solutions, but having a bounded dissemination overhead independent of the number of relays. First, we consider homogeneous relay clusters, where each cluster serves all S-D pairs and manages the multiuser interference independently from the other clusters. The diversity reduction due to the loss of coherency in between clusters is recovered through cluster and time specific phase rotations [2, 50].

As a second approach, we assign the relays into two hierarchically separate clusters. The members of the first cluster employ only local CSI to determine their gain factors, i.e., no CSI dissemination overhead, and contribute to the received signal power at the destinations. However, due to insufficient CSI information, they cause multiuser interference at the destinations. Whereas, the second cluster is responsible for multiuser interference mitigation by using only cluster-wide CSI knowledge and limited feedback from the destinations. We identify relay selection for clusters as an essential factor that boosts the sum rate of the network and provides full effective distributed diversity. In order to realize these gains without the need for global CSI, we devise an iterative distributed relay selection algorithm, which performs very close to the optimal brute-force search.

Chapter 6 From this chapter on, we change our focus of interest from AF to DF relaying for enabling distributed multiuser communications. For efficient separation of multiple incoming and outgoing signals at the relay(s), we assume a MIMO scenario by default. Moreover, as the nature of network traffic suggests in practice, we address bidirectional traffic between terminals. Specifically in this introductory chapter, we consider two MIMO terminals exchanging information via a single MIMO relay node, and present a unified overview consisting of the signaling schemes and the capacity regions of both one- and two-way relaying protocols. Our main contribution therein is to extend the two-way protocol to multiple antenna equipped nodes with a primary focus on network coding based signal combination at the relay. We propose a modified network coding based approach, which supports unbalanced relay-to-terminal rates. That is, the full MIMO two-way broadcast capacity region derived in [137] can be achieved. To do so, we apply *zero-padding per symbol* on the information sequence to be transmitted to the weak terminal, prior to exclusive-or (XOR) addition, and further we provide *a priori* decoding information about this process to the corresponding terminal.

Chapter 7 Having introduced the achievable rate expressions and capacity regions for given deterministic transmit covariance matrices in the previous chapter, in this chapter we study the impact of transmit CSI at the relay on both one- and two-way MIMO relaying protocols. We start with presenting how to find the covariances that characterize the capacity boundaries for both phases of two-way relaying, which are then combined to obtain the two-way capacity region. Afterwards, we design two SDP based transmit covariance optimization schemes, each of which respectively maximizes the sum or the minimum of the two terminal-to-terminal information rates. Moreover, within the context of this chapter, we also study the imperfect transmit CSI case for the special case of multiple-input single-output (MISO) downlink channels. Therein, we derive the worst-case broadcast capacity region, and further propose robust counterparts for the two optimization problems designed for the perfect transmit CSI case. In summary, in addition to depicting the constructive impact of transmit CSI, we also emphasize the substantial spectral efficiency improvement of two-way relaying compared to its one-way counterpart independent of the CSIT assumptions.

Chapter 8 We extend the previous discussions in Chapters II and 7 to the case of simultaneous MIMO communication of multiple S-D pairs via a single DF relay. For both one- and two-way protocols, either a multiple access or a broadcast channel is established for each transmission towards the relay and from the relay, respectively. The relay is equipped with multiple antennas to resolve the MIMO multiple access channel spatially and also to manage the interference through the downlink. The role of such a MIMO capable relay is played by an access point in a WLAN scenario or a relay/base station in a cellular network. Through the broadcast phase of multiuser one-way relaying, we apply conventional ZF beamforming based MIMO broadcasting techniques. However, for the two-way case, we propose a novel two-level terminal separation for broadcasting: bit-level and spatial-level. Consequently, we design a modular relay transmit covariance optimization scheme that aims at maximizing the sum rate of all terminal-to-terminal rates optimally over the two phases.

Chapter 9 Finally, we draw our conclusions and present an outlook on future research directions of wireless multiuser networks.

Appendices We first present a fairly inclusive overview of convex optimization theory, which in general follows an engineer's view and summarizes the findings in various references such as the convex optimization book of Boyd *et al.* [26] and the robust methodology of Ben-Tal *et al.* [9–14]. In the sequel of the appendices, we present various depictive examples of designed protocols and some further derivations, which are dropped from the corresponding main text for the clarity of information flow.

Part I.

Coherent Multiuser Amplify-and-Forward Relaying

Chapter 2

Fundamentals of Coherent Multiuser Amplify-and-Forward Relaying

This chapter takes a fundamental look at the problem of achieving distributed spatial multiplexing through coherent AF relaying. After a general chronological survey of coherent relaying, we present the system and signal models of the two-hop multiuser relaying setup we consider. This setup will serve as the theoretical playground for all discussions on coherent AF relaying to be covered in this thesis. Subsequently, we elaborate on distributing beamforming, where we expose the sufficient conditions and required channel information set per relay for efficient interference mitigation at the destinations. The corresponding complex relay gain decisions are derived for different levels of channel state information at the relays and for different types of interference sources.

2.1. Introduction

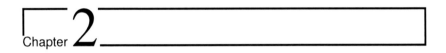

We consider a multiuser interference channel scenario, where N S-D pairs want to communicate concurrently over the same physical channel without interfering with each other. Such multiuser multiplexing can be realized in a distributed manner through a two-hop half-duplex coherent AF relaying protocol [133]. The AF relays choose their complex relay gains such that the two-hop equivalent channels of interfering streams at the destinations are suppressed and non-zero received signal powers are attained for all intended signals. Here and in the sequel, we refer to this approach as *coherent multiuser AF relaying*. In the related multiuser literature [23,80,85], the term "coherence" translates to perfect CSI knowledge at the relays. That is in the sense that the anticipated channel knowledge at the relays is the same as the actual channel realization, and it is further independent of the impact of different RF chains at the terminals and relays. In other words, as the relay gain factors are adjusted according to the phase information of CSI, there needs to be either a global phase reference or alternatively the channel estimation should be performed such that the global phase reference can be avoided [17,20]. For the sake of the clarity of the exposition throughout the thesis, we adapt the following definition for coherent AF relaying: Any arbitrary variation at the phase offset of any relay in the network is tracked and its impact on destinations' received signals is correspondingly corrected either through the nature itself or through the signal processing defined by the gain allocation scheme.

The most significant overhead coming from relaying-based SM is the stringent requirement for perfect *global CSI* at each relay. That is to say, the calculation of the relay gain factors requires knowledge of the channel coefficients from all terminal nodes to all relays. Based on the training sequences in the transmit packets of the terminal nodes, each relay can estimate the subset of channel coefficients from all terminals to itself, i.e. its *local CSI*, without overhead. Whereas, for the aforementioned complex gain factor calculation, this local CSI has to be disseminated to all other relays. Such a CSI dissemination can be performed over the same (physical) communication channel, where the data transmission takes place, and the corresponding necessary channel resources may be counted within the training period of the protocol. However, the overhead involved may be prohibitive, in particular if the number of relays is large and/or the channel is fast fading. As an alternative approach, assuming that the relay nodes are in close proximity, we propose that some other independent short-range wireless media are employed for CSI dissemination purposes. For instance, Bluetooth and ultra-wideband systems, which are well known short-range and low-cost solutions, can be potent candidates. Furthermore, the wireline infrastructure of power line communications can be as well deployed [69].

The multiuser coherent AF relaying requires $N^2 - N + 1$ distributed relay antennas [133] for simultaneously multiplexing N terminal pairs each equipped with a single-antenna. While this requirement appears to be daunting at first glance, it is put into perspective by the remarkably simple structure and low power consumption of the AF relay nodes. In a frequency division duplex (FDD) implementation each relay just performs an amplification and a frequency translation of the received signal, which can be fully implemented in the analog domain, i.e., they do not require analog-to-digital conversion and digital baseband processing. An equivalent time division duplex

(TDD) implementation involves a store and forward (delay) operation, which is typically implemented in digital baseband (random access memory). Nevertheless also for TDD, the relays have very low complexity. In essence, the relay nodes in the network have a significantly lower complexity compared to the terminal nodes. This fact allows for the implementation of heterogeneous networks, where a set of terminal nodes coexists with a far larger set of low-cost relay nodes.

Prior to moving on the technical details of distributed beamforming through coherent AF relaying, in the next section we present a brief literature survey for coherent relaying and explore how it has evolved and adapted through time.

State-of-the-Art The conventional coherent relaying setup of a single S-D pair with multiple relays, has been considered in several publications with different aspects [53, 65,73,97]. Reference [73] is one of the first works that considered coherent combining for the conventional multi-relay channel. While aiming at the optimum power allocation through the relays, the complex relay gains are chosen such that the superposed relay signals' phase are aligned to result in maximal received signal-to-noise ratio (SNR) at the destination. Considering a similar scenario, the optimal relay gain allocation is derived in [53]. In both of the previous references, a sum (also called aggregate) relay transmit power was assumed. However, in [97] both sum and per-relay transmit power constraints are addressed and efficient conic programming based solutions have been proposed to maximize the achievable rate, which is a monotonically equivalent problem to the maximization of received SNR for single-antenna terminal pairs. Moreover, in [65] the objective of complex relay gain decision has been chosen to minimize the mean-square error (MSE) between the source's transmit and the destination's receive symbol.

The previous works on AF relaying are carried over to multiple antenna equipped terminals and/or relays in [84, 141], where now the maximum mutual information between terminals is targeted. In [84], a single MIMO AF relay channel was considered with different levels of CSI knowledge at the relay, e.g., different combinations of the MIMO channels belonging to the first-hop, the second-hop and the direct source-to-destination link. Multiple MIMO relays are studied in [141], where consequently, transmit CSI was assumed at the source terminal. Therein, each data stream of the source was matched to a specific relay node, based on the conventional eigen-mode transmission.

In [23], the sum-capacity scaling of the multiuser relay networks has been derived in the number of relays. There, it was assumed that a set of relays is assigned to each S-D terminal pair, where each relay has the local CSI knowledge for its assigned S-D pair.

The relays perform match filtering with respect to their local channels. Whereas, in [37], each relay in the network serves for all S-D pairs and is assumed to know global CSI within the network. Correspondingly, the power efficiency scaling was derived in the number of terminal pairs. Reference [80] has generalized these two results and provided the sum-capacity scaling of the network in both the number of S-D terminal pairs and the number of relays. It is further shown that if the number of relays grows fast enough as a function of number of terminal pairs, the network decouples in the sense that the capacity of each S-D terminal pair is strictly positive.

Reference [133] was the first to consider distributed spatial multiplexing for finite number of single-antenna relays, where a nullspace projection method has been proposed to zero-force the multiuser interference terms at the destination terminals. This scheme is later referred in [102] as orthogonalize-and-forward relaying, as so each S-D pair communicates through orthogonalized channels. A similar setting of [133] has been also studied in [37,85] with different focus. References [17–19,69,134] address the zero-forcing gain allocation from different perspectives. In [134], the case of excess number of relays was considered, where there are more relays available than the reported minimum number of relays for ZF in [133]. Moreover, it was assumed that the relays could cooperate in order to exchange the received signals at the end of first hop transmissions. It has been presented that both excess relays and relay cooperation could be employed to optimize the zero-forcing relay gain factors in order to attain some other MIMO gains besides spatial multiplexing. A simple choice of maximizing the minimum received signal power at the destinations has been shown to improve the outage rates, which was there translated to an effective diversity gain. In [17], a demonstrator for coherent multiuser AF relaying was presented. The crucial aspect of attaining coherency between relays through global carrier phase synchronization has been addressed in [18,19]. In [19] a method to achieve global carrier phase synchronization was proposed, whereas in [18], the impact of noisy phase synchronization on the performance of zero-forcing relaying, was investigated. Further, in [69], the power line was assumed to be employed to initialize and synchronize the AF relays and to disseminate information in between the relays in such a multiuser zero-forcing relaying scenario.

The zero-forcing constraint that is enforced to manage the interference between multiple S-D pairs, has been removed in [21]. Instead, the complex relay gains were designed such that the MSE of the signal at all destinations is jointly minimized. A generalization of the multiuser relaying technique to multiple-antenna terminal and relay nodes has been provided in [1]. Further, adapting multiuser relaying to a cellular relaying structure, the direct link between the base station (source) and the mobile station (destination) is also incorporated through the distributed linear precoding in [135].

Contribution and Outline of this Chapter In this chapter we present an extensive overview of coherent multiuser AF relaying, where we aim to provide the big picture without avoiding any possible details. Moreover, we supplement the content with recent emerging applications of coherent relaying.

We first present a general system model for the considered multiuser relaying scenario in Section 2.2. This will serve as the basis for all related sections throughout this thesis. In Section 2.3, we present the two-hop traffic pattern, where we considered two different transmission modes for the second hop. The fundamentals of distributed beamforming through AF relays is given in Section 2.4. First, we summarize the conditions for efficient multiuser interference cancellation and the corresponding complex relay gain vector allocation. Therein, we also address partial relay cooperation, in which relays share their received signals in the first hop with other relays. Next, assuming that only local CSI is available per relay, we present a benchmark relay gain decision which is asymptotically optimal for infinitely many relays. Finally in the same section, we consider a scenario, where the multiuser communications within the primary cluster of S-D terminal pairs, is subject to out-of-cluster interference. We assume multiple unauthorized sources, which transmit through the same physical channel with the primary cluster at the focus of interest, and hence, cause interference either at the relays, the destinations, or both. Following the zero-forcing framework, we derive the conditions for interference-free multiuser communication within the primary cluster. In the last Section 2.5, we discuss and compare the performance of different distributed beamforming schemes and provide insight into the proposed out-of-cluster interference cancellation approach.

2.2. System Model

We consider a distributed wireless network consisting of N source, N destination and N_r AF relay terminals. Each source terminal constitutes a pair with a corresponding destination such that N distinct S-D terminal pairs communicate concurrently over the same physical channel, i.e., no FDD or TDD, while N_r relay terminals assist this communication in a half-duplex scheme without private benefit. Each relay node in the network is equipped with multiple antennas, where the kth relay has $M_{\mathsf{r},k}$ antennas, and $k = 1, \dots, N_\mathsf{r}$. Without loss of generality, we assume that sources and destinations are single-antenna terminals. The extension to the case of S-D pairs with multiple antennas is quite trivial and dropped here for the sake of simplicity; however, the detailed expressions can be found in [16].

The propagation channel between any two antennas, each belonging to a different terminal within the network, is assumed to be frequency flat and independent. We model the channels as Rayleigh fading such that each coefficient is drawn from a zero-mean circularly symmetric complex Gaussian distribution. Moreover, we assume a slowly varying block-fading environment, where the channels stay constant within at least one transmission cycle duration, and the succeeding realizations of the propagation channels are statistically independent.

We denote the uplink channel between the ith source and the kth relay terminals with

$$\mathbf{h}_{k,i} \in \mathbb{C}^{M_{\mathsf{r},k}} = \left[h_{k,i}^{(1)} \ \cdots \ h_{k,i}^{(M_{\mathsf{r},k})}\right]^T,$$

the downlink channel between the kth relay and the ith destination terminals with

$$\mathbf{f}_{i,k} \in \mathbb{C}^{M_{\mathsf{r},k}} = \left[f_{i,k}^{(1)} \ \cdots \ f_{i,k}^{(M_{\mathsf{r},k})}\right]^T,$$

and the direct-link channel matrix between the source and destination terminals with $\mathbf{D} \in \mathbb{C}^{N \times N}$. As they are extensively used throughout Part I, we define the following channel matrices explicitly, which represent various combinations of $\mathbf{h}_{k,i}$ and $\mathbf{f}_{i,k}$:

- $\mathbf{H}_k \in \mathbb{C}^{M_{\mathsf{r},k} \times N} \triangleq \left[\mathbf{h}_{k,1} \ \cdots \ \mathbf{h}_{k,N}\right]$: the uplink channel matrix from all N sources to the kth relay,
- $\mathbf{H} \in \mathbb{C}^{M_{\mathsf{r}} \times N} \triangleq \left[\mathbf{H}_1^T \ \cdots \ \mathbf{H}_{N_{\mathsf{r}}}^T\right]^T$: the compound uplink channel matrix from all N sources to all N_{r} relays,
- $\mathbf{F}_k \in \mathbb{C}^{N \times M_{\mathsf{r},k}} \triangleq \left[\mathbf{f}_{1,k} \ \cdots \ \mathbf{f}_{N,k}\right]^T$: the downlink channel matrix from the kth relay to all N destinations,
- $\mathbf{F} \in \mathbb{C}^{N \times M_{\mathsf{r}}} \triangleq \left[\mathbf{F}_1 \ \cdots \ \mathbf{F}_{N_{\mathsf{r}}}\right]$: the compound downlink channel matrix from all N_{r} relays to all N destinations,

where $M_{\mathsf{r}} = \sum_{k=1}^{N_{\mathsf{r}}} M_{\mathsf{r},k}$ is the total number of relay antennas.

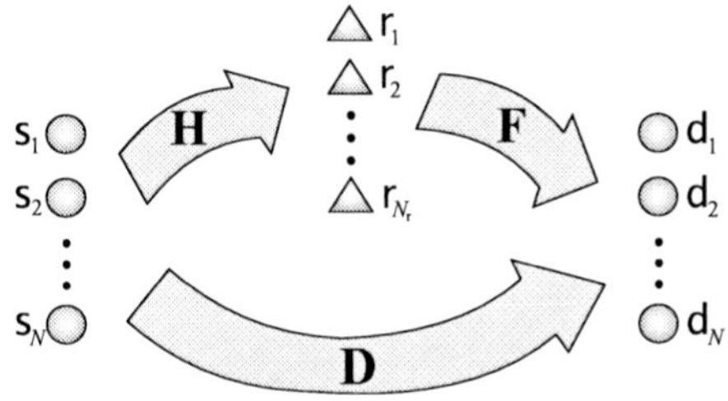

Figure 2.1.: Two-hop network configuration with N S-D pairs and N_r half-duplex relays.

Moreover, for the special case of single-antenna relays, i.e., $M_\mathsf{r} = N_\mathsf{r}$, we further define

- $\mathbf{h}_{\mathsf{s}k} \triangleq \begin{bmatrix} h_{k,1} & \cdots & h_{k,N} \end{bmatrix}^T$: $N \times 1$ vector of channel coefficients from all sources to the kth relay,
- $\mathbf{f}_{k\mathsf{d}} \triangleq \begin{bmatrix} f_{1,k} & \cdots & f_{N,k} \end{bmatrix}^T$: $N \times 1$ vector of channel coefficients from the kth relay to all destinations,
- $\mathbf{h}_{i\mathsf{r}} \triangleq \begin{bmatrix} h_{1,i} & \cdots & h_{N_\mathsf{r},i} \end{bmatrix}^T$: $N_\mathsf{r} \times 1$ vector of channel coefficients from the ith source to all relays,
- $\mathbf{f}_{\mathsf{r}i} \triangleq \begin{bmatrix} f_{i,1} & \cdots & f_{i,N_\mathsf{r}} \end{bmatrix}^T$: $N_\mathsf{r} \times 1$ vector of channel coefficients from all relays to the ith destination.

Note that we dropped the superscript at each single channel coefficient that indicates the index of relay antenna. The two-hop network configuration is illustrated in Fig. 2.1.

Every terminal in the network employs a free running LO. As aforementioned, phase synchronization through different LOs is a crucial issue in coherent AF relaying in order to satisfy coherent combination at the destination terminals. The phase offsets of the LO belonging to the ith source, the ith destination, and the kth relay, with respect to a global phase reference at time instant t are denoted by $\phi_{\mathsf{s}_i}^{(t)}$, $\phi_{\mathsf{d}_i}^{(t)}$, and $\phi_{\mathsf{r}_k}^{(t)}$, respectively. We model the LO phase offset as

$$\phi_\mathsf{x}^{(t)} = 2\pi f_\mathsf{x} t + \tilde{\theta}_\mathsf{x}^{(t)}, \qquad \mathsf{x} \in \{\mathsf{s}_i, \mathsf{d}_i, \mathsf{r}_k\}, \tag{2.1}$$

where f_x is the carrier frequency and $\tilde{\theta}_\mathsf{x}^{(t)}$ represents the random offset induced by the phase noise [17, 40]. This random term is assumed to be a Levy-Wiener process, which models the phase variations of a free running LO disturbed by white noise. Hence, assuming $t - t_0 = \Delta t$, $\tilde{\theta}_{\mathsf{x}}^{(t)}$ is modeled as a Gaussian random variable with mean $\bar{\theta}_{\mathsf{x}}^{(t_0)}$ and variance $\Delta t \sigma_p^2$ as $\tilde{\theta}_{\mathsf{x}}^{(t)} = \bar{\theta}_{\mathsf{x}}^{(t_0)} + \theta_{\mathsf{x}}^{(\Delta t)}$, where $\theta_{\mathsf{x}}^{(\Delta t)} \sim \mathcal{N}(0, \Delta t \sigma_p^2)$. Note that since each antenna of a single relay is connected to the same LO, we assume equal phase offsets for all antennas within a relay node.

2.3. Two-Hop Relaying Protocol and Signal Model

The coherent relaying protocol is composed of two hops (phases), where a single stage of AF relays aids the source-to-destination communication by listening to the sources in the first hop, and then, forwarding the processed versions of these received signals to the destinations in the second hop. We consider two options for the second-hop. In the primary operation mode, the sources are assumed to be silent in the second hop, which is motivated by a range extension scenario via relays. However, for the entirety of the presentation and the network model, we also present a second optional mode for the second hop, where the source terminals also contribute to the received signal power at the destinations. Note that in any case, we neglect any possible direct-link transmission from the sources to the destinations in the first hop. Although we set this constraint for the sake of the simplicity of the analysis, this can also be motivated by the assumption that the destinations do not have the ability for joint decoding of symbols in two time/frequency slots.

2.3.1. The First Hop

In the first hop, all sources transmit concurrently to the relays, while the destination terminals are in idle mode. Each source transmits with full transmit power $\mathbb{E}\{|s_i|^2\} = P_{\mathsf{s}}$ (also accounting for path loss and shadowing), where the transmit symbol of the ith source terminal is denoted by $s_i \in \mathbb{C}$, and $\mathbb{E}\{s_i s_j^H\} = 0, \forall i \neq j, i, j = 1, \ldots, N$. Stacking these symbols in a compound transmit symbol vector $\mathbf{s} \in \mathbb{C}^N$, the received signal vector $\mathbf{y}_{\mathsf{r}} \in \mathbb{C}^{M_{\mathsf{r}}}$ observed at the relays is written as

$$\mathbf{y}_{\mathsf{r}} = \mathbf{\Phi}_{\mathsf{r}}^H \mathbf{H} \mathbf{\Phi}_{\mathsf{s}} \mathbf{s} + \mathbf{n}_{\mathsf{r}}, \tag{2.2}$$

where

$$\mathbf{\Phi}_{\mathsf{x}} \in \mathbb{C}^{N \times N} = \mathrm{diag}\Big\{\big[e^{j\phi_{\mathsf{x},1}} \; e^{j\phi_{\mathsf{x},2}} \; \cdots \; e^{j\phi_{\mathsf{x},N}}\big]\Big\}, \; \mathsf{x} \in \{\mathsf{s}, \mathsf{d}\},$$

$$\mathbf{\Phi}_{\mathsf{r}} \in \mathbb{C}^{M_{\mathsf{r}} \times M_{\mathsf{r}}} = \mathrm{bdiag}\Big\{e^{j\phi_{\mathsf{r},1}} \cdot \mathbf{I}_{M_{\mathsf{r},1}}, \; e^{j\phi_{\mathsf{r},2}} \cdot \mathbf{I}_{M_{\mathsf{r},2}}, \; \ldots, \; e^{j\phi_{\mathsf{r},N_{\mathsf{r}}}} \cdot \mathbf{I}_{M_{\mathsf{r},N_{\mathsf{r}}}}\big]\Big\},$$

denote the phase offset matrices for sources/destinations and relays, respectively. The $\mathbf{n}_\mathsf{r}$ is an $M_\mathsf{r} \times 1$ spatio-temporally white circularly symmetric complex Gaussian noise vector sequence with zero-mean and covariance matrix $\mathbb{E}\{\mathbf{n}_\mathsf{r}\mathbf{n}_\mathsf{r}^H\} = \sigma_{n_\mathsf{r}}^2 \mathbf{I}_{N_\mathsf{r}}$.

Each relay terminal k linearly processes its received signal by multiplying it with an $M_{\mathsf{r},k} \times M_{\mathsf{r},k}$ complex diagonal relay gain matrix $\mathbf{G}_k$ in equivalent baseband, which induces amplification and phase-rotation on the received signal. These relay gains are dependent on the realizations of the propagation channels of sources/relays/destinations. Further information regarding their computation and the corresponding necessary CSI knowledge at each relay will be given in details in the following sections.

2.3.2. The Second Hop

We consider two possible options for the second hop: a primary and an optional protocol. For both, the relays simultaneously transmit the compound vector $\mathbf{r} \in \mathbb{C}_{\mathsf{r},k}^M = \mathbf{G}\mathbf{y}_\mathsf{r}$ to the destinations through the downlink channel $\mathbf{F}$, where $\mathbf{G}$ is an $M_\mathsf{r} \times M_\mathsf{r}$ block-diagonal matrix, and is defined as $\mathbf{G} \triangleq \mathrm{bdiag}\{\mathbf{G}_1, \ldots, \mathbf{G}_{N_\mathsf{r}}\}$. Moreover, we impose an instantaneous (in terms of channel realization) sum transmit power constraint P_r on $\mathbf{r}$ such that

$$\mathrm{Tr}\Big(\mathbf{G}(P_\mathsf{s}\mathbf{H}\mathbf{H}^H + \sigma_{n_\mathsf{r}}^2 \mathbf{I}_{M_\mathsf{r}})\mathbf{G}^H\Big) \leq P_\mathsf{r}. \tag{2.3}$$

The difference of the two options for the second hop originates from the action that is taken by the sources during this phase: either sources are kept silent or they transmit concurrently with the relays to the destinations through $\mathbf{D}$.

2.3.2.1. Primary Second Hop Protocol - Silent Sources

Motivated by a scenario of range extension through relays and/or weak source-to-destination links due to path loss and shadowing, the sources are silent in the second hop. Hence, the received signal vector $\mathbf{y}_\mathsf{d} \in \mathbb{C}^N$ observed at the destination antennas is

$$\mathbf{y}_\mathsf{d} = \mathbf{\Phi}_\mathsf{d}^H \mathbf{F} \mathbf{\Phi}_\mathsf{r} \mathbf{\Theta}_\mathsf{r} \mathbf{G} \mathbf{y}_\mathsf{r} + \mathbf{n}_\mathsf{d}, \tag{2.4}$$

where $\mathbf{n}_\mathsf{d}$ is an $N \times 1$ spatio-temporally white circularly symmetric complex Gaussian noise vector sequence with zero-mean and covariance matrix $\mathbb{E}\{\mathbf{n}_\mathsf{d}\mathbf{n}_\mathsf{d}^H\} = \sigma_{n_\mathsf{d}}^2 \mathbf{I}_N$. The M_r dimensional diagonal matrix $\mathbf{\Theta}_\mathsf{r}$ is defined similarly to $\mathbf{\Phi}_\mathsf{r}$, and represents the change in the phases of LO at the relays due to phase noise triggered by the time-difference

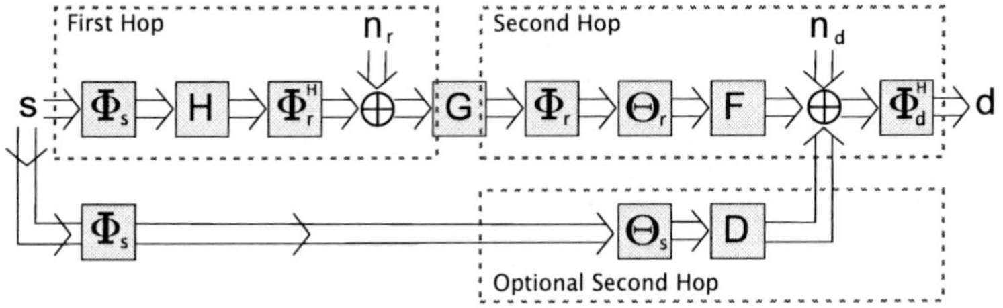

Figure 2.2.: Block diagram for the two-hop signaling protocol.

between reception and forwarding, i.e., the random $\tilde{\theta}_\mathsf{r}^{(t)}$ term in (2.1). If we write the received destination signal $\mathbf{y}_\mathsf{d}$ in terms of the source transmit signals $\mathbf{s}$, we have

$$\mathbf{y}_\mathsf{d} = \mathbf{\Phi}_\mathsf{d}^H \mathbf{F} \mathbf{\Phi}_\mathsf{r} \mathbf{\Theta}_\mathsf{r} \mathbf{G} \mathbf{\Phi}_\mathsf{r}^H \mathbf{H} \mathbf{\Phi}_\mathsf{s} \mathbf{s} + \mathbf{\Phi}_\mathsf{d}^H \mathbf{F} \mathbf{\Phi}_\mathsf{r} \mathbf{\Theta}_\mathsf{r} \mathbf{G} \mathbf{n}_\mathsf{r} + \mathbf{n}_\mathsf{d}, \tag{2.5}$$

$$= \mathbf{\Phi}_\mathsf{d}^H \mathbf{F} \mathbf{\Phi}_\mathsf{r} \mathbf{G} \mathbf{\Phi}_\mathsf{r}^H \mathbf{H} \mathbf{\Phi}_\mathsf{s} \mathbf{s} + \mathbf{\Phi}_\mathsf{d}^H \mathbf{F} \mathbf{\Phi}_\mathsf{r} \mathbf{G} \mathbf{n}_\mathsf{r} + \mathbf{n}_\mathsf{d}, \tag{2.6}$$

$$= \mathbf{\Phi}_\mathsf{d}^H \underbrace{\mathbf{F}\mathbf{G}\mathbf{H}} \mathbf{\Phi}_\mathsf{s} \mathbf{s} + \underbrace{\mathbf{\Phi}_\mathsf{d}^H \mathbf{F} \mathbf{\Phi}_\mathsf{r} \mathbf{G} \mathbf{n}_\mathsf{r} + \mathbf{n}_\mathsf{d}}, \tag{2.7}$$

$$= \mathbf{\Phi}_\mathsf{d}^H \cdot \mathbf{H}_\mathsf{srd} \cdot \mathbf{\Phi}_\mathsf{s} \mathbf{s} \quad + \quad \mathbf{n}_\mathsf{e}. \tag{2.8}$$

In (2.6) we assume for simplicity that the processing time at the relays is negligible, i.e., $\mathbf{\Theta}_\mathsf{r} = \mathbf{I}_{M_\mathsf{r}}$. A detailed analysis of the effect of $\mathbf{\Theta}_\mathsf{r}$ on the system performance and a corresponding robust gain allocation scheme will be presented in Section 4.2. Moreover, (2.7) follows from the self-cancellation of relay offsets through $\mathbf{\Phi}_\mathsf{r} \mathbf{G} \mathbf{\Phi}_\mathsf{r}^H = \mathbf{G} \mathbf{\Phi}_\mathsf{r} \mathbf{\Phi}_\mathsf{r}^H = \mathbf{G}$, and finally in (2.8) we define a *equivalent two-hop channel* matrix $\mathbf{H}_\mathsf{srd} \in \mathbb{C}^{N \times N} \triangleq \mathbf{F}\mathbf{G}\mathbf{H}$ and an *equivalent two-hop noise* vector $\mathbf{n}_\mathsf{e} \triangleq \mathbf{\Phi}_\mathsf{d}^H \mathbf{F} \mathbf{\Phi}_\mathsf{r} \mathbf{G} \mathbf{n}_\mathsf{r} + \mathbf{n}_\mathsf{d}$.

Lastly, we remark that for the sake of fair transmit power consumption at both hops, unless otherwise stated, we set the sum relay transmit power as $P_\mathsf{r} = N P_\mathsf{s}$ when this primary protocol is employed in the second hop.

2.3.2.2. Optional Second Hop Protocol - Transmitting Sources

In the case that the source terminals are in the communication range of the destination terminals, and hence, source-to-destination direct links are rather strong, the sources can transmit a scaled version of $\mathbf{s}$ through $\mathbf{D}$ to the destinations, while the relays are still transmitting $\mathbf{r}$ through $\mathbf{F}$. With the additional transmissions from the sources, the received signal vector in (2.8) modifies to

$$\mathbf{y}_\mathsf{d} = \mathbf{\Phi}_\mathsf{d}^H (\mathbf{H}_\mathsf{srd} + \varrho \cdot \mathbf{D} \mathbf{\Theta}_\mathsf{s}) \mathbf{\Phi}_\mathsf{s} \mathbf{s} + \mathbf{n}_\mathsf{e}, \tag{2.9}$$

where $\varrho \in \mathbb{R}_+$ is the scaling applied on $\mathbf{s}$ in the second hop, and the N-dimensional diagonal matrix $\mathbf{\Theta}_\mathsf{s}$ represents the change in the phases of LO at the sources due to phase noise triggered by the time-difference between the first and the second hop transmissions. Note that we do not consider any power allocation over the source transmissions in the second hop, but simply assume that each transmits with equal power of $\varrho^2 P_\mathsf{s}$. As we assume that the total network power consumption in the second hop is also upper bounded by NP_s, the scaling parameter ϱ and the sum relay transmit power P_r should be appropriately chosen such that

$$P_\mathsf{r} + \varrho^2 N P_\mathsf{s} \leq N P_\mathsf{s} \quad \rightarrow \quad \frac{P_\mathsf{r}}{1-\varrho^2} \leq N P_\mathsf{s} \tag{2.10}$$

holds.

Fig. 2.2 depicts the full picture of the two-hop communication pattern between source and destination terminals by summarizing it with a block diagram for both variations of the second hop.

Remark 2.3.1: Our main focus of interest throughout the Part I is the primary second hop protocol. Hence, unless otherwise is explicitly stated, the reader can assume that all given expressions in the sequel are related with the primary option.

2.4. Distributed Beamforming

Coherent AF relaying has been proposed to *supersede* the classical point-to-point MIMO transmit beamforming in a distributed fashion. Hence, the relay gains should be appropriately chosen such that simultaneous beams are formed between each S-D pair. In other words, the relay gains should cancel (or at least suppress) the multiuser interference, and hence, provide sufficiently high received S(I)NRs at all destination terminals.

In this section, we present an overview of how to design the distributed beamforming matrix $\mathbf{G}$. To this end, we want to modify (2.8) so that we have an explicit expression for the ith received symbol $y_{\mathsf{d},i} \in \mathbb{C}$ which requires the following algebraic manipulations and definitions. Firstly, we define an auxiliary matrix $\tilde{\mathbf{H}}_\mathsf{srd} \triangleq \mathbf{\Phi}_\mathsf{d}^H \mathbf{H}_\mathsf{srd} \mathbf{\Phi}_\mathsf{s}$ to denote the *effective equivalent two-hop channel* subjecting to the transmit signal vector $\mathbf{s}$. Then, using the matrix relation

$$\mathrm{vec}(\mathbf{X}_1\mathbf{X}_2\mathbf{C}\mathbf{Y}_1\mathbf{Y}_2) = (\mathbf{Y}_2^T \otimes \mathbf{X}_1)(\mathbf{Y}_1^T \otimes \mathbf{X}_2)\mathrm{vec}(\mathbf{C}),$$

where $\mathbf{X}_1, \mathbf{X}_2, \mathbf{C}, \mathbf{Y}_1, \mathbf{Y}_2$ are any full matrices with appropriate sizes, we vectorize $\tilde{\mathbf{H}}_{\mathsf{srd}}$ to $\tilde{\mathbf{h}}_{\mathsf{srd}} \in \mathbb{C}^{N^2}$ as

$$\begin{aligned}\tilde{\mathbf{h}}_{\mathsf{srd}} = \mathrm{vec}(\tilde{\mathbf{H}}_{\mathsf{srd}}) &= \mathrm{vec}(\mathbf{\Phi}_{\mathsf{d}}^H \mathbf{F}\mathbf{G}\mathbf{H}\mathbf{\Phi}_{\mathsf{s}}), \\ &= (\mathbf{\Phi}_{\mathsf{s}}^T \otimes \mathbf{\Phi}_{\mathsf{d}}^H)(\mathbf{H}^T \otimes \mathbf{F}) \cdot \mathrm{vec}(\mathbf{G}), \\ &\triangleq \mathbf{\Phi}_{\mathsf{sd}} \tilde{\mathbf{Z}} \tilde{\mathbf{g}},\end{aligned}$$

where we defined $\mathbf{\Phi}_{\mathsf{sd}} \in \mathbb{C}^{N^2 \times N^2} \triangleq \mathbf{\Phi}_{\mathsf{s}}^T \otimes \mathbf{\Phi}_{\mathsf{d}}^H$, $\tilde{\mathbf{Z}} \in \mathbb{C}^{N^2 \times M_{\mathsf{r}}^2} \triangleq \mathbf{H}^T \otimes \mathbf{F}$, and $\tilde{\mathbf{g}} \triangleq \mathrm{vec}(\mathbf{G})$ for the brevity of the following derivations. Since the $M_{\mathsf{r}}^2 \times 1$ vector $\tilde{\mathbf{g}}$ contains some ineffective zero elements, we drop these and obtain a full vector $\mathbf{g} \in \mathbb{C}^{N_{\mathsf{c}}}$, where $N_{\mathsf{c}} = \sum_{k=1}^{N_{\mathsf{r}}} M_{\mathsf{r},k}^2$. Furthermore, the corresponding columns of the compound channel matrix $\tilde{\mathbf{Z}}$ related with the dropped elements from $\tilde{\mathbf{g}}$, can also be deleted from $\tilde{\mathbf{Z}}$. Hence, this results in a new reduced $N^2 \times N_{\mathsf{c}}$ matrix $\mathbf{Z}$. To sum up, the equivalent two-hop channel vector becomes $\tilde{\mathbf{h}}_{\mathsf{srd}} = \mathbf{\Phi}_{\mathsf{sd}} \mathbf{Z}\mathbf{g}$.

Secondly, we modify the relay noise related term $\mathbf{\Phi}_{\mathsf{d}}^H \mathbf{F} \mathbf{\Phi}_{\mathsf{r}} \mathbf{G} \mathbf{n}_{\mathsf{r}}$ such that it can be read in terms of $\mathbf{g}$. To this mean, we define an auxiliary relay noise matrix $\mathbf{N}_{\mathsf{r}} \in \mathbb{C}^{M_{\mathsf{r}} \times N_{\mathsf{c}}}$ such that $\mathbf{N}_{\mathsf{r}}\mathbf{g} = \mathbf{G}\mathbf{n}_{\mathsf{r}}$ holds. The linear construction of $\mathbf{N}_{\mathsf{r}}$ is explained in Appendix A.3 in detail. Moreover, we define two more auxiliary matrices $\tilde{\mathbf{\Lambda}} \triangleq \mathbf{\Phi}_{\mathsf{d}}^H \mathbf{F} \mathbf{\Phi}_{\mathsf{r}}$ and $\mathbf{\Lambda} \triangleq \tilde{\mathbf{\Lambda}} \mathbf{N}_{\mathsf{r}}$ for notational convenience in the following derivations.

Finally, assuming that the primary second hop protocol is employed, we can write the received signal $y_{\mathsf{d},i}$ at the ith destination as

$$y_{\mathsf{d},i} = \underbrace{\mathbf{\Phi}_{\mathsf{sd}}[\hat{i},\hat{i}] \cdot \mathbf{Z}[\hat{i},:] \cdot \mathbf{g} \cdot s_i}_{\text{intended signal}} + \underbrace{\sum_{j \in \mathcal{Z}_i} \mathbf{\Phi}_{\mathsf{sd}}[j,j] \cdot \mathbf{Z}[j,:] \cdot \mathbf{g} \cdot s_{\hat{j}}}_{\text{interference}} + \underbrace{\mathbf{\Lambda}[i,:]\mathbf{g} + n_{\mathsf{d},i}}_{\text{noise}}, \quad (2.11)$$

where $\hat{i} \triangleq N(i-1)+i$, $\hat{j} \triangleq \lceil j/N \rceil$, $n_{\mathsf{d},i}$ is the ith element of $\mathbf{n}_{\mathsf{d}}$. Further, $\mathcal{Z}_i$ is the index set of the rows of both $\mathbf{\Phi}_{\mathsf{sd}}$ and $\mathbf{Z}$ represents the interference terms associated with ith S-D link, and is defined as

$$\mathcal{Z}_i = \{i, N+i, 2N+i, \ldots, N^2 - N + i\} \backslash \{\hat{i}\}.$$

Hence, the received SINR at the ith destination is written through (2.11) as

$$\begin{aligned}\mathrm{SINR}_i &= \frac{P_{\mathsf{s}} \cdot |\mathbf{Z}[\hat{i},:]\mathbf{g}|^2}{P_{\mathsf{s}} \cdot \sum_{j \in \mathcal{Z}_i} |\mathbf{Z}[j,:]\mathbf{g}|^2 + \sigma_{n_{\mathsf{r}}}^2 \mathbf{g}^H \mathbf{R}_{\mathbf{\Lambda},i} \mathbf{g} + \sigma_{n_{\mathsf{d}}}^2}, \\ &= \frac{\mathbf{g}^H \Big(P_{\mathsf{s}} \cdot \mathbf{Z}[\hat{i},:]^H \mathbf{Z}[\hat{i},:] \Big) \mathbf{g}}{\mathbf{g}^H \Big(P_{\mathsf{s}} \cdot \sum_{j \in \mathcal{Z}_i} \mathbf{Z}[j,:]^H \mathbf{Z}[j,:] + \sigma_{n_{\mathsf{r}}}^2 \mathbf{R}_{\mathbf{\Lambda},i} \Big) \mathbf{g} + \sigma_{n_{\mathsf{d}}}^2},\end{aligned}$$

$$\triangleq \frac{\mathbf{g}^H \mathbf{A}_i \mathbf{g}}{\mathbf{g}^H \mathbf{B}_i \mathbf{g} + \sigma_{n_\mathsf{d}}^2}, \tag{2.12}$$

where in the last line we define the matrices

$$\mathbf{A}_i \triangleq P_\mathsf{s} \cdot \mathbf{Z}[\hat{i},:]^H \mathbf{Z}[\hat{i},:],$$
$$\mathbf{B}_i \triangleq P_\mathsf{s} \cdot \sum_{j \in \mathcal{Z}_i} \mathbf{Z}[j,:]^H \mathbf{Z}[j,:] + \sigma_{n_r}^2 \mathbf{R}_{\mathbf{\Lambda},i},$$

for notational convenience in the remainder of Part I. Moreover, $\mathbf{R}_{\mathbf{\Lambda},i}$ is a modified covariance matrix for $\mathbf{\Lambda}[i,:]$ and its form is given in Appendix A.3 in details.

Correspondingly, assuming that i.i.d. Gaussian symbols are transmitted, the information rate between the ith S-D pair is

$$\mathrm{R}_i = \frac{1}{2} \log_2(1 + \mathrm{SINR}_i), \tag{2.13}$$

where the pre-log factor 1/2 follows because of the half-duplex mode of the AF relays, and two channel uses are needed to transmit the information from the source to the destination.

Optional Second-Hop Protocol In order to extend the received signal model (2.11) to the case of optional second-hop protocol, we also need to vectorize the contribution from the sources during the second hop. That is,

$$\mathrm{vec}(\varrho \mathbf{\Phi}_\mathsf{d}^H \mathbf{D} \mathbf{\Theta}_\mathsf{s} \mathbf{\Phi}_\mathsf{s}) = \varrho(\mathbf{\Phi}_\mathsf{s}^T \otimes \mathbf{\Phi}_\mathsf{d}^H)\mathbf{d} = \varrho \mathbf{\Phi}_\mathsf{sd} \mathbf{d}, \tag{2.14}$$

where $\mathbf{d} \in \mathbb{C}^{N^2} \triangleq \mathrm{vec}(\mathbf{D})$. Note that we neglect the phase noise uncertainty induced by the time difference between two transmissions of the sources as for the relays, i.e., $\mathbf{\Theta}_\mathsf{s} = \mathbf{I}_N$. Incorporating (2.14) with (2.11), we obtain

$$y_{\mathsf{d},i}^{\text{wi-dl}} = \mathbf{\Phi}_\mathsf{sd}[\hat{i},\hat{i}] \cdot \Big(\mathbf{Z}[\hat{i},:] \cdot \mathbf{g} + \underbrace{\varrho \cdot d_{\hat{i}}}_{\substack{\text{direct-link signal}\\ \text{contribution}}}\Big) \cdot s_i + \sum_{j \in \mathcal{Z}_i} \mathbf{\Phi}_\mathsf{sd}[j,j] \cdot \Big(\mathbf{Z}[j,:] \cdot \mathbf{g} + \underbrace{\varrho \cdot d_j}_{\substack{\text{direct-link interference}\\ \text{contribution}}}\Big) \cdot s_{\hat{j}}$$
$$+ \mathbf{\Lambda}[i,:]\mathbf{g} + n_{\mathsf{d},i}, \tag{2.15}$$

where d_j is the jth element of $\mathbf{d}$ and the superscript "wi-dl" stands for "with direct-link". With (2.15), the $\mathrm{SINR}_i^{\text{wi-dl}}$ is given by

$$\mathrm{SINR}_i^{\text{wi-dl}} = \frac{P_\mathsf{s} \cdot |\mathbf{Z}[\hat{i},:]\mathbf{g} + \varrho d_{\hat{i}}|^2}{P_\mathsf{s} \cdot \sum_{j \in \mathcal{Z}_i} |\mathbf{Z}[j,:]\mathbf{g} + \varrho d_j|^2 + \sigma_{n_r}^2 \mathbf{g}^H \mathbf{R}_{\mathbf{\Lambda},i} \mathbf{g} + \sigma_{n_\mathsf{d}}^2}. \tag{2.16}$$

After explicitly defining the SINR expressions in terms of the distributed beamforming vector $\mathbf{g}$, the next question to answer is how to design $\mathbf{g}$ for efficient and reliable multiuser communications. There are various beamforming approaches for the broadcast channel and MIMO communications in the literature which aim at different figures of merit and correspondingly requiring different design complexities. Some (but not all) of the approaches that can be adapted to our distributed scenario are listed as follows: Beamforming matrix/vector optimization with or without a zero-forcing constraint for maximizing the minimum SINR under given power constraints; providing max-min or proportional fairness in between SINR/rate per link; minimizing (some) power constraint while providing certain QoS for each S-D link in terms of SINR/rate, maximizing the sum rate; minimizing the minimum MSE (MMSE) between the terminals of each pair.

In this chapter, we particularly address the adaptation of zero-forcing beamforming from a sufficiency perspective without further optimization. Subsequently, in Chapter 3, we touch on most of the above mentioned beamforming methods, and investigate them for the favor of several different figures of merit, which are in general based on SINR/rate expressions. However, specifically for MMSE based distributed multiuser beamforming methods, we refer the interested reader to [21].

We would like to emphasize here, that as stated in the introduction, our main concern in coherent multiuser relaying, independent from the employed beamforming method, is to provide *full* distributed spatial multiplexing gain N which is determined by the rank of the *effective equivalent two-hop channel matrix* $\tilde{\mathbf{H}}_{\mathsf{srd}}$:

$$\operatorname{rank}(\tilde{\mathbf{H}}_{\mathsf{srd}}) \leq \min\left\{\operatorname{rank}(\mathbf{\Phi}_{\mathsf{d}}^{H}), \operatorname{rank}(\mathbf{F}), \operatorname{rank}(\mathbf{G}), \operatorname{rank}(\mathbf{H}), \operatorname{rank}(\mathbf{\Phi}_{\mathsf{s}})\right\}. \tag{2.17}$$

Since all matrices except $\mathbf{G}$ in (2.17) satisfy the rank constraint, the distributed beamforming matrix $\mathbf{G}$ should have at least a rank of N.

However, besides choosing the beamforming method to apply and achieving spatial multiplexing gain, there are some other crucial issues regarding the practicality of the distributed system proposal. These are the required *CSI knowledge per relay* and *number of relays/relay antennas* for efficient multiplexing and beamforming gains. Henceforth, an important portion of this thesis is devoted to the problem of designing practical and efficient multiuser relaying protocols. Here and in the sequel, we mainly distinguish between *global* and *local* CSI knowledge, which are stated as follows:

- **Global CSI:** When a relay is assumed to have global CSI, then it has perfect knowledge of the compound sources-to-relays and relays-to-destinations channel coefficients, i.e. $\mathbf{H}$ and $\mathbf{F}$.

- **Local CSI:** When the k relay r_k is assumed to have local CSI, then it has perfect knowledge of the channel coefficients from all sources to itself, and the channel coefficients from itself to all destinations, i.e., $\mathbf{H}_k$ and $\mathbf{F}_k$, respectively.

In the following, we present two fundamental gain allocation schemes, which are commonly cited in the literature. Both are based on low-complexity zero-forcing beamforming, but they differ in their necessary conditions for zero-forcing:

1) ZF beamforming with global CSI knowledge [102, 133],

2) Asymptotic ZF beamforming with only local CSI knowledge [23, 133].

Subsequently, we investigate further implications of coherent relaying and propose *out-of-cluster interference* mitigation.

2.4.1. Distributed Zero-Forcing Beamforming

We want to zero-force all $N(N-1)$ inter-user interference terms, i.e.,

$$\mathbf{Z}[j,:] \cdot \mathbf{g} = 0, \quad \text{for all } \; j \in \mathcal{Z}_i, i = 1, \ldots, N, \tag{2.18}$$

such that the equivalent two-hop channel matrix $\mathbf{H}_{\mathsf{srd}}$ is diagonal. Hence, defining a *compound interference* matrix $\mathbf{Z}_{\mathcal{I}} \in \mathbb{C}^{N(N-1) \times N_{\mathrm{c}}}$ whose rows are composed by the $\mathbf{Z}[j,:]$, $\forall j \in \mathcal{Z}_i, i = 1, \ldots, N$, the *ZF condition* is written as

$$\mathbf{Z}_{\mathcal{I}} \cdot \mathbf{g} = \mathbf{0}. \tag{2.19}$$

Note that $\mathbf{\Phi}_{\mathsf{sd}}$ can also be introduced into the ZF condition; namely, $\mathbf{\Phi}_{\mathsf{sd},\mathcal{I}} \cdot \mathbf{Z}_{\mathcal{I}} \cdot \mathbf{g} = \mathbf{0}$, where the $N(N-1)$ dimensional diagonal matrix $\mathbf{\Phi}_{\mathsf{sd},\mathcal{I}}$ is reduced from $\mathbf{\Phi}_{\mathsf{sd}}$ by deleting the unchosen rows according to $\mathbf{Z}_{\mathcal{I}}$. However, since any vector $\mathbf{g}$ that fulfills the ZF condition (2.19), also fulfills the condition $\mathbf{\Phi}_{\mathsf{sd},\mathcal{I}} \cdot \mathbf{Z}_{\mathcal{I}} \cdot \mathbf{g} = \mathbf{0}$ as long as $\mathbf{\Phi}_{\mathsf{sd}}$ is non-singular, in the sequel we disregard $\mathbf{\Phi}_{\mathsf{sd},\mathcal{I}}$ from the zero-forcing gain vector calculation.

A simple choice for choosing $\mathbf{g}$ to satisfy the *ZF condition* is to search it within the nullspace of $\mathbf{Z}_{\mathcal{I}}$. Hence, we define the nullspace of $\mathbf{Z}_{\mathcal{I}}$ as

$$\mathbf{V} \triangleq \mathrm{null}(\mathbf{Z}_{\mathcal{I}}),$$

which can be computed through the singular value decomposition (SVD) of $\mathbf{Z}_{\mathcal{I}}$; namely, the null space is spanned by the right singular vectors corresponding to vanishing singular values of $\mathbf{Z}_{\mathcal{I}}$. The nullspace $\mathbf{V}$ is of dimension $N_c - \text{rank}(\mathbf{Z}_{\mathcal{I}})$, where $\text{rank}(\mathbf{Z}_{\mathcal{I}}) \leq \min\{N(N-1), N_c\}$ by definition. Hence, in order not to have a non-empty nullspace, i.e., $N_c - \text{rank}(\mathbf{Z}_{\mathcal{I}}) > 0$, we need

$$N_c > \min\{N(N-1), N_c\} \;\rightarrow\; N_c > N(N-1). \tag{2.20}$$

In other words, we need at least $N(N-1)+1$ non-zero entries in $\mathbf{G}$, or equivalently $\mathbf{g}$ should be at least length of $N(N-1)+1$ for cancelling multi-user interference at each destination terminal. As N_c plays a crucial role on the efficiency of beamforming, and will be more effective on further optimizations (see Chapter 3), in the sequel of Part I we refer it as the *number of degrees of freedom*.

If $N_c = N(N-1)+1$, then $\mathbf{V}$ boils down to be a vector of size N_c, and we use it as the unique relay gain vector $\mathbf{g}_{\text{zf}}$ for zero-forcing up to a constant factor, which is determined according to the sum relay transmit power constraint (2.3). We refer this case as *minimum relay configuration for ZF* in the following.

However, if $N_c > N(N-1)+1$, then any vector $\mathbf{g}_{\text{v}}$ of size $N_c - N(N-1)$ lying in the nullspace $\mathbf{V} \in \mathbb{C}^{N_c \times N_c - N(N-1)}$ is a candidate for distributed ZF beamforming vector, i.e.,

$$\mathbf{g}_{\text{zf}} = \mathbf{V}\mathbf{g}_{\text{v}}. \tag{2.21}$$

The search of $\mathbf{g}_{\text{v}}$ within $\mathbf{V}$ can be based on different figures of merit, and the corresponding optimization formulations will be given in Chapter 3.

Substituting (2.21) in the SINR expression of (2.12), the SINR boils down to be the SNR

$$\begin{aligned} \text{SNR}_i &= \frac{\mathbf{g}_{\text{v}}^H \mathbf{V}^H \mathbf{A}_i \mathbf{V} \mathbf{g}_{\text{v}}}{\sigma_{n_r}^2 \mathbf{g}_{\text{v}}^H \mathbf{V}^H \mathbf{R}_{\Lambda,i} \mathbf{V} \mathbf{g}_{\text{v}} + \sigma_{n_d}^2}, \\ &\triangleq \frac{\mathbf{g}_{\text{v}}^H \tilde{\mathbf{A}}_i \mathbf{g}_{\text{v}}}{\mathbf{g}_{\text{v}}^H \tilde{\mathbf{B}}_i \mathbf{g}_{\text{v}} + \sigma_{n_d}^2}. \end{aligned} \tag{2.22}$$

where in the second line we embedded $\mathbf{V}$ within $\mathbf{A}_i$ and $\mathbf{B}_i$, and defined $\tilde{\mathbf{A}}_i \triangleq \mathbf{V}^H \mathbf{A}_i \mathbf{V}$ and $\tilde{\mathbf{B}}_i \triangleq \sigma_{n_r}^2 \mathbf{V}^H \mathbf{R}_{\Lambda,i} \mathbf{V}$. Note that in order to compute $\mathbf{g}_{\text{zf}}$, each relay needs to know $\mathbf{V}$, which requires the aforementioned global CSI knowledge per relay. Alternatively, there can be a master node which has global CSI knowledge, computes $\mathbf{g}_{\text{zf}}$, and distributes the corresponding gain factors to the relays.

Although zero-forcing beamforming through nullspace projection is an efficient means to achieve distributed multiplexing gain, it comes with an unavoidable dissemination overhead of local CSI within the network. Depending on the size of the network and the dissemination medium, this can as well result in some loss in spectral efficiency. Moreover, the scheme needs a certain amount of N_c (degrees of freedom), which improves quadratically with N and hence, can be prohibitive on the system for large N.

In general, zero-forcing based methods make all links independent from each other and provides perfect separation up to the associated power constraint. For instance, looking at (2.22), it may be thought that each user can adjust its transmit power independently, so as its rate, without causing a penalty on the others, e.g., P_s is present only in the numerator within $\tilde{\mathbf{A}}_i$. However, recall that any change in one of the source's transmit power will induce a corresponding change in relay sum transmit power. As we assume the sum power to be fixed, this triggers a scaling in gain vectors, which does not cancel out in the SNR expression due to the presence of destination noise. For the special case of low noise levels at the destinations, such a scaling would not change the anticipated SNR. In summary, ZF based relay gain allocation enables perfect separation of S-D links as $\sigma_{n_\mathrm{d}} \rightarrow 0$.

Optional Second-Hop Protocol In the presence of direct links between S-D pairs through the second hop, the ZF condition (2.19) modifies to

$$\mathbf{Z}_\mathcal{I} \cdot \mathbf{g} + \mathbf{d}_\mathcal{I} = \mathbf{0}, \tag{2.23}$$

where the vector $\mathbf{d}_\mathcal{I} \in \mathbb{C}^{N(N-1)}$ is the reduced version of $\mathbf{d}$ defined in (2.15), and collects all corresponding interference terms at the destinations. Hence, the zero-forcing gain vector is computed immediately as

$$\mathbf{g}_{\mathrm{zf-dl}} = -\mathbf{Z}_\mathcal{I}^{\dagger} \cdot \mathbf{d}_\mathcal{I}, \tag{2.24}$$

where the subscript "zf − dl"stands for "zero-forcing with direct link". The minimum number of relays required for ZF here is $N(N-1)$, which is one relay less than that of the case without direct links. The reason for this follows simply from the rank of $\mathbf{Z}_\mathcal{I}$ and that there is no trivial solution of all-zero gain vector here. Note that the ZF condition (2.23) requires further that on top of global CSI knowledge of all relay-to-terminal links, the relays have the CSI of direct links as well.

We have mentioned in Section 2.3.2.2 that the total network power consumption in the second hop is upper bounded by NP_s regardless of the presence of direct-links or not.

Hence, in order to fulfill this requirement, we need to determine an appropriate scaling parameter ϱ for the transmit power of sources through the second hop. In addition to the related bound on ϱ in (2.10), the chosen $\mathbf{g}_{\mathsf{zf-dl}}$, in fact, defines a certain and fixed ϱ for the $N_{\mathsf{r}} = N(N-1)$ case. This is stated with the following theorem.

Theorem 2.4.1: Given that $N_{\mathsf{r}} = N(N-1)$, and the total network power consumption in the second hop is fixed to NP_{s}, the scaling factor ϱ is given by

$$\varrho = \sqrt{\frac{N}{N+\alpha}}$$

where

$$\alpha = \left(\mathbf{Z}_{\mathcal{I}}^{\dagger} \cdot \mathbf{d}_{\mathcal{I}}\right)^{H} \cdot \left(\sigma_{n_{\mathsf{r}}}^{2} \mathbf{I}_{N_{\mathsf{r}}} + P_{\mathsf{s}}(\mathbf{H}\mathbf{H}^{H}) \odot \mathbf{I}_{N_{\mathsf{r}}}\right) \cdot \left(\mathbf{Z}_{\mathcal{I}}^{\dagger} \cdot \mathbf{d}_{\mathcal{I}}\right),$$

and $\mathbf{d}_{\mathcal{I}}$ is computed for a unit transmit power for each source.

Proof: If each source transmits with a power of $\varrho^2 P_{\mathsf{s}}$ in the second hop, the remaining total power to the relays is $(1-\varrho^2)NP_{\mathsf{s}}$. Hence, the proof follows immediately from the definition of the instantaneous transmission power at the relays:

$$\begin{aligned}(1-\varrho^2)NP_{\mathsf{s}} &= \mathbf{g}_{\mathsf{zf-dl}}^{H}\left(\sigma_{n_{\mathsf{r}}}^{2} \mathbf{I}_{N_{\mathsf{r}}} + P_{\mathsf{s}}(\mathbf{H}\mathbf{H}^{H}) \odot \mathbf{I}_{N_{\mathsf{r}}}\right)\mathbf{g}_{\mathsf{zf-dl}} \\ &= \varrho^2 P_{\mathsf{s}} \cdot \left(\mathbf{Z}_{\mathcal{I}}^{\dagger} \cdot \mathbf{d}_{\mathcal{I}}\right)^{H} \cdot \left(\sigma_{n_{\mathsf{r}}}^{2} \mathbf{I}_{N_{\mathsf{r}}} + P_{\mathsf{s}}(\mathbf{H}\mathbf{H}^{H}) \odot \mathbf{I}_{N_{\mathsf{r}}}\right) \cdot \left(\mathbf{Z}_{\mathcal{I}}^{\dagger} \cdot \mathbf{d}_{\mathcal{I}}\right) \\ &= \alpha\varrho^2 P_{\mathsf{s}}, \end{aligned} \tag{2.25}$$

where in the second line we inserted the definition of $\mathbf{g}_{\mathsf{zf-dl}}$. After some trivial algebraic manipulations on (2.25), we find that $\varrho = \sqrt{\frac{N}{N+\alpha}}$ and conclude the proof. □

If $N_{\mathsf{r}} > N(N-1)$, then there are more than one possible gain vector allocations, so as the possible scaling ϱ. However, for the brevity of this chapter, we drop further details here and refer to Section 5.3.2, from which the related mathematical derivations can be adapted here as a special case.

2.4.1.1. Partial Relay Cooperation

In the previous discussions we assume that the relays are non-cooperating, i.e., they do not share (exchange) their received signals. However, assuming a reliable communication backbone throughout the network, which can be either wireless, e.g., Bluetooth, ultra-wideband communication, or wired, e.g., power-line communication, the relays can exchange some (or all) of their received signals. Any share of received signal (per antenna) adds one non-zero entry to the gain matrix $\mathbf{G}$, and hence, increases the number of degrees of freedom by one. The signal-shares do not have to be symmetric; in other words $\mathbf{G}$ does not need to be symmetric in terms of the non-zero entries, e.g., one relay can broadcast its received signal to the network. Please see Appendix A.4 for some simple relay cooperation examples.

An extreme case of a such scenario is the so-called linear distributed antenna system (LDAS), where all relays are connected with each other. Hence, the relay gain matrix $\mathbf{G}$ is a full $M_\mathsf{r} \times M_\mathsf{r}$ matrix. This is referred to either *full relay cooperation* or a single MIMO relay node with M_r antennas. Nevertheless, the zero-forcing gain design is independent from the amount of relay cooperation, and directly follows from the definition of $\mathbf{g}$. However, a special attention should now be paid to the phase synchronization at the relays. In case the relays have *direct* access to the *exact* received signals (down-converted) of each other, a phase synchronization procedure is needed since the offsets do no cancel as in (2.7). On the other hand, if up- and down-conversion processes are made once again during the signal-shares, then the phase offsets continue to cancel each other. The former case can be typical for a wired scenario, whereas the latter can be observed in a wireless link in between. Phase synchronization for distributed systems is out of the scope of this thesis, we refer the interested reader to [16, 19].

2.4.2. Asymptotic Zero-Forcing Beamforming

It has been recently shown that ZF beamforming can also be achieved with only local CSI knowledge per relay in the asymptotic limit of N_r. We present here only the case of single-antenna equipped relays [133], i.e., $M_{\mathsf{r},\mathsf{k}} = 1\ \forall k$, and refer the interested reader to [23, 85] for an extension to arbitrary $M_{\mathsf{r},\mathsf{k}}$.

As $M_\mathsf{r} = N_\mathsf{r}$, $\mathbf{G}$ becomes an $N_\mathsf{r} \times N_\mathsf{r}$ diagonal matrix. Likewise, the uplink and downlink channels of the kth relay boil down to be a vector of size N, i.e., $\mathbf{h}_{\mathsf{s}k} \in \mathbb{C}^{N\times 1}$ and $\mathbf{f}_{k\mathsf{d}} \in \mathbb{C}^{N\times 1}$ as defined in Section 2.2. Moreover, in contrary with the previous coherent multiuser ZF relaying scheme, it is assumed that each relay employs a per-node transmit power constraint, and transmit with a full power of $P_{\mathsf{r},k}$, where $\sum_k P_{\mathsf{r},k} = P_\mathsf{r}$.

In [133], it has been proposed that each relay chooses its gain factor $g_k \triangleq \mathbf{G}[k,k]$ based only on $\mathbf{h}_{\mathsf{s}k}$ and $\mathbf{f}_{k\mathsf{d}}$, and appropriately matches its gain to the uplink and downlink channels. That is to say,

$$g_{\mathsf{azf},k} = \xi \cdot \mathbf{h}_{\mathsf{s}k}^H \cdot \mathbf{f}_{k\mathsf{d}}^*, \tag{2.26}$$

where $\xi \in \mathbb{R}_+$ is a scaling to meet the transmit power constraint. Such a choice results in maximum ratio combining (MRC) of intended signal terms at the destinations, but does not suppress the inter-user interference. However, it has been shown in [85, 133] that, the choice of (2.26) orthogonalizes the N S-D links for $N_\mathsf{r} \to \infty$ in the sense that the network capacity scales as $\frac{N}{2}\log(N_\mathsf{r}) + \mathcal{O}(1)$.

2.4.3. Out-of-cluster Interference Mitigation

So far, we have assumed that our network is composed of a single cluster of N S-D pairs, and the main concern of coherent multiuser relaying was to cancel the multiuser interference within this cluster. Consider now that there are out-of-cluster signals whose intended destinations are out of the considered cluster and hence, they act as interference to all of the N destination terminals within the primary cluster. Having sufficient number of degrees of freedom and sufficient CSI knowledge, the proposed interference management scheme can also eliminate such out-of-cluster interference.

The interference can enter the system either in the first hop at the relays, in the second hop at the destinations or in both hops. Since we assume to employ no receive beamforming at the destinations, the relays can only manage the cases of either interference only at the first hop or the same kind of interference at the both hops. For the sake of generality we consider the latter case. A typical scenario can be two neighbouring clusters each simultaneously employing coherent multiuser AF relaying. One of the two is assumed to be the primary cluster and is subject to interference from the other secondary cluster. Hence, while the sources of the secondary cluster are transmitting, their signals appear as interference at the relays of the primary cluster. Likewise, in the second hop, as the relays of the secondary cluster are forwarding signals, these act as interference at the destinations of the primary cluster.

Without loss of generality, we assume for simplicity that there are N_i out-of-cluster interference sources each with a single-antenna at both hops, and that each interferer transmits with a power of P_i at each hop. In the presence of interference, the received signal vector at the relays of the primary cluster becomes

$$\mathbf{y}_\mathsf{r} = \mathbf{\Phi}_\mathsf{r}^H \mathbf{H} \mathbf{\Phi}_\mathsf{s} \mathbf{s} + \underbrace{\mathbf{\Phi}_\mathsf{r}^H \mathbf{Q}_\mathsf{ir} \mathbf{\Phi}_\mathsf{i}^{(1)} \mathbf{u}}_{\substack{\text{out-of-cluster}\\ \text{interference}}} + \mathbf{n}_\mathsf{r}, \tag{2.27}$$

where $\mathbf{Q}_\mathsf{ir} \in \mathbb{C}^{M_\mathsf{r} \times N_\mathsf{i}}$ is the fading channel between the interferer and the relays, $\mathbf{\Phi}_\mathsf{i}^{(1)}$

is an N_i dimensional diagonal phase-offset matrix for the interferers in the first hop, $\mathbf{u} \in \mathbb{C}^{N_\mathsf{i}}$ is the interference signal observed in the first hop, and $\mathbb{E}\{\mathbf{u}_\mathsf{r}\mathbf{u}_\mathsf{r}^H\} = P_\mathsf{i} \cdot \mathbf{I}_{N_\mathsf{i}}$. Upon receiving $\mathbf{y}_\mathsf{r}$, the relays multiply their received signals with the corresponding element of the gain matrix $\mathbf{G}$, and simultaneously transmit to the destinations in the second hop.

Next, we write the received signal observed at the destinations of the primary cluster in the presence of interference (typically from the relays of the secondary cluster) as

$$\mathbf{y}_\mathsf{d} = \mathbf{\Phi}_\mathsf{d}^H \mathbf{H}_\mathsf{srd} \mathbf{\Phi}_\mathsf{s} \mathbf{s} + \underbrace{\mathbf{\Phi}_\mathsf{d}^H \mathbf{F} \mathbf{G} \mathbf{Q}_\mathsf{ir} \mathbf{\Phi}_\mathsf{i}^{(1)} \mathbf{u}}_{\substack{1^\text{st} \text{ hop out-of-cluster} \\ \text{interference}}} + \underbrace{\mathbf{\Phi}_\mathsf{d}^H \mathbf{Q}_\mathsf{id} \mathbf{\Phi}_\mathsf{i}^{(2)} \mathbf{u}}_{\substack{2^\text{nd} \text{ hop out-of-cluster} \\ \text{interference}}} + \mathbf{n}_\mathsf{e}, \tag{2.28}$$

where $\mathbf{Q}_\mathsf{id} \in \mathbb{C}^{N \times N_\mathsf{i}}$ is the fading channel between the interferers and the destinations, and $\mathbf{\Phi}_\mathsf{i}^{(2)}$ is an N_i dimensional diagonal phase-offset matrix for the interferers in the second hop. Although we have assumed that the same interference sources are transmitting at both hops, the corresponding LO phase offsets of these sources are different due to random phase terms affected by different transmission instants, i.e., $\mathbf{\Phi}_\mathsf{i}^{(1)} \neq \mathbf{\Phi}_\mathsf{i}^{(2)}$.

Since now the received signals at the destinations are subject to two types of interference, i.e., multiuser and out-of-cluster, the designed gain vector allocation should cancel both within the primary cluster. To this end, in the following, we write the ZF condition for the additional interference terms, and then combine it with the ZF condition for multiuser interference (2.19).

At first we vectorize the equivalent two-hop channel of the interference $\mathbf{u}$ entered at the relays during the first hop, i.e., $\mathbf{q}_\mathsf{ir} \triangleq \mathrm{vec}\left(\mathbf{\Phi}_\mathsf{d}^H \mathbf{F} \mathbf{G} \mathbf{Q}_\mathsf{ir} \mathbf{\Phi}_\mathsf{i}^{(1)}\right)$. Dropping the details here, we follow the previous derivation steps to obtain $\tilde{\mathbf{h}}_\mathsf{srd}$, and find

$$\mathbf{q}_\mathsf{ir} \triangleq \mathrm{vec}\left(\mathbf{\Phi}_\mathsf{d}^H \mathbf{F} \mathbf{G} \mathbf{Q}_\mathsf{ir} \mathbf{\Phi}_\mathsf{i}^{(1)}\right) = \mathbf{\Phi}_\mathsf{id}^{(1)} \mathbf{Z}_\mathsf{i} \mathbf{g}, \tag{2.29}$$

where $\mathbf{\Phi}_\mathsf{id}^{(1)} \in \mathbb{C}^{N_\mathsf{i}N \times N_\mathsf{i}N} \triangleq (\mathbf{\Phi}_\mathsf{i}^{(1)})^T \otimes \mathbf{\Phi}_\mathsf{d}^H$ and $\mathbf{Z}_\mathsf{i} \in \mathbb{C}^{N_\mathsf{i}N \times N_\mathsf{c}}$ represents the reduced version of the Kronecker matrix product $\mathbf{Q}_\mathsf{ir}^T \otimes \mathbf{F}$, such that the columns associated with the dropped non-zero elements of $\tilde{\mathbf{g}}$ are deleted.

Secondly, we apply the same vectorization process to the channel seen by the interference $\mathbf{u}$ at the destinations, i.e., $\mathbf{q}_\mathsf{id} \triangleq \mathrm{vec}\left(\mathbf{\Phi}_\mathsf{d}^H \mathbf{Q}_\mathsf{id} \mathbf{\Phi}_\mathsf{i}^{(2)}\right) = \mathbf{\Phi}_\mathsf{id}^{(2)} \cdot \mathrm{vec}(\mathbf{Q}_\mathsf{id})$, where $\mathbf{\Phi}_\mathsf{id}^{(2)} \in \mathbb{C}^{N_\mathsf{i}N \times N_\mathsf{i}N} \triangleq (\mathbf{\Phi}_\mathsf{i}^{(2)})^T \otimes \mathbf{\Phi}_\mathsf{d}^H$. Hence, the ZF condition for the out-of-cluster interference is written as

$$\mathbf{q}_\mathsf{ir} + \mathbf{q}_\mathsf{id} = \mathbf{0} \;\rightarrow\; \mathbf{\Phi}_\mathsf{id}^{(1)} \mathbf{Z}_\mathsf{i} \mathbf{g} + \mathbf{\Phi}_\mathsf{id}^{(2)} \mathrm{vec}(\mathbf{Q}_\mathsf{id}) = \mathbf{0}. \tag{2.30}$$

When combined with the ZF condition for multiuser interference, we obtain the overall interference cancellation condition:

$$\begin{bmatrix} \mathbf{\Phi}_{\mathsf{sd},\mathcal{I}} \mathbf{Z}_{\mathcal{I}} \\ \mathbf{\Phi}_{\mathsf{id}}^{(1)} \mathbf{Z}_{\mathsf{i}} \end{bmatrix} \mathbf{g} + \begin{bmatrix} 0 \\ \mathbf{\Phi}_{\mathsf{id}}^{(2)} \mathrm{vec}(\mathbf{Q}_{\mathsf{id}}) \end{bmatrix} = \mathbf{0}. \tag{2.31}$$

Specifically, the solution of (2.31) is

$$\mathbf{g}_{\mathsf{oci}} = - \begin{bmatrix} \mathbf{\Phi}_{\mathsf{sd},\mathcal{I}} \mathbf{Z}_{\mathcal{I}} \\ \mathbf{\Phi}_{\mathsf{id}}^{(1)} \mathbf{Z}_{\mathsf{i}} \end{bmatrix}^{\dagger} \cdot \begin{bmatrix} 0 \\ \mathbf{\Phi}_{\mathsf{id}}^{(2)} \mathrm{vec}(\mathbf{Q}_{\mathsf{id}}) \end{bmatrix}, \tag{2.32}$$

where the subscript "oci"of $\mathbf{g}$ stands for "out-of-cluster interference".

The next question to answer is the minimum number of degrees of freedom N_{c} to fulfill (2.31). Its answer is immediately noticed from the rank of the pseudoinversed matrix in (2.32), which is $N(N_{\mathsf{i}}+N-1)$. As a final note, we recall that such a choice of *distributed beamforming plus interference cancellation* vector requires each relay in the primary cluster to have global CSI for both clusters.

Relay Transmit Power Employing $\mathbf{g}_{\mathsf{oci}}$ at the AF relays, we can both suppress out-of-cluster interference and provide distributed spatial multiplexing to the terminal pairs in the primary cluster. We now shift our focus to the sum transmit power consumption at the relays. We have assumed that each source terminal in the primary cluster and each interference source transmit with a fixed full power of P_{s} and P_{i}, respectively. Hence, the relay sum transmit power defined in (2.3) modifies to

$$P_{\mathsf{r}} = \mathrm{Tr}\Big(\mathbf{G}_{\mathsf{oci}}(P_{\mathsf{s}}\mathbf{H}\mathbf{H}^{H} + P_{\mathsf{i}}\mathbf{Q}_{\mathsf{ir}}\mathbf{Q}_{\mathsf{ir}}^{H} + \sigma_{n_{\mathsf{r}}}^{2}\mathbf{I}_{M_{\mathsf{r}}})\mathbf{G}_{\mathsf{oci}}^{H}\Big), \tag{2.33}$$

where $\mathbf{G}_{\mathsf{oci}} = \mathrm{diag}\{\mathbf{g}_{\mathsf{oci}}\}$. Previously in Section 2.4.1, where the only target was multiuser interference mitigation, the relay sum transmit power could be scaled to any desired value without affecting the zero-forcing condition and changing P_{s}, i.e., $\mathbf{g}_{\mathsf{v}}$ still lies in the nullspace $\mathbf{V}$ even if it is scaled with a real scalar. However, in the current case of out-of-cluster mitigation, if we scale $\mathbf{g}_{\mathsf{oci}}$ with a real value, the ZF condition is not fulfilled anymore. Specifically, say c is the scalar, although $c \cdot \mathbf{\Phi}_{\mathsf{sd},\mathcal{I}}\mathbf{Z}_{\mathcal{I}}\mathbf{g}_{\mathsf{oci}} = \mathbf{0}$ still holds, we face the inequality $c \cdot \mathbf{\Phi}_{\mathsf{id}}^{(1)}\mathbf{Z}_{\mathsf{i}}\mathbf{g}_{\mathsf{oci}} \neq \mathbf{\Phi}_{\mathsf{id}}^{(2)}\mathrm{vec}(\mathbf{Q}_{\mathsf{id}})$. Thus, in order not to disturb the ZF condition, we should also scale $\mathrm{vec}(\mathbf{Q}_{\mathsf{id}})$ with c. This, in return, means that we need to change the transmit power of interference sources P_{i}. However, as we do not have control on the interference sources in practice, instead we need to adapt the relay sum transmit power according to the effect of interference. So to say, the penalty to pay for cancelling the out-of-cluster interference is to require higher transmit power in total at the relays. Note that the impact of the transmit power is highly dependent on the number of interference sources, and the proximity of these to the primary cluster, i.e. path loss effect.

2.5. Performance Results

Simulation Setup In this section we study the performance of distributed multiuser beamforming through Monte Carlo simulations. Each element of the channel matrices $\mathbf{H}$, $\mathbf{F}$ and $\mathbf{D}$ is i.i.d. Rayleigh fading coefficient with zero-mean and variance of $\sigma_{\mathbf{H}}^2 = 1$, $\sigma_{\mathbf{F}}^2 = 1$, $\sigma_{\mathbf{D}}^2$, respectively. These propagation channel matrices are assumed to remain the same over at least one transmission cycle (two-hops). Unless otherwise stated, $P = P_{\mathsf{s}} = P_{\mathsf{i}} = 1$ and $\sigma^2 \triangleq \sigma_{n_{\mathsf{r}}}^2 = \sigma_{n_{\mathsf{d}}}^2$. Here and in the sequel, we define the *average signal-to-noise ratio* as SNR $= P/\sigma^2$. For a fair performance comparison between different values of N, we set $P_{\mathsf{r}} = NP$. Moreover, besides perfect frequency synchronization, we assume that there is a global phase reference available for all nodes in the system (please refer to [16, 17, 19] for further details on how to achieve global phase reference). Hence, all phase offset indicating matrices are assumed to be an identity matrix, i.e., $\mathbf{\Phi}_{\mathsf{s}} = \mathbf{\Phi}_{\mathsf{r}} = \mathbf{\Phi}_{\mathsf{d}} = \mathbf{\Phi}_{\mathsf{i}}^{(1)} = \mathbf{\Phi}_{\mathsf{i}}^{(2)} = \mathbf{\Theta}_{\mathsf{s}} = \mathbf{\Theta}_{\mathsf{r}} = \mathbf{I}$. The impact of phase offsets are briefly covered in Section 4.2 from a robustness perspective, whereas a more detailed analysis can be found in [16].

Simulation Results

Distributed Spatial Multiplexing: We initially focus on the sum rate performance of distributed multiuser beamforming without direct-links in the second hop and on the corresponding spatial multiplexing capability. We show the provided distributed spatial multiplexing gain in Fig. 2.3 for $N = 2, 3, 4, 5, 6$ S-D pairs. For each choice of N, we employ as many relays as minimum number of relays required for perfect zero-forcing, i.e., $N(N-1)+1$. We read from the figure that the distributed ZF approach provides, as expected, a spatial multiplexing gain of half the order of number of pairs, e.g., $1, \frac{3}{2}, 2, \frac{5}{2}, 3$, for $N = 2, 3, 4, 5, 6$, respectively. The half-order effect is due the two channel-uses required for the two-hop communication between the source and the destination via relays. Note that changing N_{r} for each chosen N would result in nothing but parallel shifts in the curves, whereas the slopes of these curves would also

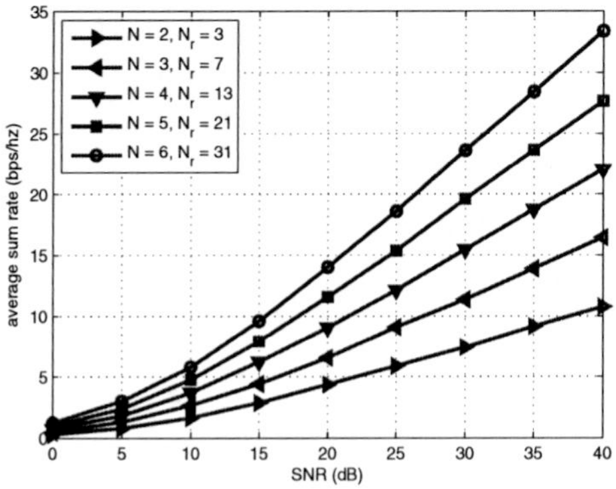

Figure 2.3.: Average sum rates vs. average SNR for $N = 2, 3, 4, 5, 6$ S-D pairs and minimum relay configuration for zero-forcing.

change when N varies further.

In Figures 2.4 and 2.5, we compare three different choices of relay gain allocation for a two-hop AF relaying scenario without direct-links:

1. We assume that each relay has only local CSI and hence, uses the asymptotic zero forcing approach introduced in Section 2.4.2. Each relay transmits with a full power of P_r/N_r. We label the corresponding curves with "AZF".

2. We assume full global CSI per each relay. Hence, the relay gain vector can be computed through (2.21) with which multiuser interference is completely cancelled for each instantaneous channel realization. If $N_r > N(N-1)+1$, then the nullspace $\mathbf{V}$ is multi-dimensional which results in more than one possible zero-forcing gain vector. We choose the relay gain vector to be a random vector out of $\mathbf{V}$, say the last column vector, and assume a sum relay transmit power constraint (see (2.3)). We label this gain allocation choice with "ZF-1".

3. Alternatively, assuming full global CSI per relay and a fixed sum relay transmit power constraint, we combine the first two gain choices. That is, we project the AZF gain vector, whose kth element is defined in (2.26), onto the nullspace $\mathbf{V}$. Such a combinatorial gain vector choice is labelled with "ZF-2".

In Fig. 2.4 we consider $N_r = 20$ relays either serving $N = 2$ or $N = 4$ S-D pairs. Thus, as the minimum relay configurations are 3 and 13 for the $N = 2$ and $N = 4$ cases, we

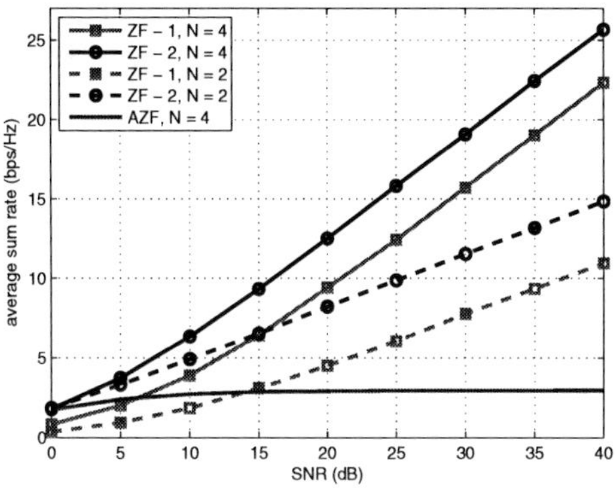

Figure 2.4.: Average sum rates vs. average SNR for $N_r = 20$ relays.

have 17 and 7 excess degrees of freedom, respectively. We immediately observe that AZF fails to provide distributed spatial multiplexing with such a finite number of relays. The multiuser interference is dominant in the high SNR regime, and prevents the sum rates to improve with SNR. In spite of the provided full multiplexing gain, unfortunately the random vector choice of ZF-1 can not exploit the available additional degrees of freedom, and performs very poorly when compared with ZF-2. The substantial sum rate advantage of the smart gain choice of ZF-2 motivates us to investigate more on efficient gain vector searches within the nullspace. Corresponding optimization issues will be addressed in the next Chapter 3.

Fig. 2.5 depicts the sum rate performance of these three gain allocation choices for increasing number of relays N_r. We fix the average SNR to 20 dB and simulate for $N = 2$ and $N = 4$ S-D pairs. Observe that the curves of ZF-1 and ZF-2 start from 3 and 7 relays, respectively, for 2 and 4 pairs, as they correspond to the minimum relay configuration for efficient zero-forcing. Unlike its counterparts, the ZF-1 approach can not enjoy the high number of relays, and behaves indifferent to the increase in N_r. On the other hand, the sum rate of ZF-2 improves monotonically with N_r. This is a result of the fact that each additional relay contributes to the coherent addition at the destination terminals, and hence provides an array gain. Recall from Section 2.4.2 that the vector projected onto the nullspace implements a MRC combining at the destinations (but without caring to cause interference). However, with nullspace projection, the multiuser interference terms are anyhow removed. The impact of interference-cancellation can also

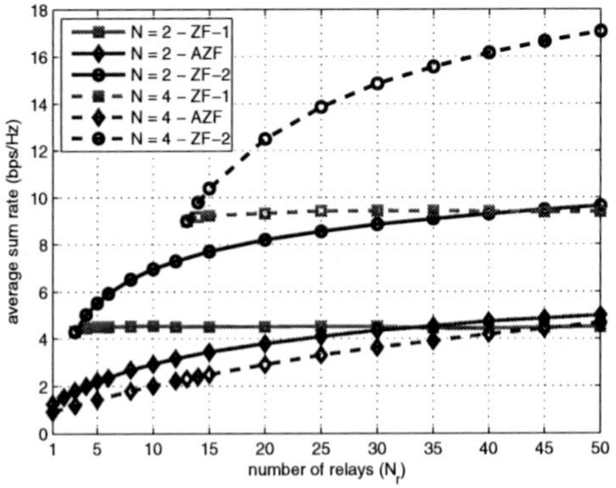

Figure 2.5.: Average sum rates vs. number of relays (N_r) for average $\mathrm{SNR} = 20$ dB.

be appreciated through the comparison of the curves of AZF and ZF-2. In principle, the same gain vector is employed in both with a difference for ZF-2 that it is projected onto the nullspace of compound interference matrix $\mathbf{Z}_{\mathcal{I}}$, i.e., it cancels all interference terms. Although the sum rate improves monotonically with N_r, the achieved rates are incomparable with that of ZF-2. Moreover, note the not-so-typical behaviour of AZF for high N values. The sum rate for $N = 4$ S-D pairs is lower than that of $N = 2$ S-D pairs, due to the augmented multiuser interference effect.

Optional Second-Hop with Transmitting Sources: Next in Fig. 2.6, we study the impact of direct source-to-destination links $\mathbf{D}$ in the second-hop transmission. The corresponding zero-forcing gain vector is computed through (2.24), and we denote the corresponding curves with "ZF-1 with direct links" As one of the fundamental motivations for relaying application is the range extension, we expect that source-to-destination link $\mathbf{D}$ is weaker with respect to both source-to-relay $\mathbf{H}$ and relay-to-destination $\mathbf{F}$ links. In order to introduce this effect, we introduce the ratio $\rho_{\mathrm{dl}} = \frac{1}{\sigma^2_{\mathbf{D}}}$ to decide on $\sigma^2_{\mathbf{D}}$ correspondingly (recall that $\sigma^2_{\mathbf{H}} = \sigma^2_{\mathbf{F}} = 1$). In Fig. 2.6, we employ $N_r = 20$ relays for $N = 4$ S-D pairs, which is much more than the minimum relay configuration of both zero-forcing with and without direct links. We plot ZF-1 with direct links for ρ_{dl} ranging from 0 to 18 dB, where, for instance, 0 dB indicates equal channel quality for the direct-links and the links via the relays. We observe that until ρ_{dl} is around 5 dB (corresponds to a $\sigma^2_{\mathbf{H}}$ of about 0.316), the direct-links help to improve the rates of ZF-1. Beyond this value, the transmit power spent to null the direct-link related interference, out of a total

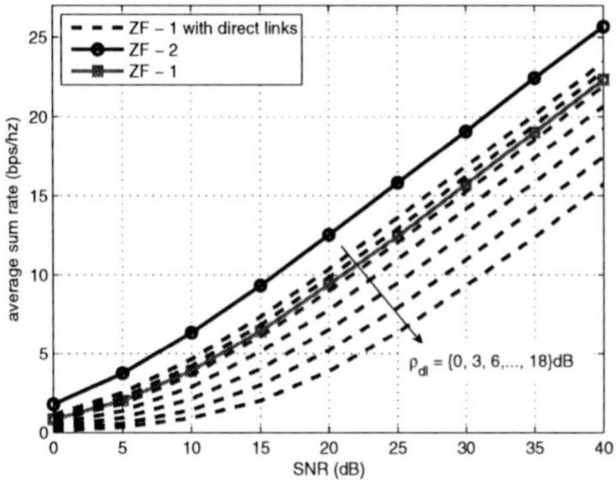

Figure 2.6.: Impact of direct source-to-destination links on average sum rates for $N = 4$ S-D pairs and $N_r = 20$ relays.

power of NP_s, reduces the attained array gain at the destinations. However, note that it may not be that fair to draw a strict conclusion out of this figure, as both ZF-1 gain allocations are choosing a random vector out of the nullspace. For example, comparison with a smarter choice of ZF-2, shows that ZF-1 with direct links performs relatively very poor for all ρ_{dl} values. Summing up, in order to fairly answer the question whether the direct links are beneficial, we should rather optimize both cases with and without direct-links for a chosen figure-of-merit and compare the resulting performances.

Out-of-cluster Interference Mitigation: We have shown that coherent multiuser relaying techniques are also capable of managing the interference originated from out of the main cluster of terminals at the focus of interest. In Fig. 2.7, we investigate the negative impact of this out-of-cluster interference depending on the number of interference sources and the corresponding strength of the interference-to-relay and interference-to-destinations links. We assume that each interference source transmit with unit power, and each element of the interference channels $\mathbf{Q}_{ir}$ and $\mathbf{Q}_{id}$ is an i.i.d. Rayleigh fading coefficient with zero-mean and variance of $\sigma_{\mathbf{Q}}^2$. Similar to the previous direct-link considerations, in order to introduce the effect of the strength of the interference channels, we introduce the ratio $\rho_{int} = \frac{1}{\sigma_{\mathbf{Q}}^2}$ to decide on $\sigma_{\mathbf{Q}}^2$ correspondingly. We consider both relay gain vector choices of ZF-1 and ZF-2 for $N = 2$ S-D pairs, and simulate the sum rate performances against increasing ρ_{int} for five sets of curves. Each set corresponds to $N_i = 1, 3, 5, 7, 10$ interferers, respectively. Note further that we take the interference

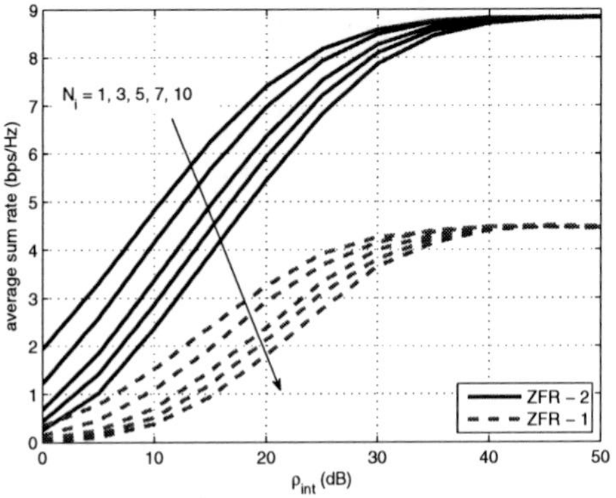

Figure 2.7.: Impact of out-of-cluster interference on sum rate of coherent multiuser AF relaying with $N = 2$ S-D pairs, $N_r = 30$ relays, and $\text{SNR} = 20$ dB.

power into account while scaling the gain vectors to fulfill the relay sum transmit power constraint.

We observe for both ZF-1 and ZF-2 that even a single interference source is prohibitive for coherent relaying until $\rho_{\text{int}} = 25-30$ dB, whereas beyond these values the sum rate is marginally affected. We can immediately conclude that the strength of the interference channels emphasize itself more than the number of interference sources does. Let us specifically focus on ZF-2. For $\rho_{\text{int}} = 10$ dB, we see that coherent relaying suffers almost a half bits/sec/Hz for each additional interferer. Likewise, from an another perspective, as ρ_{int} is decremented 1 dB in the regime of 0-to-30 dB, more than a half bits/sec/Hz penalty is paid.

We next focus on the out-of-cluster interference cancellation satisfied through the gain vector choice of (2.32). As reported in the corresponding section, the required relay sum transmit power is identified to be the primary performance measure for fixed source and interference transmit powers. In Fig. 2.8, we plot the average required relay sum transmit power for $N = 2$ S-D pairs in the presence of out-of-cluster interference. We simulate various number of interference sources N_i (1 to 10) and different ρ_{int} parameter (0 to 30 dB), which indicates the relative strength of the interference channels. As the minimum number of relays required for perfect interference cancellation is given by $N(N_i + N - 1)$, and the maximum assumed number of interferers is $N_i = 10$, we set $N_r = 30$ so that for any simulated N_i value, zero-forcing is perfectly performed, i.e.,

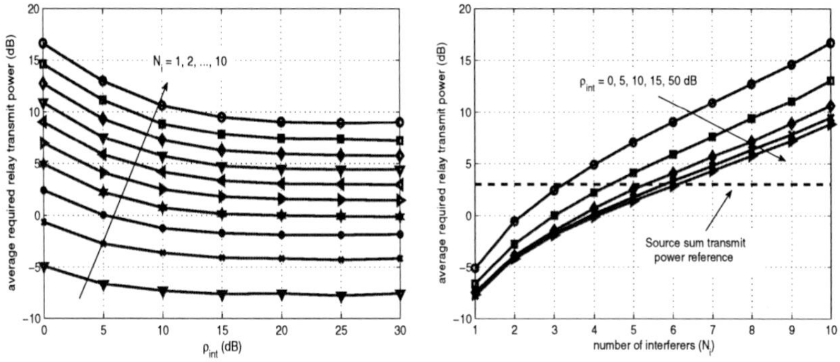

(a) (b)

Figure 2.8.: Average required relay sum transmit power in the presence of out-of-cluster interference: (a) Over varying ρ_{int}, (b) Over varying number of interferers N_{i} ($N = 2$, $N_{\text{r}} = 30$, $\text{SNR} = 20$dB).

$2(N_{\text{i}} + 1) < 30$.

It is depicted in Fig. 2.8(a) that $\rho_{\text{int}} > 10$dB has a marginal impact on the relay transmit power. However, the number of receivers plays the major role for determining the relay transmit power, where almost 20 dB power difference is observed from $N_{\text{i}} = 1$ to $N_{\text{i}} = 10$. These statements are also confirmed by Fig. 2.8(b). Therein, we also plot the sum transmit power of sources as a reference which is equal to 2 (3 dB). Recall that this power level was used to be allocated to the relays as a sum transmit power in the absence of out-of-cluster interference. A related comparison with the other curves reveals the additional power penalty paid for cancelling out-of-cluster interference on top of multiuser interference.

We conclude that out-of-cluster interference mitigation can be an efficient means to enable multiple networks to coexist within the same physical channel without causing interference to each other. However, the challenging issue is that interferer's channels and LO phase offsets should be available to the primary cluster.

2.6. Concluding Remarks

In this chapter, we aimed to introduce the fundamentals on coherent multiuser AF relaying, which will be further investigated in the following four chapters. To this end, we reviewed the corresponding system and signal models in a detailed generic setup. Considering two different relaying protocols (with and without direct source-to-destination links), we have presented the sufficiency conditions for perfect multiuser interference cancellation at the terminals, i.e., full spatial multiplexing gain. Within this context, we have elaborated on the minimum required number of relays for distributed orthogonalization and on the amount of channel knowledge required at each relay. We have reported that smart gain choices (optimizations) are compulsory for efficiently employing the excess degrees of freedom. In a last part, we have adapted the zero-forcing gain allocation to additionally mitigate with the out-of-cluster interference. Hence, the multiuser interference cancellation framework has been extended to a more general interference cancellation structure.

Chapter 3

Relay Gain Optimization for Distributed Networks with Single-Relay-Cluster

In the previous chapter we have introduced the concept of distributed orthogonalization through AF relays, and generalized the state-of-the-art relay gain allocation schemes for general interference mitigation purposes. In this chapter, we seek answers to the natural questions of how to design optimal distributed beamforming relay gains and what kind of measures to base these designs on. To address these questions, we first introduce essential figures of merit employed for beamforming in general. While designing the relay gains, we esteem the "practicality" as the main concern. Hence, we pay special attentions on the computational complexities of the corresponding optimization algorithms, the required CSI knowledge per relay, the power consumption of the network, and the necessary number necessary degrees of freedom for fulfilling fundamental distributed multiplexing gains.

3.1. Introduction

We have learned from the previous chapter that in the presence of sufficient number of relays, the multiuser interference can be efficiently mitigated and full spatial multiplexing gain can be achieved. However, in the presence of high (excess) number of relays, the simulation results in Section 2.5 highlight the fact that we need smart relay gain decisions in order to fully utilize the additional spatial degrees. Specifically for ZF beamforming, when the nullspace is multi-dimensional, although any vector lying in there cancels the interference, not all of them result in good sum or outage rate performances. So to say, besides interference-cancellation, we can as well attain additional

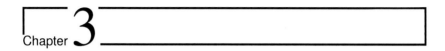

MIMO gains in a distributed manner through efficient relay gain vector optimizations. To this end, we devote the first part of this chapter to find answers for the questions of what kind of other gains can be achieved besides SM and how we can design distributed beamformers to achieve these gains.

The most essential requirement for coherent relaying is to have perfect global CSI knowledge at each relay or at a master node in the proximity of all relays. Hence, each relay needs to estimate its local CSI perfectly and to distribute it to all other relays within the network. In general, the assumption of perfect local CSI availability per relay, is commonly approved by the related references [23,37,80,85,133,134]. For the clarity of exposition, we consider a unidirectional traffic pattern from a set of source nodes to a set of destination nodes. In practice all terminal nodes in the network will transmit and receive packets, leading to a bidirectional traffic pattern. We assume that the packets transmitted by different terminal nodes comprise orthogonal training sequences. Thus, the relays can estimate the local CSI without additional overhead. However, the time and/or frequency resources allocated for the dissemination of these local CSI throughout the network can be a major performance-limiting factor in terms of spectral efficiency. Moreover, it can also be the case that the topological conditions of the network do not let the relays to communicate.

The second part of this chapter, addresses this global CSI dissemination problem. We focus on gradient based algorithms and aim to develop *distributed* schemes which do not require global CSI knowledge anymore. However, we still intend to compute exactly the same gain allocation vector that would be obtained with global CSI knowledge requiring optimizations.

Considering rather different setups than ours, similar distributed designs have been studied in several related publications with different focus. For instance, reference [46] focused on a single S-D pair scenario and devised an heuristic iterative algorithm, which does not need global CSI but benefits from multiple one-bit feedbacks to perform distributed beamforming. In [118], the feasibility of distributed SM has been proved for a multiuser setup, where the sources are partitioned into multiple clusters, and every cluster is dedicated to transmit a common message to a unique destination. Therein, 1-bit feedbacks from the destinations are employed to design the distributed multiuser beamformers. On the other hand, there are known methods that incorporate feedback to circumvent the perfect CSI requirements in some other *point-to-point* communication scenarios. For instance, in [5], a gradient adaptation is proposed to iteratively design the complex baseband weights of a transmit antenna array, where stochastic vector perturbations and feedback from the destination are employed.

Contribution of this Chapter In the first part, we assume perfect global CSI per relay and study two kinds of optimization frameworks for relay gain allocation. We either maximize the sum rate of all S-D links or maximize the minimum link rate. For the sake of generality, we address both ZF beamforming and general beamforming without ZF constraint. The corresponding performance and complexity differences are further discussed. Through sum rate maximization, the array gains per destination are adjusted such that the resultant individual link rates add up to the maximal sum rate of the multiuser network. On the other hand, maximization of the minimum link rate introduces, by definition, *max-min fairness* in between the S-D pairs, and further improves the outage rate performance per S-D link. We finally show that each additional relay on top of the minimum number of relays required for ZF adds one order of effective diversity gain. In order to achieve this, we either employ relay selection for zero-forcing or optimize the relay gains using max-min type optimizations.

In the second part, we devise a gradient based method which allows for designing the relay gain factors based on a substantially reduced CSI dissemination overhead. In particular, our method renders the amount of CSI to be disseminated over the network independent of the number of relays as opposed to a linear scaling in the conventional approach. Our approach is based on the key insight that the derivative of the SINR for a given terminal pair with respect to a specific relay gain coefficient can be determined on the basis of local CSI, if the terminal nodes provide some very limited feedback, which is now independent of the number of relays. We develop our scheme starting out with the maximization of the SINR for a single S-D pair in the presence of interfering sources. In a second step, we generalize our results to multiple S-D pairs and spatial multiplexing. It is well-known for gradient based schemes that fast convergence requires a careful choice of the step-size and the starting point. We derive two protocols for the adaptive distributed calculation of the optimal step-size. The centralized step-size selection protocol is most efficient during the initial phase of the gradient algorithm (acquisition), whereas the decentralized protocol is preferable in the tracking phase. Furthermore, we design heuristic but efficient starting gain factor computation methods, which are also based on available local information at the relays. We provide an experimental analysis of the convergence speed in terms of both outer and inner (step-size) iterations, and elaborate on the initialization of the algorithm.

Outline of this Chapter The rest of this chapter is organized as follows. Before proceeding to the contribution sections, in the following we summarize the common figure-of-merits employed for similar optimization frameworks in the literature. In Section 3.2 we focus on relay gain vector optimization with global CSI knowledge. Specifically, we maximize the sum rate in Section 3.2.1 and the minimum link rate in Section **??**. In both sections, we address the cases with and without ZF constraint. In section 3.2.4, we discuss the corresponding simulation results of the designed beamforming methods. Finally, in the last Section 3.2.3 of the first part, we investigate the achievable diversity gains through relay selection and max-min type optimization. In Section 3.3, we introduce the distributed gradient method, where we explain the main principles for a single S-D pair in Section 3.3.1 and extend these findings to a coherent multiuser relaying setup in Section 3.3.2. We discuss key issues in algorithm implementation in Section 3.3.3. The simulation results and the discussions on convergence speed and optimization parameters are presented in Section 3.3.4, which is then followed by a comparative CSI dissemination analysis in Section 3.3.5. After presenting some further application areas for distributed gain allocation in Section 3.3.6, we conclude this chapter with conclusive remarks in Section 3.4.

Figures of Merit for Distributed Multiuser Beamforming Although we are primarily aiming at providing *distributed spatial multiplexing* gain with coherent multiuser AF relaying, further gains can be achieved depending on the number of degrees of freedom available within the system, i.e. N_c. Specifically, when zero-forcing is enforced on relay gains (although not necessary), a certain amount of N_c is spent for nulling the interference; however the rest can be still employed to obtain the following listed gains:

- **Array gain:** Joint interference cancellation and coherent combination at the destinations results in a certain *array gain* (also called *power gain*) for the intended signal per terminal. However, additional degrees of freedom (refering to additional antennas in a MIMO system) may enable the relays to choose their gain factors such that the formed beams between each S-D paid are narrower and correspondingly with higher gains.

- **Outage gain:** The coherent multiuser relaying scheme is said to be in an outage when the information rate of at least one of its S-D links is below a predefined outage rate $\mathrm{R}_{\mathsf{outage}}$. Hence, for a given set of source-destination terminals, the outage probability at a rate of R_{outage} is defined as

$$P_{\text{outage}}(\mathcal{H},\mathcal{R},\mathbf{g}) = \Pr\Big(\min_{1\le i\le N} R_i(\mathcal{H},\mathcal{R},\mathbf{g}) \le R_{\text{outage}}\Big), \tag{3.1}$$

where $\mathcal{R}$ and $\mathcal{H} = \{\mathbf{H},\mathbf{F}\}$ are given sets of relay and corresponding channel coefficients, respectively; and $\mathbf{g}$ stands for the assigned relay gain vector. A smart beamforming strategy with or without ZF is to choose relay gain vector such that $R_{\text{min}} = \min_{1\le i\le N} R_i$ is maximized within the given channel and transmit power conditions. Consequently this results in either a reduction in outage probability for a fixed R_{outage}, or improvement on R_{outage} for a fixed outage probability. We refer this as *outage gain* in the sequel, and translate it as the *distributed diversity gain* (without proof) from a practical point of view.

- **Sum Rate:** Sum rate is a practical and commonly preferred performance measure for most of the multiuser multiplexing setups, as it provides an efficient means of comparison with the basic TDD or FDD systems. Sum rate of N S-D pairs is defined as

$$R_{\text{sum}} = \sum_{i=1}^{N} R_i. \tag{3.2}$$

Beamforming approaches targeting at maximum sum rate inherently provide array gain to the system, since they correspondingly improve the SINRs at the destinations (for the considered single-antenna S-D case). However, for the sake of maximum sum rate, this gain may not be uniform over different S-D links, i.e., not fair for each S-D pair.

- **Fairness and QoS:** As noted above, improving sum rate oriented performance measures may favor the strong links while punishing the weak ones especially when path loss and shadowing are taken into account. This kind of gain allocation is known to be rather unfair as there may be links that are only provided with a small information rate. To avoid this, fairness can be introduced to the system by choosing the relay gain correspondingly. There exist several approaches to provide fairness such as *max-min* or *proportional* fairness [22]. Here and in the sequel, we focus on *max-min* fairness, which aims at maximizing R_{min} with the constraint that an increment in a S-D link's rate does not lead to the reduction of some other link's rate which was already smaller than that of the same link.

Another way of satisfying the needs of each S-D link is to set QoS constraints on transmission rate or SINR. Thinking of each S-D pair as a provider-consumer pair of a voice over IP and/or IP-TV service, each link requires a certain and fixed amount of data transmission. Hence, with QoS assurance, the capacity of the network can be utilized much more efficiently.

Here and in the sequel, we mainly concentrate on maximizing two objective functions while designing the relay gain vector $\mathbf{g}$: Sum rate and minimum link rate. While the former provides a comparable system-wide perspective in terms of spectral efficiency, the latter proposes max-min fairness and outage gain. Note that both are performed within the transmit power constraints of sources and relays stated in Section 2.3.

3.2. Optimization with Global CSI Knowledge

In this section, we assume either that each relay has perfect global CSI knowledge to compute its gain factor, or that there is central node which has perfect global CSI knowledge, calculates the gain factors, and then distributes these to the corresponding relays.

3.2.1. Maximizing the Sum Rate

3.2.1.1. General Beamforming without ZF

The sum rate maximization problem for coherent multiuser relaying with N S-D pairs is formulated as

$$\mathcal{P}_{\mathsf{sum}}: \quad \max_{\mathbf{g}} \sum_{i=1}^{N} \log_2(1+\mathrm{SINR}_i) \qquad \textbf{subject to} \qquad \mathbf{g}^H\mathbf{M}\mathbf{g} \le P_{\mathsf{r}}, \tag{3.3}$$

where SINR_i is as defined in (2.12), and $\mathbf{M} \in \mathbb{C}^{N_{\mathsf{c}} \times N_{\mathsf{c}}}$ is adapted from $(P_{\mathsf{s}}\mathbf{H}\mathbf{H}^H + \sigma_{n_{\mathsf{r}}}^2\mathbf{I}_{M_{\mathsf{r}}})$ such that the power constraint is expressed in terms of $\mathbf{g}$ instead of $\mathbf{G}$, i.e.

$$\mathbf{M} \triangleq \mathbf{g}^H\left(P_{\mathsf{s}}(\mathbf{H}\mathbf{H}^H \odot \mathbf{I}_{M_{\mathsf{r}}}) + \sigma_{n_{\mathsf{r}}}^2\mathbf{I}_{M_{\mathsf{r}}}\right)\mathbf{g}.$$

Note that here and in the sequel, we have dropped the pre-log factor 1/2 since it does not affect the optimality argument of the solution. The problem $\mathcal{P}_{\mathsf{sum}}$ does not have any closed-form solution to the best of our knowledge. Moreover, the sum rate maximization problem is not necessarily convex, since the objective function relates with sum of quasi-convex functions, i.e., SINRs, which may not be (quasi-)convex. Nevertheless, with smart starting vector decisions, the sum maximization problem can be solved with a gradient based algorithm in combination with penalty or barrier method [26]. In the following, we present the adaptation to the barrier method, and drop the details for the penalty method, as it can be trivially extended from the findings below.

First, we convert the inequality constrained problem $\mathcal{P}_{\mathsf{sum}}$ to an unconstrained problem by making the corresponding power constraint implicit in the objective. To this end, we use the logarithmic barrier method [26], which conceptually inserts a barrier to the problem such that the solution stays always within the boundary of the feasible set (see [26] or Appendix A.1). The logarithmic barrier function for the power constraint is written as

$$\varphi(\mathbf{g}) = -\log_e(P_{\mathsf{r}} - \mathbf{g}^H\mathbf{M}\mathbf{g}).$$

Associating this with the objective function to be minimized

$$-\sum_{i=1}^{N} \log_2(1 + \mathrm{SINR}_i),$$

we obtain an unconstrained problem

$$\mathbf{g}^{\star}(\nu) = \underset{\mathbf{g}}{\mathbf{argmin}} \; -\nu \cdot \sum_{i=1}^{N} \log_2(1 + \mathrm{SINR}_i) - \log_e(P_{\mathsf{r}} - \mathbf{g}^H\mathbf{M}\mathbf{g}), \tag{3.4}$$

where $\nu \in \mathbb{R}_+$ is an accuracy parameter. The minimization in (3.4) can now be basically solved with any gradient based method for a given ν [26]. During this minimization, we need to calculate the gradient, i.e., search direction, at each iteration. Denoting the objective function in (3.4) with $f_{\mathsf{sum}}(\mathbf{g})$, the gradient is written as

$$\begin{aligned} \nabla_{\mathbf{g}} f_{\mathsf{sum}}(\mathbf{g}) = {} & \frac{-\nu}{\log_e(2)} \cdot \sum_{i=1}^{N} \frac{1}{1 + \mathrm{SINR}_i} \cdot \left(\frac{\mathbf{A}_i\mathbf{g}}{\mathbf{g}^H\mathbf{B}_i\mathbf{g} + \sigma_{n_{\mathsf{d}}}^2} - \frac{\mathbf{B}_i\mathbf{g}}{\mathbf{g}^H\mathbf{B}_i\mathbf{g} + \sigma_{n_{\mathsf{d}}}^2} \mathrm{SINR}_i \right) \\ & + \frac{1}{P_{\mathsf{r}} - \mathbf{g}^H\mathbf{M}\mathbf{g}} \cdot \mathbf{M}\mathbf{g}. \end{aligned} \tag{3.5}$$

We remark here that conjugate search directions can be incorporated with the minimization of $\nabla_{\mathbf{g}} f_{\mathsf{sum}}(\mathbf{g})$ in order to accelerate convergence [112].

Summing up, we compute $\mathbf{g}^{\star}(\nu)$ for an increasing sequence of ν values, until an ϵ-optimal solution of $\mathcal{P}_{\mathsf{sum}}$ is obtained, i.e., $\nu \leq 1/\epsilon$, where $\epsilon \in \mathbb{R}_+$ denotes the tolerance level. The overall sum rate maximization algorithm is summarized with **Algorithm 1**.

Algorithm 1 Sum Rate Maximization Algorithm for Coherent Multiuser Relaying

initiate $\epsilon > 0, \iota > 1$, $\nu \Leftarrow \nu_{(0)}$, feasible starting vector $\mathbf{g}$.

repeat

Step 1. Minimize (3.4), obtain $\mathbf{g}^{\star}(\nu)$.

Step 2. Update. $\mathbf{g} \Leftarrow \mathbf{g}^{\star}(\nu)$.

Step 3. Check and **quit** if $1/\nu < \epsilon$.

Step 4. Increase. $\nu \Leftarrow \iota \cdot \nu$.

3.2.1.2. ZF Beamforming

When the ZF constraint is enforced on the relay gain factors, the sum rate maximization problem becomes

$$\mathcal{P}_{\text{sum-zf}}: \quad \max_{\mathbf{g}_{\text{v}}} \sum_{i=1}^{N} \log_2(1+\text{SNR}_i) \qquad \textbf{subject to} \qquad \mathbf{g}_{\text{v}}^H \tilde{\mathbf{M}} \mathbf{g}_{\text{v}} \leq P_{\text{r}}, \quad (3.6)$$

where SNR is defined in (2.22) and $\tilde{\mathbf{M}} = \mathbf{V}^H \mathbf{M} \mathbf{V}$. The mathematical formulation of $\mathcal{P}_{\text{sum,zf}}$ is same as that of $\mathcal{P}_{\text{sum}}$, and hence, it can be solved through the previously introduced barrier method. For the sake of brevity, we drop further details, which can be trivially extracted from Section 3.2.1.1, and only write the unconstrained version of the problem:

$$\mathbf{g}_{\text{v}}^{\star}(\nu) = \underset{\mathbf{g}_{\text{v}}}{\textbf{argmin}} \; -\nu \cdot \sum_{i=1}^{N} \log_2(1+\text{SNR}_i) - \log_e(P_{\text{r}} - \mathbf{g}_{\text{v}}^H \tilde{\mathbf{M}} \mathbf{g}_{\text{v}}). \quad (3.7)$$

The related gradient can be calculated through the corresponding definitions given in Appendix A.2 for generalized quadratic expressions. To summarize, $\mathcal{P}_{\text{sum,zf}}$ is solved with **Algorithm 1** by just substituting $\mathbf{g}_{\text{v}}$ instead of $\mathbf{g}$, and solving (3.7) instead of (3.4) in the first step.

Remark 3.2.1: Both $\mathcal{P}_{\text{sum}}$ and $\mathcal{P}_{\text{sum-zf}}$ can be extended to have some QoS constraints on any S-D link rate. For instance, assume that the ith link has the following QoS constraint:

$$\text{R}_i = \log_2(1+\text{SINR}_i) \geq \text{R}_{\text{QoS},i}, \quad (3.8)$$

where $\text{R}_{\text{QoS},i}$ represents the QoS constraint for the rate. After some algebraic manipulations, we express (3.8) as the following quadratic function of $\mathbf{g}$

$$\mathcal{C}_{\text{QoS}}^{(i)}: \quad \mathbf{g}^H \Big(\mathbf{A}_i - (2^{\text{R}_{\text{QoS},i}} - 1) \mathbf{B}_i \Big) \mathbf{g} \geq \sigma_{n_{\text{d}}}^2 (2^{\text{R}_{\text{QoS},i}} - 1), \quad (3.9)$$

and add it to the corresponding problem definition. Adaptation of (3.9) to the barrier method is in the same way as the power constraint.

3.2.2. Maximizing the Minimum Link Rate

Depending on the instantaneous channel realizations and/or relative distances of the terminals to the relays, sum rate maximization may lead some nodes to have a very low transmission rate, which may be unfair but chosen for the sake of sum rate over the network. In order to prevent such situations, the S-D link rate can be chosen such that *max-min fairness* is applied over the whole network.

3.2.2.1. General Beamforming without ZF

First we remark that the minimum link rate maximization problem is equivalent to the minimum SINR maximization problem, due to the monotonic equivalence of the logarithm function. Hence, in the following we stick up to the SINR expressions. The problem of maximizing the minimum SINR for coherent multiuser relaying with N S-D pairs is formulated as

$$\mathcal{P}_{\text{mm}}: \quad \max_{\mathbf{g}} \min_{1\leq i\leq N} \ \text{SINR}_i \qquad \textbf{subject to} \qquad \mathbf{g}^H\mathbf{M}\mathbf{g} \leq P_{\mathsf{r}}, \tag{3.10}$$

where the subscript "mm" denotes the initials of "max-min". Introducing a dummy variable $\tau \geq 0$ and lower bounding each SINR constraint, $\mathcal{P}_{\text{mm}}$ is equivalently written as

$$\begin{aligned} \max_{\mathbf{g},\tau\geq 0} \ \tau \qquad \textbf{subject to} \qquad & \text{SINR}_i \geq \tau, \ i \in \{1,\ldots,N\} \\ & \mathbf{g}^H\mathbf{M}\mathbf{g} \leq P_{\mathsf{r}}. \end{aligned} \tag{3.11}$$

This problem formulation is in a well-known NP-hard problem structure as proved in [63, 113] which states that it is not solvable optimally in polynomial time. Morover, the problem is in fact a quadratically constrained quadratic programming (QCQP), and non-convex because of its SINR constraints, i.e., this is easy to see by simply converting the SINR constraint to a lower bounded quadratic form. Further, note that although an upper bound constraint is enforced for relay sum transmit power, the optimal result will be met with equality; otherwise, it will contradict with the optimality of the resultant maximal minimum SINR.

In the following, we relax the problem formulation in (3.11) appropriately to a quasi-convex problem structure. The ith SINR constraint in (3.11) can be expressed as

$$\mathcal{C}_{\text{SINR}}^{(i)}: \quad \mathbf{g}^H\Big(\mathbf{A}_i - \tau\mathbf{B}_i\Big)\mathbf{g} \geq \tau\sigma_{n_\mathsf{d}}^2. \tag{3.12}$$

Defining $\hat{\mathbf{G}} \triangleq \mathbf{g}\mathbf{g}^H$, we use the relation

$$\mathbf{g}^H\Big(\mathbf{A}_i - \tau\mathbf{B}_i\Big)\mathbf{g} = \mathrm{Tr}\Big((\mathbf{A}_i - \tau\mathbf{B}_i)\mathbf{g}\mathbf{g}^H\Big) = \mathrm{Tr}\Big((\mathbf{A}_i - \tau\mathbf{B}_i)\hat{\mathbf{G}}\Big) \tag{3.13}$$

in (3.12), and obtain

$$\mathcal{C}_{\text{SINR}}^{(i)}: \quad \mathrm{Tr}\Big((\mathbf{A}_i - \tau\mathbf{B}_i)\hat{\mathbf{G}}\Big) \geq \tau\sigma_{n_\mathsf{d}}^2. \tag{3.14}$$

After applying a similar "trace-relation" for the power constraint, we substitute all in $\mathcal{P}_{\text{mm}}$, and obtain

$$\begin{aligned} \max_{\hat{\mathbf{G}},\tau\geq 0} \ \tau \qquad \textbf{subject to} \qquad & \mathrm{Tr}\Big((\mathbf{A}_i - \tau\mathbf{B}_i)\hat{\mathbf{G}}\Big) \geq \tau\sigma_{n_\mathsf{d}}^2, \ i \in \{1,\ldots,N\} \\ & \mathrm{Tr}(\mathbf{M}\hat{\mathbf{G}}) \leq P_\mathsf{r}, \ \hat{\mathbf{G}} \succeq 0, \ \mathrm{rank}(\hat{\mathbf{G}}) = 1, \end{aligned} \tag{3.15}$$

where the optimization variable has changed to be $\hat{\mathbf{G}}$ instead of $\mathbf{g}$. However, the problem in (3.15) is not convex due to the rank constraint. Applying semidefinite relaxation (SDR) and hence, dropping the rank constraint [7, 41, 58, 113, 123], the problem (3.15) is relaxed to a quasi-convex SDP problem :

$$\begin{aligned} \tilde{\mathcal{P}}_{\text{mm}}: \quad \max_{\hat{\mathbf{G}},\tau\geq 0} \ \tau \qquad \textbf{subject to} \qquad & \mathrm{Tr}\Big((\mathbf{A}_i - \tau\mathbf{B}_i)\hat{\mathbf{G}}\Big) \geq \tau\sigma_{n_\mathsf{d}}^2, \ i \in \{1,\ldots,N\} \\ & \mathrm{Tr}(\mathbf{M}\hat{\mathbf{G}}) \leq P_\mathsf{r}, \ \hat{\mathbf{G}} \succeq 0, \end{aligned} \tag{3.16}$$

where all constraints are linear in the semidefinite matrix $\hat{\mathbf{G}}$. Consequently, the $\tilde{\mathcal{P}}_{\text{mm}}$ is equivalent to an *SDP feasibility check* problem when τ is given a priori, and can be efficiently solved by any interior-point method based SDP tool, e.g. YALMIP [77] or SeDuMi [116]. Henceforth, in order to solve $\tilde{\mathcal{P}}_{\text{mm}}$ optimally, we employ the bisection method [26] combined with a feasibility check at each iteration.

Since we drop the rank constraint from the original problem in (3.16), the optimal $\hat{\mathbf{G}}^\star$ obtained by solving $\tilde{\mathcal{P}}_{\text{mm}}$ may or may be a rank-1 matrix. If it is by any chance, then its eigenvector that corresponds to the only non-zero eigenvalue will be the optimal solution $\mathbf{g}^\star$ to the original problem $\mathcal{P}_{\text{mm}}$. In any case, we need to extract a well-performing $\mathbf{g}$ out of the (most probably) non-rank-1 matrix $\hat{\mathbf{G}}^\star$. To this end, we use the recently developed *randomization* techniques [113], which generate a set of candidate gain vectors using $\hat{\mathbf{G}}^{\star}$, from which the "best-performing" solution is chosen. In [113], three different randomization techniques have been proposed, whereas here we only employ one of those, i.e., randA as called in [113]. In this technique, the eigenvalue decomposition of $\hat{\mathbf{G}}^{\star} = \mathbf{U}\boldsymbol{\Sigma}\mathbf{U}^H$ is computed, and a candidate vector is chosen such that $\mathbf{g} = \mathbf{U}\boldsymbol{\Sigma}^{\frac{1}{2}}\mathbf{e}$, where the vth element of $\mathbf{e}$ is $e^{j\xi_v}$ and ξ_v's are independently and uniformly distributed on $[0, 2\pi)$. Independent of any realization of ξ_v, it is assured that $\mathbf{g}^H\mathbf{g} = \mathrm{Tr}(\hat{\mathbf{G}}^{\star})$. After generating each candidate, any violation of constraints should be checked and the candidate vector should be scaled to satisfy the most violated constraint accordingly. Finally, we choose the best vector out of the set of candidates, and assign it to be the sub-optimal result $\mathbf{g}^{\star}$. Note that although there are no theoretical bounds indicating the performance of this randomization technique, the simulation results are very encouraging for small-to-medium number of S-D pairs [113].

The **Algorithm 2** summarizes the minimum SINR/link rate maximization algorithm in details. Note that there we choose τ_{max} sufficiently large, and adapt it to the average operation SNR value to avoid excess number of iterations.

Algorithm 2 Minimum SINR Maximization Algorithm for Coherent Multiuser Relaying

---

initiate $\epsilon > 0, \tau_{\text{max}} > 0, \tau_{\text{min}} = 0.$

repeat

Step 1. Set. $\tau \Leftarrow (\tau_{\text{max}} + \tau_{\text{min}})/2.$

Step 2. Check the feasibility $\mathcal{F}_{\text{mm}}(\tau)$.

$$\mathcal{F}_{\text{mm}}(\tau): \begin{cases} \textbf{check } \hat{\mathbf{G}}, \tau \quad \textbf{subject to} & \mathrm{Tr}\Big((\mathbf{A}_i - \tau\mathbf{B}_i)\hat{\mathbf{G}}\Big) \geq \tau\sigma_{n_\text{d}}^2,\ i \in \{1, \ldots, N\} \\ & \mathrm{Tr}(\mathbf{M}\hat{\mathbf{G}}) \leq P_\text{r},\ \hat{\mathbf{G}} \succeq 0. \end{cases}$$

Step 3. **if** feasible

Set. $\tau_{\text{min}} \Leftarrow \tau.$

else

Set. $\tau_{\text{max}} \Leftarrow \tau.$

Step 4. Check and **quit** if $\tau_{\text{max}} - \tau_{\text{min}} < \epsilon.$

apply randomization on $\hat{\mathbf{G}}^{\star}$.

3.2.2.2. ZF Beamforming

As it was for the sum rate maximization, also for the max-min optimization, the ZF constraint does only change the optimization variable and the input data of the problem, but not the general structure of the employed optimization algorithm. Hence, dropping the details we briefly outline the required modifications with respect to the case without the ZF constraint. The problem of maximizing the minimum SNR (or link rate) for coherent multiuser relaying with zero-forced interference is formulated as

$$\mathcal{P}_{\text{mm-zf}}: \quad \max_{\mathbf{g}_\mathsf{v}} \min_{1\leq i\leq N} \ \text{SNR}_i \qquad \textbf{subject to} \qquad \mathbf{g}_\mathsf{v}^H \tilde{\mathbf{M}} \mathbf{g}_\mathsf{v} \leq P_\mathsf{r}. \tag{3.17}$$

Defining $\mathbf{g}_\mathsf{v}\mathbf{g}_\mathsf{v}^H = \hat{\mathbf{G}}_\mathsf{v}$ and following similar steps as in the previous case, we conclude that the relaxed problem can be solved by the bisection method where at each iteration the following SDP feasibility check is performed:

$$\mathcal{F}_{\text{mm-zf}}(\tau): \begin{cases} \textbf{check } \hat{\mathbf{G}}_\mathsf{v}, \tau \quad \textbf{subject to} & \text{Tr}\Big((\tilde{\mathbf{A}}_i - \tau\tilde{\mathbf{B}}_i)\hat{\mathbf{G}}_\mathsf{v}\Big) \geq \tau\sigma_{n_\mathsf{d}}^2, \\ & \text{Tr}(\tilde{\mathbf{M}}\hat{\mathbf{G}}_\mathsf{v}) \leq P_\mathsf{r}, \ \hat{\mathbf{G}}_\mathsf{v} \succeq 0. \end{cases}$$

The **Algorithm 2** can be efficiently adapted for the case with ZF constraint, by just exchanging $\mathcal{F}_{\text{mm}}(\tau)$ and $\hat{\mathbf{G}}$ with $\mathcal{F}_{\text{mm-zf}}(\tau)$ and $\hat{\mathbf{G}}_\mathsf{v}$, respectively.

3.2.2.3. Minimizing Relay Transmit Power with QoS Assurance

A related problem with max-min beamforming is to minimize the relay sum transmit power while maintaining certain SINR (or rate) to each link depending on the chosen QoS constraint. In the following, we only consider the general beamforming case for brevity; but as it is anticipated from the previous sum rate and max-min beamforming discussions, it is trivial to extend it for the case with the ZF constraint.

We refer such a distributed beamforming design as *minimum relay transmit power* problem, and write it as

$$\mathcal{P}_{\text{mp}}: \quad \min_{\mathbf{g}} \ \mathbf{g}^H\mathbf{M}\mathbf{g} \qquad \textbf{subject to} \qquad \text{SINR}_i \geq \gamma_i, \ \ i \in \{1, \ldots, N\}. \tag{3.18}$$

where $\gamma_i \in \mathbb{R}_+$ represents the minimum QoS requirement on the SINR of the ith S-D link and the subscript "mp" denotes the initials of "minimum power". Adapting the $\hat{\mathbf{G}}$ formulation instead of $\mathbf{g}$ from the previous subsections, and assuming SDR for the rank constraint on $\hat{\mathbf{G}}$, we can relax the $\mathcal{P}_{\text{mp}}$ to

$$\tilde{\mathcal{P}}_{\text{mp}}: \quad \min_{\hat{\mathbf{G}}} \ \text{Tr}(\mathbf{M}\hat{\mathbf{G}}) \quad \textbf{subject to} \quad \text{Tr}\Big((\mathbf{A}_i - \gamma_i \cdot \mathbf{B}_i)\hat{\mathbf{G}}\Big) \geq \sigma_{n_\mathsf{d}}^2\gamma_i$$
$$\hat{\mathbf{G}} \succeq 0, \ i \in \{1, \ldots, N\}, \tag{3.19}$$

where the cost function is linear in $\hat{\mathbf{G}}$ and the constraints are linear inequalities in $\hat{\mathbf{G}}$. Hence, the readily SDP-formulated problem $\tilde{\mathcal{P}}_{\mathrm{mp}}$ can be efficiently solved. However, as we omitted the rank constraint that should have been imposed on $\hat{\mathbf{G}}$, we still need to apply the randomization technique on the optimal $\hat{\mathbf{G}}^{\star}$ obtained by solving $\tilde{\mathcal{P}}_{\mathrm{mp}}$.

There is a strong relation between the max-min and the minimum relay transmit power beamforming approaches such that $\mathcal{P}_{\mathrm{mm}}$ can be solved via successive $\mathcal{P}_{\mathrm{mp}}$'s. To this end, we set equal QoS constraints for each S-D link, i.e., $\gamma = \gamma_i\ \forall i$, and employ a bisection method on γ. At each iteration we check if the minimum power obtained by solving $\mathcal{P}_{\mathrm{mp}}$ is lower than the upper bound value set for power constraint in $\mathcal{P}_{\mathrm{mm}}$. However, due to the relaxation and randomization processes we do not expect identical $\mathbf{g}^{\star}$ from the direct solution of $\mathcal{P}_{\mathrm{mm}}$ and the successive solution via $\mathcal{P}_{\mathrm{mp}}$'s, but approximately equal maximal minimum SINR.

3.2.3. Performance Results

Simulation Setup In this section we study the performance of different optimization techniques through Monte Carlo simulations. We used the MATLAB based semidefinite tool YALMIP [77] to solve the designed semidefinite problems. Each element of the channel matrices $\mathbf{H}$ and $\mathbf{F}$ is i.i.d. Rayleigh fading coefficient with zero-mean and unit-variance. These propagation channel matrices are assumed to remain the same over at least one transmission cycle (two-hops). Unless otherwise stated, $P_{\mathsf{s}} = 1$, $P_{\mathsf{r}} = NP_{\mathsf{s}}$, and $\sigma^2 = \sigma^2_{n_{\mathsf{r}}} = \sigma^2_{n_{\mathsf{d}}}$. Correspondingly the average SNR is assumed to be SNR $= P_{\mathsf{s}}/\sigma^2$. Moreover, we assume perfectly frequency and phase synchronous relays. All relays considered throughout the following simulations are equipped with only a single-antenna, i.e., $N_{\mathsf{c}} = N_{\mathsf{r}}$.

As benchmarks, we consider the relay gain choices ZF-1 and ZF-2 explained in Section 2.5. Summarizing briefly, we choose the last column vector of $\mathbf{V}$ as the relay gain vector in ZF-1, whereas we project the AZF gain vector presented in Section 2.4.2 onto the nullspace $\mathbf{V}$ so that we implement a maximum ratio combining at the destinations (see Section 2.4.2 for further details).

Simulation Results We are first interested in finding the maximum achievable sum rate with coherent multiuser AF relaying. To this end, we plot average sum rate versus average SNR in Fig. 3.1 for $N_{\mathsf{r}} = 20$ relays. The complex relay gains are computed through the general beamforming problem $\mathcal{P}_{\mathsf{sum}}$ (employing **Algorithm 1**) so that

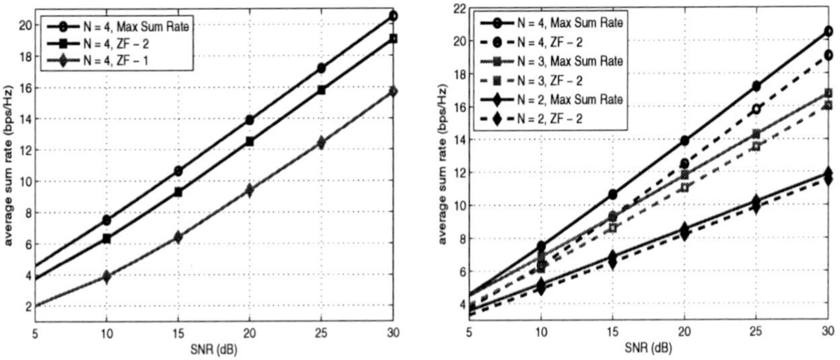

(a) (b)

Figure 3.1.: Average sum rate vs. average SNR for $N_\mathsf{r} = 20$ relay and $N = 2, 3, 4$ S-D pairs.

the sum of all S-D link rates is maximized. We denote the corresponding curves with Max Sum Rate. We observe from Fig. 3.1(a) that sum rate maximization provides a substantial improvement over the random vector choice of ZF-1 independent of the average SNR. Moreover, for the considered case of $N = 4$ pairs, there is almost 1 bps/Hz sum rate gain achieved on top of the smart choice of ZF-2. In Fig. 3.1(b), we consider three sets of curves, where each corresponds to $N = 2, 3$, and 4 S-D pairs. Comparing the corresponding curves, we notice that as N increases, the gain of sum rate maximization with respect to ZF-2 improves. Note that we do not enforce a ZF constraint here for optimization. However, even in that case, $N_\mathsf{r} = 20$ would be sufficient for perfect interference cancellation and would provide 17, 13, 7 excess relays for $N = 2, 3, 4$ pairs, respectively.

It may seem unfair at first glance to compare the sum rate performances of zero-forcing based benchmarks (ZF-1 and ZF-2) with the general beamforming problem $\mathcal{P}_{\text{sum}}$ without ZF constraint. As the optimization space shrinks with the ZF constraint, it is known that the general beamforming serves as an upper bound for ZF beamforming. This suspicion triggers the question of how the performances of sum rate maximization methods with and without the ZF constraint compare. Hence, we focus on the impact of the ZF constraint on the sum rate optimization in Fig. 3.2. We consider two scenarios of $N = 2$ and $N = 4$ S-D pairs and compare the average sum rate performances of the general beamforming $\mathcal{P}_{\text{sum}}$ and the ZF-beamforming $\mathcal{P}_{\text{sum-zf}}$. As the minimum number

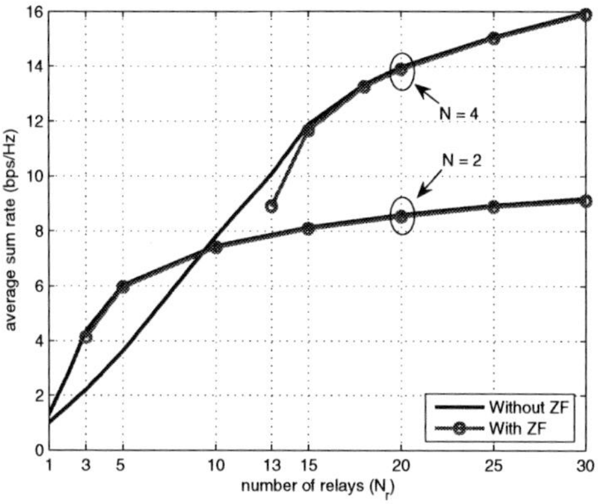

Figure 3.2.: Average sum rate comparison of sum rate maximization optimization with and without ZF constraint for $N = 2$ S-D pairs.

of relays for a non-empty nullspace is 3 for $N = 2$ and 13 for $N = 4$, the curves of ZF-beamforming start from these values. It is realized that, in essence, the performance reduction due to the ZF constraint is almost negligible except for the minimum relay configuration case. There is a certain sum rate performance difference at $N_r = 3$ for $N = 2$ and $N_r = 13$ for $N = 4$, which is much more significant for the four S-D pairs case. The reason for this simply arises from the enforced ZF constraint in combination with being in shortage of degrees of freedom. The ZF constrained beamforming employs all spatial degrees of freedom for complete interference cancellation without caring the sum rate. In essence, since the corresponding gain vector is uniquely given through the single column of the nullspace, the optimization can only scale this vector to fulfill the power constraint. On the other hand, when the ZF constraint is removed, the spatial degrees are employed for the sake of sum rate but not nulling interference, which result in an improvement in sum rate. As N_r grows larger, the additional degrees of freedom support $\mathcal{P}_{\text{sum-zf}}$ to catch up with the performance of $\mathcal{P}_{\text{sum}}$.

We observe that at the expense of almost no performance penalty, ZF-beamforming provides two major advantages: First, the optimization complexity is reduced since the searched gain vector size is reduced from N_r to $N_r - N(N-1)$ complex dimensions. Second, we provide some sort of independence to the individual S-D links in the sense that the transmit power variation of a single source influences the other links only through the enforced relay power adjustment.

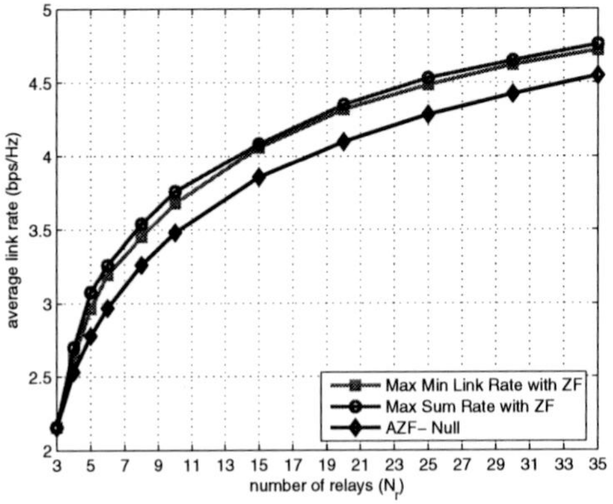

Figure 3.3.: Average link rate comparison between the optimizations of maximizing sum rate and maximizing minimum link rate for $N = 2$ S-D pairs and $\mathrm{SNR} = 20\mathrm{dB}$.

We now shift our focus to the average performance of individual S-D links. Sum rate oriented optimizations disregard the performance of individual links for the sake of overall network performance. However, this may lead to some unfair rate allocations for specific channel realizations. It is known *by definition* that max-min type optimizations provide as fair as possible rates, in the sense of equivalence, to all links. At this point, we wonder how much performance penalty per link rate should be paid for the supplied fairness. In Fig. 3.3, we plot the average S-D link rate for a scenario of zero-forcing $N = 2$ S-D pairs, where the gain allocations is performed through either sum rate maximizing $\mathcal{P}_{\mathrm{sum-zf}}$ or minimum link rate maximizing $\mathcal{P}_{\mathrm{mm-zf}}$. As a reference we also show the corresponding performance of ZF-2. It is shown that the performance difference per link is limited by less than a rate of 0.1 bps/Hz, which is quite an affordable penalty for the sake of fairness. Moreover, this behaviour is continuous over N_{r}. On the other hand, it is also interesting to note that the growing rates of all three curves are very similar to each other, where the difference between the schemes Max Sum Rate and ZF-2 can be identified to be an array gain.

We have already mentioned in Section 3.2.2 that max-min type gain allocations not only provide fairness, but also outage (rate) gain. We have defined outage as that at least one of the S-D links has an information rate below the pre-determined outage rate. Since, for each channel realization, we are forcing the individual rates to be

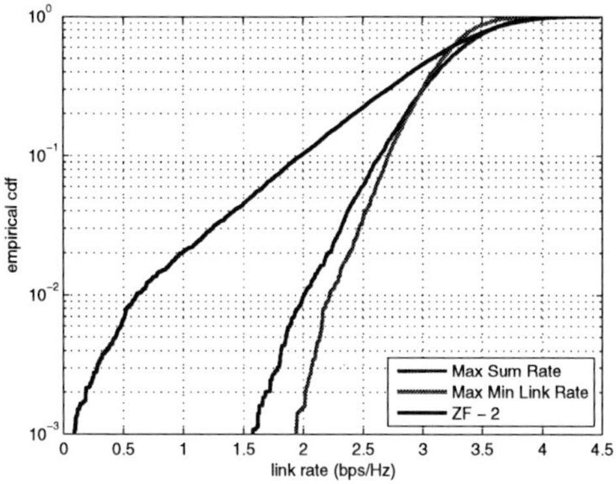

Figure 3.4.: Empirical CDFs of single S-D link rate with different relay gain optimizations for $N = 2$ S-D pairs, $N_\mathsf{r} = 6$ relays, $\mathrm{SNR} = 20$ dB.

as high as possible by $\mathcal{P}_{\mathrm{mm}}$ or $\mathcal{P}_{\mathrm{mm-zf}}$, we achieve, by definition, the lowest outage probability that can be attained through gain optimization. In Fig. 3.4 we investigate this aforementioned outage improvement through max-min optimization. We consider $N = 2$ pairs and $N_\mathsf{r} = 6$ relays and plot the cumulative distribution function (CDF) of any one S-D link for three different zero-forcing gain allocation choices considered for the previous figure. First of all, we immediately notice that the ZF-2 performs very poor with respect to its counterparts, e.g. there is almost 1.5 bps/Hz rate loss at an outage probability of 1×10^{-2} when compared with the other two. On the other hand, the provided outage gain of $\mathcal{P}_{\mathrm{mm-zf}}$ with respect to the outage performance of $\mathcal{P}_{\mathrm{sum-zf}}$ is confirmed within the figure. For instance, say at an outage probability of 1×10^{-2}, outage rate is improved by 0.25 bps/Hz for the given system parameters. Note that the number of relays plays an important role on the outage improvement, as each additional relay provides a surplus degree of freedom. However, we do not conjecture that the outage rate difference between the optimization schemes $\mathcal{P}_{\mathrm{mm-zf}}$ and $\mathcal{P}_{\mathrm{sum-zf}}$ will monotonically increase with each additional N_r for a given outage probability.

An alternative approach of studying the outage improvement is to compare the "minimum" link rate performance over multiple pairs. In Fig. 3.5 we plot the corresponding zero-forcing rate performances of Max Sum Rate, Max Min Link Rate, and ZF-2 for $N = 2, 3$ S-D pairs and increasing number of relays. As expected, we see that Max Min Link Rate provides a continuous improvement with respect to Max Sum Rate over all N_r values (except of the minimum relay configuration case, where all designs

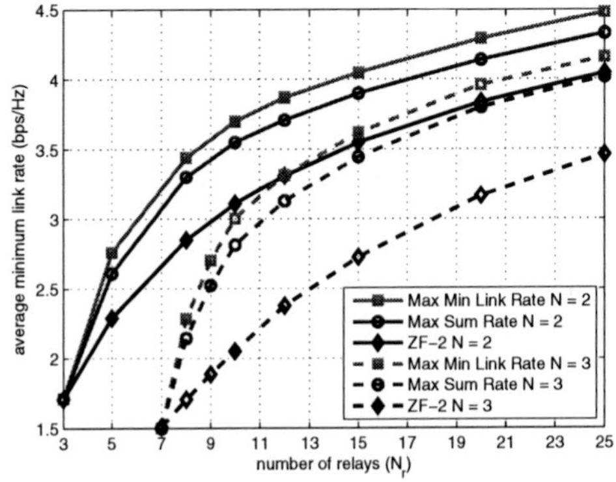

Figure 3.5.: Average minimum link rate comparison between the optimizations of maximizing the sum rate and maximizing the minimum link rate for $N = 2$ and $N = 3$ S-D pairs and $\mathrm{SNR} = 20\mathrm{dB}$.

achieve the same performance). However, for both N choices we note on average that maximizing the sum rate does not substantially penalize the weak link, i.e., there is around 0.1 bps/Hz loss with respect to max-min optimization. This is primarily because we have assumed similar path losses for all links and the weakness does only arise from the fading. On the other hand, the poor performance of ZF-2 is also confirmed with the associated curves, where the performance loss with respect to its counterparts is emphasized more for $N = 3$ than for $N = 2$. This impact can be explained as follows. The AZF relay gain decision results in both coherent and non-coherent additions at the destinations, where the growing non-coherent part ruins the coherent additions more severely as N grows.

While concluding this section, we would like to highlight the following findings:

1. The ZF constraint serves its advantages almost for free in terms of performance-penalty to pay.
2. Max-min type optimizations not only provide fairness and outage rate improvement, but also provides sufficient sum/link rate performances, which are almost as good as that of sum rate oriented optimizations.
3. The smart gain choice of ZF-2 performs surprisingly well for average oriented performance measures for no optimization cost.

4. Assuming that full global CSI is available, relay gain optimization is compulsory for utilizing the available resources (relay diversity and transmit power) for maximal sum/link rate, outage gain and fairness.

3.2.4. Diversity through Relay Selection or Max-Min Optimization?

In the previous section, we shortly touched on the outage gain that can be achieved through relay gain optimization. Moreover, in the beginning of this chapter, we have claimed that this outage gain can be translated to diversity gain at the destinations, which, in practice, can be read as the slope of outage probability curves. We refer to this gain as the *effective distributed diversity*. In this section, we aim at elaborating more on this issue, providing empirical validations and answering the questions of how to achieve, how much to achieve and under which conditions to achieve this diversity gain.

Recall that the outage probability at a rate of $\mathrm{R}_{\text{outage}}$ is defined as

$$P_{\text{outage}}(\mathcal{H}, \mathcal{R}, \mathbf{g}) = \Pr\Big(\min_{1 \le i \le N} \mathrm{R}_i(\mathcal{H}, \mathcal{R}, \mathbf{g}) \le \mathrm{R}_{\text{outage}}\Big).$$

The S-D link rates R_i, so as the P_{outage}, are functions of instantaneous channel realization, employed relays, and the designed gain allocation vector. Hence, for a given instantaneous channel realization set $\mathcal{H}$, we can influence P_{outage} by maximizing the minimum R_i through either relay gain optimization or relay selection.

Max-Min Optimization We first assume that the relay set $\mathcal{R}$ is fixed for a given set of channel realizations. We then optimize $\mathbf{g}$ through one of the proposed max-min type optimizations, e.g. $\mathcal{P}_{\text{mm}}$ or $\mathcal{P}_{\text{mm-zf}}$. Correspondingly, the definition of outage probability modifies to

$$P^{\text{mm}}_{\text{outage}}(\mathcal{H}, \mathcal{R}) = \Pr\Big(\max_{\mathbf{g}} \min_{1 \le i \le N} \mathrm{R}_i(\mathcal{H}, \mathcal{R}, \mathbf{g}) \le \mathrm{R}_{\text{outage}}\Big),$$

where $\mathbf{g}$ must satisfy a pre-assumed relay sum transmit power constraint. With this max-min optimization for $\mathbf{g}$, we maximize the worst link rate for given sets of $\mathcal{H}, \mathcal{R}$, and hence minimize the "instantaneous" outage probability.

This outage probability gain would eventually translate into diversity gain. The interesting question to answer here is: What is the order of this achieved diversity? Recall that the primary aim of coherent multiuser AF relaying is to provide full spatial multiplexing gain in a distributed manner. We have already shown that we need a

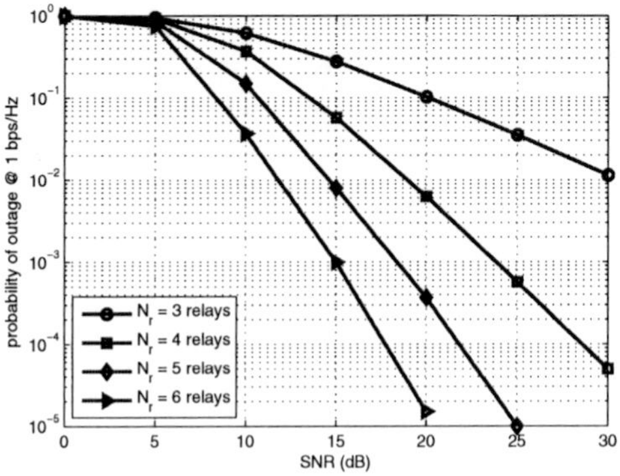

Figure 3.6.: Max-min optimization diversity: Probability of outage at 1 bps/Hz vs. average SNR for $N = 2$ S-D pairs.

certain number of relays for perfect interference mitigation, which is crucial to achieve the desired spatial multiplexing. Following this statement, we have also experienced that with less than this required minimum number of relays, any optimization can not result in a diversity gain. Summing up, we claim (without any explicit spatial diversity gain proofs defined for infinite average SNR) that the maximization of minimum link rate per each set of channel realizations, can fully convert the additional spatial degrees into an effective distributed diversity. The order of this diversity gain is conjectured to be as much as the remaining number of degrees of freedom left after interference mitigation. Specifically, say we have $|\mathcal{R}| = N_r$ relays serving N S-D pairs. Then the maximum achievable effective diversity gain is $N_r - N(N-1)$. Note that our maximal effective diversity conjecture is also in line with the achievable diversity-multiplexing trade-off derived in [2, 4] for relay-channels.

Now, we perform computer simulations in order to numerically validate our conjecture. The Fig. 3.6 shows the outage probability performance of max-min optimization for $N = 2$ S-D pairs and various N_r choices. We read the figure as that a diversity order of (say) two corresponds to two orders of magnitude change in 10 dB scale. Following our full diversity gain proposal of $N_r - N(N-1) = N_r - 2$, max-min optimization provides an additional diversity gain for each relay introduced to the system on top of the minimum relay configuration, i.e., 1, 2, 3, 4 orders of diversity respectively for the corresponding curve in the figure. Note that we have enforced zero-forcing constraint for the max-min results in the figure, however, removing it would result in nothing but an array gain at each curve, i.e., a parallel shift to the left.

Relay Selection Let us now assume that we have N S-D pairs and a set of relays $\mathcal{R}$, whose cardinality is strictly larger than $N(N-1)+1$. That is, we have more relays than sufficient number of relays for perfect multiuser interference cancellation. However, we restrict our scenario such, that the N pairs are served by only $N_\mathsf{r} = N(N-1)+1$ relays. For such a number of relays, the relay gain vector $\mathbf{g}$ is unique and immediately given by (2.21). Correspondingly, the link rates are not anymore function of $\mathbf{g}$. We have already shown in the previous section, that employing such a fixed set with minimal sufficient number of relays, does not result in any effective diversity gain, i.e., diversity of 1. The basic reason is that all available spatial degrees of freedom is spent to null the multiuser interference.

Previously, we proposed to increase the number of relays and employed a max-min optimization in order to attain some outage gain. Instead, we propose here to choose the best set of relays $\tilde{\mathcal{R}}$ (with cardinality $N(N-1)+1$) out of the global set $\mathcal{R}$ so that the the minimum link rate is maximized over all possible $\tilde{\mathcal{R}} \subset \mathcal{R}$. The outage probability definition consequently modifies to

$$P^{\mathrm{rs}}_{\mathsf{outage}}(\mathcal{H}) = \Pr\Big(\max_{\tilde{\mathcal{R}} \subset \mathcal{R}} \min_{1 \le i \le N} \mathrm{R}_i(\mathcal{H}, \tilde{\mathcal{R}}) \le \mathrm{R}_{\mathsf{outage}} \Big),$$

where we did not explicitly depict $\mathbf{g}$ as it is uniquely determined for given $\mathcal{H}$ and $\tilde{\mathcal{R}}$. Note that there are in total

$$\binom{|\mathcal{R}|}{N(N-1)+1} = \frac{|\mathcal{R}|!}{(|\mathcal{R}| - N(N-1)+1)!(N(N-1)+1)!}$$

possible relay sets. Similar to the $P^{\mathrm{mm}}_{\mathsf{outage}}$, with this optimal search over possible relay sets, we minimize the instantaneous outage probability by maximizing the worst-link rate. As the maximum attainable diversity gain (independent source-to-destination links via relays) is upper bounded by the number of available relays within the network, we again conjecture that such a selection diversity can be turned into full effective distributed diversity at an order of $|\mathcal{R}| - N(N-1)$.

In Fig. 3.7 we consider a scenario of $N = 2$ S-D pairs and simulate the achievable effective diversity for four different global relay $\mathcal{R}$ set with cardinalities $|\mathcal{R}| = 3, 4, 5, 6$. Hence, we are looking for the best three relays (minimum number of relays required for $N = 2$) within each given global set. According to our conjecture above, we expect

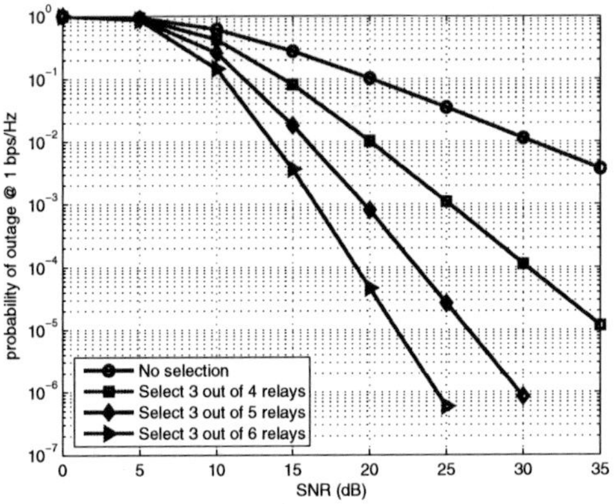

Figure 3.7.: Selection diversity: Probability of outage at 1 bps/Hz vs. average SNR for $N = 2$ S-D pairs.

the respective diversity orders of 1, 2, 3, 4, each of which can be confirmed through the corresponding curves in the figure as being the respective slope.

To sum up, using either max-min optimization or relay selection, an effective diversity gain at an order that is equal to the number of degrees of freedom left after interference mitigation can be achieved. Or, putting differently, we have proposed two different schemes that attain the maximal diversity gain that is achievable in two-hop multiuser networks while providing full spatial multiplexing [2, 4].

Alternatively, a combination of the two schemes can be also employed to achieve the same diversity gains. That is, we search for the best relay set with any cardinality larger than $N(N-1)+1$ and employ max-min optimization within this chosen set. A final remark is that although both schemes provide the same diversity gain for a given number of relays in the global set, max-min optimization based scheme results in better outage rates as it employs more relays than that of relay selection.

3.3. Distributed Optimization of Relay Gain Factors

As aforementioned in the introduction section, the assumption of perfect global CSI knowledge per relay can be an unbearable limitation to the practicality of distributed multiuser orthogonalization especially when the number of nodes within the network is very large. In this section, we take a practical look at the same problem of optimizing the distributed beamforming and ask the question: Can we obtain the *exact* gain allocation vector that is obtained by the previously proposed global CSI requiring solutions, with a reduced set of channel information at the relays? The answer is particularly important to make the scaling of local CSI dissemination overhead independent of number of relays, and hence, to enjoy the high number of relays without paying the penalty of spectral deficiency.

We will show in the following that in fact the answer of the above question is "yes". To this end, we identify gradient descent based relay gain allocation schemes as the approach to take and propose a gradient method where the relays determine their gain factors iteratively in a *distributed* manner. Moreover, we prove that to do so, each relay only needs local CSI plus limited feedback from the destinations.

In the following, we focus on the case of a single S-D pair under interference as an interim step towards the multiuser case, and present the details of distributed gain allocation for interference mitigation. Afterwards, we extend these findings to simultaneous communication of multiple S-D pairs. Moreover, we focus on algorithm implementation issues, e.g., starting vector computation, step-size optimization, etc., to make our proposal more practical to implement and also improve its performance.

3.3.1. Distributed Gradient Based Gain Allocation

We start by simplifying our network model and assume, here and in the sequel, that

- all relays within the network are equipped with a single antenna, i.e., $M_\mathsf{r} = N_\mathsf{r}$, $\mathbf{H} \in \mathbb{C}^{N_\mathsf{r} \times N}$, and $\mathbf{F} \in \mathbb{C}^{N \times N_\mathsf{r}}$,
- the phase offsets of all terminals are assumed to be perfectly known and compensated, i.e., perfect phase synchronization.

We assume that the ith S-D pair is the focus of interest in the presence of $N-1$ independent interferers. Incorporating the assumptions and definitions above with the received signal model at the ith destination in (2.11), we obtain a modified (and simplified) expression

$$y_{\mathsf{d},i} = (\mathbf{h}_{i\mathsf{r}} \odot \mathbf{f}_{\mathsf{r}i})^T \mathbf{g} s_i + \sum_{j=1, j\neq i}^{N} (\mathbf{h}_{j\mathsf{r}} \odot \mathbf{f}_{\mathsf{r}i})^T \mathbf{g} s_j + (\mathbf{f}_{\mathsf{r}i} \odot \mathbf{n}_{\mathsf{r}})^T \mathbf{g} + n_{\mathsf{d},i}. \tag{3.20}$$

Recall from Section 2.2 that $\mathbf{h}_{i\mathsf{r}}$ and $\mathbf{f}_{\mathsf{r}i}$ collect the channel coefficients from the ith source to all relays and from all relays to the ith destination, respectively. Correspondingly, the vector $(\mathbf{h}_{i\mathsf{r}} \odot \mathbf{f}_{\mathsf{r}i})$, $i, j \in \{1, \dots, N\}$ denotes the equivalent two-hop propagation channel between the ith source and the jth destination. Moreover, the complex vector

$$\mathbf{h}_{\mathsf{srd},i} \in \mathbb{C}^N \triangleq \left[(\mathbf{h}_{1\mathsf{r}} \odot \mathbf{f}_{\mathsf{r}i})^T \mathbf{g} \ \cdots \ (\mathbf{h}_{N\mathsf{r}} \odot \mathbf{f}_{\mathsf{r}i})^T \mathbf{g} \right]^T$$

collects all received signal terms at the ith destination and denotes the reduced version of the equivalent channel matrix $\mathbf{H}_{\mathsf{srd}}$ for a single S-D pair.

Without loss of generality, we define a normalized gain vector $\mathbf{w} \triangleq \mathbf{g} \odot \mathbf{f}_{\mathsf{r}i}$ for notational convenience in the sequel, which modifies (3.20) to

$$y_{\mathsf{d},i} = \underbrace{\mathbf{h}_{i\mathsf{r}}^T \mathbf{w} s_i}_{\text{intended signal}} + \underbrace{\sum_{j=1, j\neq i}^{N} \mathbf{h}_{j\mathsf{r}}^T \mathbf{w} s_j}_{\text{interference}} + \underbrace{\mathbf{w}^T \mathbf{n}_{\mathsf{r}} + \mathbf{n}_{\mathsf{d},i}}_{\text{noise}} . \tag{3.21}$$

On the other hand, with single-antenna relays, the sum relay transmit power constraint (2.3) becomes

$$P_{\mathsf{tx}}(\mathbf{g}) = \mathbf{g}^H \left(P_{\mathsf{s}} \cdot \operatorname{diag}\left\{ [\mathbf{h}_{\mathsf{s}1}^H \mathbf{h}_{\mathsf{s}1} \ \cdots \ \mathbf{h}_{\mathsf{s}N_{\mathsf{r}}}^H \mathbf{h}_{\mathsf{s}N_{\mathsf{r}}}] \right\} + \sigma_{n_{\mathsf{r}}}^2 \mathbf{I}_{N_{\mathsf{r}}} \right) \mathbf{g} = \mathbf{g}^H \mathbf{M} \mathbf{g} \le P_{\mathsf{r}}, \tag{3.22}$$

where $\mathbf{M}$ in (3.3) here becomes $\mathbf{M} = P_{\mathsf{s}} \cdot \operatorname{diag}\left\{ [\mathbf{h}_{\mathsf{s}1}^H \mathbf{h}_{\mathsf{s}1} \ \cdots \ \mathbf{h}_{\mathsf{s}N_{\mathsf{r}}}^H \mathbf{h}_{\mathsf{s}N_{\mathsf{r}}}] \right\} + \sigma_{n_{\mathsf{r}}}^2 \mathbf{I}_{N_{\mathsf{r}}}$.

Then, the SINR at the ith destination is given by

$$\mathrm{SINR}_i = \frac{\mathbf{w}^H \mathbf{h}_{i\mathsf{r}}^* \mathbf{h}_{i\mathsf{r}}^T \mathbf{w}}{\mathbf{w}^H (\mathbf{H}_{\mathrm{I},i}^* \mathbf{H}_{\mathrm{I},i}^T + \mathbf{I}_{N_{\mathsf{r}}} \sigma_{n_{\mathsf{r}}}^2) \mathbf{w} + \sigma_{n_{\mathsf{d}}}^2}, \tag{3.23}$$

where $\mathbf{H}_{\mathrm{I},i} \triangleq [\mathbf{h}_{1\mathsf{r}} \cdots \mathbf{h}_{i-1\mathsf{r}} \mathbf{h}_{i+1\mathsf{r}} \cdots \mathbf{h}_{N\mathsf{r}}]$ and P_{s} is assumed to be 1. We want to express the denominator of (3.23) as a single quadratic form in $\mathbf{w}$ so that the resultant SINR_i expression would become a *generalized Rayleigh quotient* (GRQ), from which the optimal $\mathbf{w}$ that maximizes SINR follows immediately. We incorporate the power constraint with SINR_i and scale $\mathbf{w}$ such that it fulfills (3.22) with equality, i.e., maximal SINR for given limited transmit power. The transmit power consumption for a given $\mathbf{w}$ is computed through (3.22) as

$$P_{\text{tx}}(\mathbf{w}) = \mathbf{g}^H \mathbf{M} \mathbf{g} = \mathbf{w}^H (\mathbf{\Gamma}_i^{-1})^H \mathbf{M} \mathbf{\Gamma}_i^{-1} \mathbf{w},$$

where $\mathbf{\Gamma}_i \triangleq \text{diag}\{\mathbf{f}_{\text{r}i}\}$ and we used its inverse conjugate to element-wise divide $\mathbf{w}$ with $\mathbf{f}_{\text{r}i}$, i.e., $\mathbf{g} = \mathbf{\Gamma}_i^{-1}\mathbf{w}$. Since $P_{\text{tx}}(\mathbf{w})$ can be maximally equal to P_{r}, we multiply $\mathbf{w}$ in (3.23) with the square root of $\frac{P_{\text{r}}}{\mathbf{w}^H(\mathbf{\Gamma}_i^{-1})^H \mathbf{M} \mathbf{\Gamma}_i^{-1}\mathbf{w}}$. After some algebraic manipulations, the SINR_i equivalently modifies to

$$\text{SINR}_i = \frac{\mathbf{w}^H \breve{\mathbf{A}}_i \mathbf{w}}{\mathbf{w}^H \breve{\mathbf{B}}_i \mathbf{w}}, \tag{3.24}$$

where $\breve{\mathbf{A}}_i = \mathbf{h}_{ir}^* \mathbf{h}_{ir}^T$ and $\breve{\mathbf{B}}_i = \mathbf{H}_{\text{I},i}^* \mathbf{H}_{\text{I},i}^T + \mathbf{I}_{N_\text{r}} \sigma_{n_\text{r}}^2 + (\sigma_{n_\text{d}}^2 / P_\text{r}) \cdot (\mathbf{\Gamma}_i^{-1})^H \mathbf{M} \mathbf{\Gamma}_i^{-1}$. The SINR_i expression in (3.24) is known as the GRQ. The maximum value of the GRQ, i.e., the maximum achievable SINR, and the corresponding optimal relay gain vector $\mathbf{w}^\star$ can be found through an eigenvalue problem incorporating $\breve{\mathbf{A}}_i$ and $\breve{\mathbf{B}}_i$:

$$\mathbf{w}^\star = c\,\mathbf{U}_{\breve{\text{B}}_i} \mathbf{\Sigma}_{\breve{\text{B}}_i}^{-\frac{1}{2}} \cdot \lambda_{\text{vec}}\left(\mathbf{\Sigma}_{\breve{\text{B}}_i}^{-\frac{1}{2}} \mathbf{U}_{\breve{\text{B}}_i}^H \breve{\mathbf{A}}_i \mathbf{U}_{\breve{\text{B}}_i} \mathbf{\Sigma}_{\breve{\text{B}}_i}^{-\frac{1}{2}} \right), \tag{3.25}$$

where $\mathbf{U}_{\breve{\text{B}}_i}$ and $\mathbf{\Sigma}_{\breve{\text{B}}_i}$ are obtained through the eigenvalue decomposition of $\breve{\mathbf{B}}_i$, i.e., $\breve{\mathbf{B}}_i = \mathbf{U}_{\breve{\text{B}}_i} \mathbf{\Sigma}_{\breve{\text{B}}_i} \mathbf{U}_{\breve{\text{B}}_i}^H$, $c \in \mathbb{R}_+$ is a constant.

So to speak, having global CSI knowledge related with the ith S-D pair, i.e., $\mathbf{H}$ and $\mathbf{f}_{\text{r}i}$, each relay can compute the maximum SINR_i resulting relay gain efficiently. However, collecting this global knowledge requires the aforementioned local CSI dissemination overhead. Addressing this problem, in the following we present a distributed gradient descent based scheme which does not require global CSI knowledge. Still, the resultant gain vector converges to the optimal relay gain vector obtained from the GRQ.

The gradient descent algorithm updates the gain vector at any iteration by using the vector of the previous iteration, a search direction, i.e., the gradient, and a step-size [26]. Applied to our scenario, the update equation for the normalized relay gain vector $\mathbf{w}$ in the nth iteration can be expressed as

$$\mathbf{w}_{n+1} = \mathbf{w}_n + \mu_n \cdot \nabla_{\mathbf{w},n} \text{SINR}_i, \tag{3.26}$$

where μ_n and $\nabla_{\mathbf{w},n}\text{SINR}_i$ denote the optimal step-size and the gradient of SINR_i with respect to $\mathbf{w}$ at the nth iteration, respectively. Starting from a locally-computable initial gain vector, at each *outer* iteration we compute the gradient in a distributed manner as will be explained in Section 3.3.1.1. For each gradient, we optimize the step-size iteratively. We refer to the step-size optimization cycle as *inner loop* (see Section 3.3.1.2). Finally, for the sake of rapid convergence, we incorporate conjugate directions in Section 3.3.1.3.

3.3.1.1. Local Gradient Calculation

The gradient of SINR_i with respect to $\mathbf{w}$ is given by [39]

$$\nabla_{\mathbf{w}}\text{SINR}_i = \frac{\breve{\mathbf{A}}_i\mathbf{w}}{\mathbf{w}^H\breve{\mathbf{B}}_i\mathbf{w}} - \frac{\breve{\mathbf{B}}_i\mathbf{w}}{\mathbf{w}^H\breve{\mathbf{B}}_i\mathbf{w}}\text{SINR}_i \equiv \mathcal{S}_1 + \mathcal{S}_2, \tag{3.27}$$

where we drop subscript n in this subsection for the sake of notational simplicity. We further define $\nabla_{\mathbf{w}}\text{SINR}_{i,k}$ to denote the kth element of $\nabla_{\mathbf{w}}\text{SINR}_i$ corresponding to the kth relay's gain coefficient. The following lemma states the sufficient information needed at each relay node for the local computation of the corresponding element of the gradient (3.27).

Lemma 3.3.1: Each relay is assumed to have an a priori knowledge of $\sigma^2_{n_\mathsf{d}}, \sigma^2_{n_\mathsf{r}}, P_\mathsf{r}$ and P_s. The sufficient information for the kth relay to "locally" compute the corresponding element $\nabla_{\mathbf{g}}\text{SINR}_{i,k}$ of the gradient of SINR_i with respect to $\mathbf{g}$, are the local CSI, i.e., $\mathbf{h}_{\mathsf{s}k}$ and $\mathbf{f}_{k\mathsf{d}}$, the equivalent channel $\mathbf{h}_{\mathsf{srd},i}$, and SINR_i.

Proof: We focus on the two summands $\mathcal{S}_1$ and $\mathcal{S}_2$ in (3.27) individually. First, we substitute the definition of $\breve{\mathbf{A}}_i$ from (3.24) into the numerator of $\mathcal{S}_1$ and express it as $\breve{\mathbf{A}}_i\mathbf{w} = \mathbf{h}^*_{ir}\mathbf{h}^T_{ir}\mathbf{w}$. The product $\mathbf{h}^T_{ir}\mathbf{w}$ represents the intended two-hop equivalent channel coefficient at the ith destination, and hence, the numerator of $\mathcal{S}_1$ can be re-written as

$$\breve{\mathbf{A}}_i\mathbf{w} = \mathbf{h}^*_{ir}\mathbf{h}^T_{ir}\mathbf{w} = h^{(i)}_{\mathsf{srd},i} \cdot \mathbf{h}^*_{ir}, \tag{3.28}$$

where $h^{(i)}_{\mathsf{srd},i}$ denotes the ith element of $\mathbf{h}_{\mathsf{srd},i}$. Further, we multiply $\breve{\mathbf{A}}_i\mathbf{w}$ with $\mathbf{w}^H$ from the left-side, and obtain the received intended signal power at the ith destination, i.e.,

$$\mathbf{w}^H\breve{\mathbf{A}}_i\mathbf{w} = h^{(i)}_{\mathsf{srd},i} \cdot \mathbf{w}^H\mathbf{h}^*_{ir} = h^{(i)}_{\mathsf{srd},i} \cdot h^{(i)*}_{\mathsf{srd},i}. \tag{3.29}$$

Next, we express the denominator of $\mathcal{S}_1$ as follows by incorporating the SINR_i definition in (3.24)

$$\mathbf{w}^H\breve{\mathbf{B}}_i\mathbf{w} = \frac{\mathbf{w}^H\breve{\mathbf{A}}_i\mathbf{w}}{\text{SINR}_i} = \frac{|h^{(i)}_{\mathsf{srd},i}|^2}{\text{SINR}_i}, \tag{3.30}$$

where the second equality follows from (3.29).

We now shift our focus to $\mathcal{S}_2$ of (3.27). Recalling the definition of $\breve{\mathbf{B}}_i$ from (3.24), we multiply it with $\mathbf{w}$ from the right-side such that we express the numerator of $\mathcal{S}_2$ as

$$\breve{\mathbf{B}}_i\mathbf{w} = \mathbf{H}_{\mathrm{I},i}^{*}\mathbf{H}_{\mathrm{I},i}^{T}\mathbf{w} + \left(\sigma_{n_\mathsf{r}}^2\mathbf{I}_{N_\mathsf{r}} + \frac{\sigma_{n_\mathsf{d}}^2}{P_\mathsf{r}}\cdot(\mathbf{\Gamma}_i^{-1})^H\mathbf{M}(\mathbf{\Gamma}_i^{-1})\right)\mathbf{w}. \tag{3.31}$$

Consider the product $\mathbf{H}_{\mathrm{I},i}^{T}\mathbf{w}$ in the first summand of (3.31), and write it in an open form by using definition of $\mathbf{H}_{\mathrm{I},i}$ as

$$\begin{aligned}\mathbf{H}_{\mathrm{I},i}^{T}\mathbf{w} = \left[\hat{\mathbf{h}}_{\mathrm{sr},1}\cdots\hat{\mathbf{h}}_{\mathrm{sr},i-1}\hat{\mathbf{h}}_{\mathrm{sr},i+1}\cdots\hat{\mathbf{h}}_{\mathrm{sr},N}\right]^T\mathbf{w}, &= \left[\hat{\mathbf{h}}_{\mathrm{sr},1}^{T}\mathbf{w}\ \cdots\ \hat{\mathbf{h}}_{\mathrm{sr},i-1}^{T}\mathbf{w}\ \hat{\mathbf{h}}_{\mathrm{sr},i+1}^{T}\mathbf{w}\ \cdots\ \hat{\mathbf{h}}_{\mathrm{sr},N}^{T}\mathbf{w}\right]^T, \\ &= \left[h_{\mathrm{sd},i}^{(1)}\cdots h_{\mathrm{sd},i}^{(i-1)}\ h_{\mathrm{sd},i}^{(i+1)}\cdots h_{\mathrm{sd},i}^{(N)}\right]^T \triangleq \tilde{\mathbf{h}}_{\mathrm{sd},i},\end{aligned}$$

where the auxiliary vector $\tilde{\mathbf{h}}_{\mathsf{sd},i}$ collects all two-hop equivalent interference channels for the ith destination.

Finally, we substitute the corresponding expressions for $\breve{\mathbf{A}}_i\mathbf{w}$, $\mathbf{w}^H\breve{\mathbf{B}}_i\mathbf{w}$, $\breve{\mathbf{B}}_i\mathbf{w}$, derived above in (3.28), (3.30), (3.31), respectively, into the gradient $\nabla_{\mathbf{w}}\mathrm{SINR}_i$ given by (3.27), i.e.,

$$\begin{aligned}\nabla_{\mathbf{w}}\mathrm{SINR}_i = \Bigg(h_{\mathsf{srd},i}^{(i)}\cdot\mathbf{h}_{i\mathsf{r}}^{*} - \Big(\mathbf{H}_{\mathrm{I},i}^{*}\tilde{\mathbf{h}}_{\mathsf{srd},i} + (\sigma_{n_\mathsf{r}}^2\mathbf{I}_{N_\mathsf{r}} \\ + \frac{\sigma_{n_\mathsf{d}}^2}{P_\mathsf{r}}\cdot(\mathbf{\Gamma}_i^{-1})^H\mathbf{M}(\mathbf{\Gamma}_i^{-1}))\mathbf{w}\Big)\mathrm{SINR}_i\Bigg)\frac{\mathrm{SINR}_i}{|h_{\mathsf{srd},i}^{(i)}|^2}.\end{aligned}$$

Note that $\nabla_{\mathbf{w}}\mathrm{SINR}_i$ is an $N_\mathsf{r}\times 1$ complex vector, and we are interested in identifying the kth element out of it, which corresponds to the kth relay's scaled gain coefficient w_k. Consequently, we reduce $\nabla_{\mathbf{w}}\mathrm{SINR}_i$ to

$$\begin{aligned}\nabla_{\mathbf{w}}\mathrm{SINR}_{i,k} = \Bigg(h_{k,i}^{*}\cdot h_{\mathsf{srd},i}^{(i)} - \Big(\mathbf{e}_k^T\mathbf{H}_{\mathrm{I},i}^{*}\tilde{\mathbf{h}}_{\mathsf{srd},i} + (\sigma_{n_\mathsf{r}}^2 \\ + \frac{\sigma_{n_\mathsf{d}}^2}{P_\mathsf{r}}\cdot\frac{(\mathbf{h}_{\mathsf{s}k}^H\mathbf{h}_{\mathsf{s}k} + \sigma_{n_\mathsf{r}}^2)}{|f_{i,k}|^2})w_k\Big)\mathrm{SINR}_i\Bigg)\frac{\mathrm{SINR}_i}{|h_{\mathsf{srd},i}^{(i)}|^2},\end{aligned}$$

where $h_{k,i}$ and $f_{i,k}$ denote the kth element of $\mathbf{h}_{i\mathsf{r}}$ and $\mathbf{f}_{\mathsf{r}i}$, respectively (recall from Section 2.2). As a last step, the gradient with respect to the physical gain vector $\mathbf{g}$ follows as [39]

$$\nabla_{\mathbf{g}}\mathrm{SINR}_{i,k} = \frac{\partial\mathrm{SINR}_i}{\partial w_k^*}\cdot\frac{\partial w_k^*}{\partial g_k^*} = \nabla_{\mathbf{w}}\mathrm{SINR}_{i,k}\cdot f_{i,k}.$$

In order to compute $\nabla_{\mathbf{g}}\mathrm{SINR}_{i,k}$ locally, the kth relay requires the information set

$$\mathcal{I} = \left\{h_{k,i}, f_{i,k}, \mathbf{h}_{\mathsf{srd},i}, \mathbf{e}_k^T\mathbf{H}_{\mathrm{I},i}^{*}, \mathbf{h}_{\mathsf{s}k}, \mathrm{SINR}_i\right\}.$$

However, we can immediately remove $h_{k,i}$ and $f_{i,k}$ from the set by noting that $h_{k,i}$ is the ith element of $\mathbf{h}_{\mathsf{s}k}$ and $f_{i,k}$ is the ith element of $\mathbf{f}_{k\mathsf{d}}$. Further, we realize that the elements of the kth row of $\mathbf{H}^*_{\mathrm{I},i}$, i.e.,

$$\mathbf{e}_k^T\mathbf{H}^*_{\mathrm{I},i} = \begin{bmatrix} h_{k,1} & \cdots & h_{k,i-1} & h_{k,i+1} & \cdots & h_{k,N} \end{bmatrix} \tag{3.32}$$

are subset of the elements of $\mathbf{h}_{\mathsf{s}k}$. Hence, the required information set is reduced to

$$\tilde{\mathcal{I}} = \left\{\mathbf{h}_{\mathsf{srd},i}, \mathbf{h}_{\mathsf{s}k}, \mathbf{f}_{k\mathsf{d}}, \mathrm{SINR}_i\right\},$$

and with this we conclude the proof. □

The next question to answer is how to gather these necessary information identified in *Lemma* 3.3.1. We have already mentioned in the introduction section that local CSI estimation per relay is handled without additional overhead, by assuming that the terminals' packets anyway carry orthogonal training sequences. The destination nodes in turn can use the training sequences to estimate the respective rows of the equivalent channel matrix $\mathbf{H}_{\mathsf{srd}}$ and the receive SINR. This information is subsequently fed back (broadcasted) to all relays. In the sequel, we will refer to the process of training sequence transmission and parameter estimation as the *estimate cycle*, and to the feedback of the estimated parameter values as the *feedback cycle*. Based on its local channel estimates and the feedback of the destinations, each relay can compute the corresponding element of the gradient locally and independently from the other relays in the network.

3.3.1.2. Distributed Step Size Calculation

Having decided in which direction to go along, i.e., the gradient, we need to now determine how much we should iterate on this direction, i.e., the optimal step-size. In general, the step-size can be efficiently computed with exact or backtracking line searches [26]. In other words, the following unconstrained optimization over μ_n

$$\underset{\mu_n \geq 0}{\arg\max} \frac{(\mathbf{g}_n + \mu_n \nabla_{\mathbf{g},n}\mathrm{SINR}_i)^H \breve{\mathbf{A}}_{\Gamma,i}(\mathbf{g}_n + \mu_n \nabla_{\mathbf{g},n}\mathrm{SINR}_i)}{(\mathbf{g}_n + \mu_n \nabla_{\mathbf{g},n}\mathrm{SINR}_i)^H \breve{\mathbf{B}}_{\Gamma,i}(\mathbf{g}_n + \mu_n \nabla_{\mathbf{g},n}\mathrm{SINR}_i)} \tag{3.33}$$

is performed, where $\breve{\mathbf{A}}_{\Gamma,i} \triangleq \mathbf{\Gamma}_i^H \breve{\mathbf{A}}_i \mathbf{\Gamma}_i$ and $\breve{\mathbf{B}}_{\Gamma,i} \triangleq \mathbf{\Gamma}_i^H \breve{\mathbf{B}}_i \mathbf{\Gamma}_i$.

In this subsection, we aim at designing schemes which perform (3.33) in a distributed manner without requiring global CSI. To this end, we propose a centralized and a decentralized protocol where both employ only local information at the relays and efficiently benefit from several additional *estimate* and *feedback* cycles. In the decentralized case, all relays participate to the step size search, whereas the centralized scheme assigns a master destination to compute the optimal value, which is later distributed back to the relays.

The Decentralized Protocol The decentralized protocol uses an *inner loop*, which consists of several estimate cycles and a 1-bit feedback cycle to determine the optimal step-size. Initially we define a minimum value $\mu_n^{\min}$, and a maximum value $\mu_n^{\max}$ of the step-size. Starting from $\mu_n^{\min}$ and iterating until $\mu_n^{\max}$ with an appropriate inner step-size δ, the network performs multiple estimate cycles. In the qth cycle the step-size $\mu_{n,q} \triangleq \mu_n^{\min} + (q-1)\delta$ is used. The destination measures the corresponding SINR at its antenna for each step-size choice $\mu_{n,q}$. As soon as it detects that SINR is reduced with respect to the previous choice, it feeds back a stop bit through a dedicated control channel. This terminates the inner loop and the relays use the step-size of the previous cycle as the optimal step-size for the present gradient. Note that in Section 3.3.1.2, we will introduce different variants of the step-size iteration for the sake of reducing the number of estimation cycles.

Optional Relay Power Scaling: The gain vector in this inner loop does not necessarily satisfy the instantaneous relay sum power constraint (3.22). Although this does not affect the convergence to the optimal value, it may influence the convergence speed. As an optional scheme to cope with this, we can normalize the updated gain vector at each inner loop iteration such that the relay sum transmit power is independent of the choice μ_n. For a given $\mu_{n,q}$, the sum transmit power without gain vector normalization is given by

$$\mathbf{g}_{n+1}^H \mathbf{M} \mathbf{g}_{n+1} = \|\mathbf{M}^{\frac{1}{2}} \mathbf{g}_n\|^2 + \mu_{n,q}^2 \|\mathbf{M}^{\frac{1}{2}} \cdot \nabla_{\mathbf{g},n} \mathrm{SINR}_i\|^2 + 2\mu_{n,q} \mathrm{Re}\{(\nabla_{\mathbf{g},n} \mathrm{SINR}_i)^H \mathbf{M} \mathbf{g}_n\}. \quad (3.34)$$

If the three summands in (3.34) are known at the relays, the necessary scaling can also be determined locally. Hence, we propose to measure these quantities at the destination and feed them back to the relays. Each measurement requires one channel use.

The key idea is to use *over-the-air addition* through the relay-to-destination channel to perform the required matrix multiplications in (3.34). During these *measurement cycles*, the relays do not act as forwarders but rather each transmit a specific locally computable symbol value each. Let us now explain the *over-the-air addition* for measuring $\|\mathbf{M}^{\frac{1}{2}} \mathbf{g}_n\|^2$ at the intended destination. With the available local CSI, the kth relay computes a transmit symbol $\alpha_k \in \mathbb{C}$, which is defined as

$$\alpha_k \triangleq \frac{|g_{n,k}|^2 \cdot (\mathbf{h}_{\mathsf{s}k}^H \mathbf{h}_{\mathsf{s}k} + \sigma_{n_\mathsf{r}}^2)}{f_{i,k}},$$

where the square root of the numerator represents the absolute value of the kth element of $\mathbf{M}^{\frac{1}{2}} \mathbf{g}_n$ (see (3.22)), and the denominator compensates the forward channel. As all N_r relays concurrently transmit the respective α_k, the received signal at the ith destination becomes

$$\mathbf{y}_{\mathsf{d},i} = \sum_{k=1}^{N_\mathsf{r}} f_{i,k} \alpha_k = \sum_{k=1}^{N_\mathsf{r}} |g_{n,k}|^2 \cdot (\mathbf{h}_{\mathsf{s}k}^H \mathbf{h}_{\mathsf{s}k} + \sigma_{n_\mathsf{r}}^2),$$

which perfectly computes $\|\mathbf{M}^{\frac{1}{2}} \mathbf{g}_n\|^2$. Thus, the destination measures this received signal amplitude, and feeds it back to the relays. Note that for such a *measurement cycle*, we require only one forward and one feedback channel use[1].

Likewise, in order to measure the second and the third summands in (3.34), we perform a very similar *measurement cycle*, where the kth relay transmits

$$|\nabla_{\mathbf{g},n} \mathrm{SINR}_{i,k}|^2 \cdot \frac{\mathbf{h}_{\mathsf{s}k}^H \mathbf{h}_{\mathsf{s}k} + \sigma_{n_\mathsf{r}}^2}{f_{i,k}} \quad \text{and} \quad \nabla_{\mathbf{g},n} \mathrm{SINR}_{i,k} \cdot g_{n,k}^* \cdot \frac{\mathbf{h}_{\mathsf{s}k}^H \mathbf{h}_{\mathsf{s}k} + \sigma_{n_\mathsf{r}}^2}{f_{i,k}},$$

respectively. Note that these measurement and feedback cycles are performed only once at the beginning of the step-size (line) search.

The Centralized Protocol Here instead of active participation of each relay in the step-size search, the ith destination computes the optimal μ_n through (3.33) locally, and then broadcasts it back to all relays in the network. To this end, assuming *no* global CSI, the destination needs to know the following six terms that build up (3.33):

$$\Big\{ \mathbf{g}_n^H \breve{\mathbf{A}}_{\Gamma,i} \mathbf{g}_n, \mathbf{g}_n^H \breve{\mathbf{B}}_{\Gamma,i} \mathbf{g}_n, (\nabla_{\mathbf{g},n} \mathrm{SINR}_i)^H \breve{\mathbf{A}}_{\Gamma,i} (\nabla_{\mathbf{g},n} \mathrm{SINR}_i), (\nabla_{\mathbf{g},n} \mathrm{SINR}_i)^H \breve{\mathbf{B}}_{\Gamma,i} (\nabla_{\mathbf{g},n} \mathrm{SINR}_i),$$
$$\mathrm{Re}\{\mathbf{g}_n^H \breve{\mathbf{A}}_{\Gamma,i} (\nabla_{\mathbf{g},n} \mathrm{SINR}_i)\}, \mathrm{Re}\{\mathbf{g}_n^H \breve{\mathbf{B}}_{\Gamma,i} (\nabla_{\mathbf{g},n} \mathrm{SINR}_i)\} \Big\}.$$

First, since the destination, as well as all relays, already have the knowledge of SINR_i and $\mathbf{h}_{\mathsf{srd},i}$, the first and the second terms are locally available, i.e.,

$$\mathbf{g}_n^H \breve{\mathbf{A}}_{\Gamma,i} \mathbf{g}_n = |h_{\mathsf{srd},i}^{(i)}|^2, \quad \text{and} \quad \mathbf{g}_n^H \breve{\mathbf{B}}_{\Gamma,i} \mathbf{g}_n = |h_{\mathsf{srd},i}^{(i)}|^2 / \mathrm{SINR}_i.$$

[1]We assume ideally that one channel use is enough for perfect (error-free) estimation or feedback.

Second, in order to measure the third, the fourth and the fifth terms, the network performs an additional two-hop *estimate and feedback* cycle (see Section 3.3.1.1). Through this cycle, the relays use $\nabla_{\mathbf{g},n}\mathrm{SINR}_i$ as the relay gain vector instead of $\mathbf{g}_n$. Hence, the destination estimates the corresponding received SINR and two-hop equivalent channel, which are respectively denoted by SINR_i^{∇} and $\mathbf{h}_{\mathsf{srd},i}^{\nabla}$ to depict the difference with SINR_i and $\mathbf{h}_{\mathsf{srd},i}$ obtained previously through $\mathbf{g}_n$. Consequently, with SINR_i^{∇} and $\mathbf{h}_{\mathsf{srd},i}^{\nabla}$ at the destination, the related terms can be computed as

$$\begin{aligned}(\nabla_{\mathbf{g},n}\mathrm{SINR}_i)^H \breve{\mathbf{A}}_{\Gamma,i}(\nabla_{\mathbf{g},n}\mathrm{SINR}_i) &= |h_{\mathsf{srd},i}^{\nabla(i)}|^2,\\ (\nabla_{\mathbf{g},n}\mathrm{SINR}_i)^H \breve{\mathbf{B}}_{\Gamma,i}(\nabla_{\mathbf{g},n}\mathrm{SINR}_i) &= |h_{\mathsf{srd},i}^{\nabla(i)}|^2/\mathrm{SINR}_i^{\nabla},\\ \mathrm{Re}\{\mathbf{g}_n^H \breve{\mathbf{A}}_{\Gamma,i}(\nabla_{\mathbf{g},n}\mathrm{SINR}_i)\} &= \mathrm{Re}\{h_{\mathsf{srd},i}^{(i)*} h_{\mathsf{srd},i}^{\nabla(i)}\}.\end{aligned}$$

The last term can only be partially calculated with the available information at the destination, i.e., SINR_i, SINR_i^{∇}, $\mathbf{h}_{\mathsf{srd},i}$ and $\mathbf{h}_{\mathsf{srd},i}^{\nabla}$. Taking a closer look, we have

$$\begin{aligned}\mathrm{Re}\Big\{\mathbf{g}_n^H \breve{\mathbf{B}}_{\Gamma,i}(\nabla_{\mathbf{g},n}\mathrm{SINR}_i)\Big\} &= \mathrm{Re}\Big\{\sigma_{n_\mathsf{r}}^2 \mathbf{g}_n^H \mathbf{\Gamma}_i^H \mathbf{\Gamma}_i (\nabla_{\mathbf{g},n}\mathrm{SINR}_i)\\ &+ \mathbf{g}_n^H \mathbf{\Gamma}_i^H \mathbf{H}_{\mathrm{I},i}^* \mathbf{H}_{\mathrm{I},i}^T \mathbf{\Gamma}_i (\nabla_{\mathbf{g},n}\mathrm{SINR}_i) + \frac{\sigma_{n_\mathsf{d}}^2}{P_\mathsf{r}} \cdot \mathbf{g}_n^H \mathbf{M}(\nabla_{\mathbf{g},n}\mathrm{SINR}_i)\Big\}, \quad (3.35)\end{aligned}$$

where the second summand can be determined using $\mathbf{h}_{\mathsf{srd},i}$ and $\mathbf{h}_{\mathsf{srd},i}^{\nabla}$ jointly, i.e.,

$$\mathbf{g}_n^H \mathbf{\Gamma}_i^H \mathbf{H}_{\mathrm{I},i}^* \mathbf{H}_{\mathrm{I},i}^T \mathbf{\Gamma}_i (\nabla_{\mathbf{g},n}\mathrm{SINR}_i) = \sum_{j=1, j\neq i}^{N} \mathbf{h}_{\mathsf{srd},i}^{(j)*} \cdot \mathbf{h}_{\mathsf{srd},i}^{\nabla(j)}.$$

However, to obtain the first and the third summands in (3.35), we need some additional *over-the-air addition* based *measurement* cycles similar to the ones used for computing (3.34). For the $\sigma_{n_\mathsf{r}}^2 \mathbf{g}_n^H \mathbf{\Gamma}_i \mathbf{\Gamma}_i^H (\nabla_{\mathbf{g},n}\mathrm{SINR}_i)$ term, a concurrent transmission is performed from all relays to the destination, where the kth relay transmits the symbol $\alpha_k = (g_{n,k}^* |f_{i,k}|^2 \cdot \nabla_{\mathbf{g},n}\mathrm{SINR}_{i,k})/f_{i,k}$. Lastly, the third term in (3.35) is similarly obtained as explained in the decentralized option. In summary, the destination is capable of performing the unconstrained maximization (3.33) with the available information gathered after several *measurement* cycles. As the optimal μ_n^* is found, it is broadcasted to all relays. Note that in order to cope with the monotonic increase of SINR with respect to μ_n, this scheme also requires an appropriate scaling as introduced in the previous subsection.

Discussions Comparing the two protocols we observe that both require some measurement cycles to keep the relay sum transmit power fixed, where specific relay transmit symbols are used to efficiently evaluate matrix expressions. The decentralized scheme furthermore occupies the relay-to-destination channel for multiple SINR estimations to iteratively determine the optimum step-size. In contrast, the centralized option requires estimate cycles, i.e., training sequences from all sources, to determine some constants in (3.33) at the destination. In general the decentralized protocol is preferable in terms of overhead, if few inner iterations are necessary. For this reason, we use the centralized option in the acquisition phase, and the decentralized option in the tracking phase of the gradient scheme.

At first glance the use of the relay-to-destination forward channel by the decentralized approach may be seen as a waste of resources. Actually, the step-size optimization comes almost free in terms of transmission/feedback overhead, if we concurrently transmit data. Due to the iterative nature of the gradient scheme, the objective function (SINR for the current case) monotonically improves with each outer and/or inner iteration. Hence, at any iteration and especially during the tracking phase, i.e., after the first few outer iterations, although not with the maximal achievable rate, the system can still provide data transmission up to a certain rate. Consequently, while iterating over each step-size choice $\mu_{n,q}$, we propose to benefit from these forward channel uses to transmit actual data with a fixed but lower rate with respect to the optimal rate that will be achieved at the end of gradient algorithm.

3.3.1.3. Conjugate Gradient and Convergence Issues

The conjugate gradient (CG) algorithm, where the search direction is chosen such that it is orthogonal to the previously searched directions, is a commonly preferred scheme due to its fast convergence properties with respect to conventional gradient algorithms. With CG, the update equation and the search direction are modified respectively to

$$\begin{aligned} \mathbf{g}_{n+1} &= \mathbf{g}_n + \mu_n \boldsymbol{\phi}_n, \\ \boldsymbol{\phi}_n &= \nabla_{\mathbf{g},n}\mathrm{SINR}_i + \beta_n \cdot \nabla_{\mathbf{g}n-1}\mathrm{SINR}_i. \end{aligned}$$

There are several different formulas to compute the constant $\beta_n \in \mathbb{R}_+$ such as the Fletcher-Reeves (FR) and Polak-Ribiere (PR) methods [112]:

$$\beta_{\mathrm{FR},n} = \frac{(\nabla_{\mathbf{g},n}\mathrm{SINR}_i)^H(\nabla_{\mathbf{g},n}\mathrm{SINR}_i)}{(\nabla_{\mathbf{g}n-1}\mathrm{SINR}_i)^H(\nabla_{\mathbf{g}n-1}\mathrm{SINR}_i)},$$

$$\beta_{\mathrm{PR},n} = \frac{(\nabla_{\mathbf{g},n}\mathrm{SINR}_i)^H\Big((\nabla_{\mathbf{g},n}\mathrm{SINR}_i) - (\nabla_{\mathbf{g},n-1}\mathrm{SINR}_i)\Big)}{(\nabla_{\mathbf{g}n-1}\mathrm{SINR}_i)^H(\nabla_{\mathbf{g}n-1}\mathrm{SINR}_i)}.$$

Of the two methods, we choose to use the FR method. In order to determine the $\beta_{\mathrm{FR},n}$ at each iteration, the relay nodes need to know the norm square of the gradient vector, i.e., $\|\nabla_{\mathbf{g},n}\mathrm{SINR}_i\|^2$. This information can be obtained through a single *measure and feedback* cycle, where a concurrent transmission is performed from all relays to the ith destination, and the kth relay transmits the symbol $|\nabla_{\mathbf{g},n}\mathrm{SINR}_{i,k}|^2/f_{i,k}$. For further details of the CG algorithm, we refer the interested reader to [112].

The CG methods for the solution of *(generalized) Rayleigh quotient* problems have been well-studied in the literature (see [45, 140] and references therein), where several different algorithms with various complexity are proposed and the corresponding convergence analyses are performed. Henceforth, we employ the CG scheme in order to accelerate the convergence speed of our distributed scheme. We emphasize here that the gradient computation and the step-sizes obtained by the proposed distributed gradient algorithm are not approximations but exactly equivalent to the ones that would be obtained by a central node having global CSI. Thus, we drop the proof of convergence of the conjugate gradient adaptation for the sake of brevity and refer the aforementioned references. However, in order to gain insight on the convergence behaviour of our specific problem, in the following we present a brief convergence analysis for the case of conventional gradient search *without* conjugate directions.

Convergence Proof Defining

$$\bar{\mathbf{g}} \triangleq (\mathbf{\Sigma}_{\breve{\mathrm{B}}_i})^{\frac{1}{2}} \cdot \mathbf{U}_{\breve{\mathrm{B}}_i}^H \cdot \mathbf{w},$$
$$\mathbf{T}_i \triangleq (\mathbf{\Sigma}_{\breve{\mathrm{B}}_i})^{\frac{-1}{2}} \cdot \mathbf{U}_{\breve{\mathrm{B}}_i}^H \cdot \breve{\mathbf{A}}_i \cdot \mathbf{U}_{\breve{\mathrm{B}}_i} \cdot (\mathbf{\Sigma}_{\breve{\mathrm{B}}_i})^{\frac{-1}{2}},$$

the SINR_i in (3.24) and the corresponding gradient with respect to $\bar{\mathbf{g}}$ are written as

$$\mathrm{SINR}_i = \frac{\bar{\mathbf{g}}^H\mathbf{T}_i\bar{\mathbf{g}}}{\bar{\mathbf{g}}^H\bar{\mathbf{g}}}, \quad \text{and} \quad \nabla_{\bar{\mathbf{g}}}\mathrm{SINR}_i = \frac{\mathbf{T}_i\bar{\mathbf{g}}}{\bar{\mathbf{g}}^H\bar{\mathbf{g}}} - \frac{\bar{\mathbf{g}}^H\mathbf{T}_i\bar{\mathbf{g}}}{(\bar{\mathbf{g}}^H\bar{\mathbf{g}})^2} \cdot \bar{\mathbf{g}}. \tag{3.36}$$

Let $\mathbf{T}_i = \mathbf{U}_{\mathrm{T}_i}\mathbf{\Sigma}_{\mathrm{T}_i}\mathbf{U}_{\mathrm{T}_i}^H$, and further define $\hat{\mathbf{g}} \triangleq \mathbf{U}_{\mathrm{T}_i}^H\bar{\mathbf{g}}$ such that we transform the coordinates, which results in an optimal real-valued gain vector $\hat{\mathbf{g}}^\star = [1\ 0\ \cdots\ 0]^T$ within this transformed optimization space. After some algebraic manipulations and incorporating $\hat{\mathbf{g}}$, the update equation at the nth iteration of the gradient algorithm can be equivalently written as

$$\hat{\mathbf{g}}_{n+1} = \left(1 - \mu_n\frac{\hat{\mathbf{g}}_n^H\mathbf{\Sigma}_{\mathrm{T}_i}\hat{\mathbf{g}}_n}{\|\hat{\mathbf{g}}_n\|^4}\right)\hat{\mathbf{g}}_n + \frac{\mu_n}{\|\hat{\mathbf{g}}_n\|^2}\mathbf{\Sigma}_{\mathrm{T}_i}\hat{\mathbf{g}}_n. \tag{3.37}$$

Noting that $\breve{\mathbf{A}}_i$ is a rank-1 matrix, and hence, $\mathbf{T}_i$ has only one nonzero eigenvalue, and further assuming that $\boldsymbol{\Sigma}_{\mathrm{T}_i}$ has the eigenvalues in a decreasing order, (3.37) can be modified to

$$\hat{\mathbf{g}}_{n+1} = \left(1 - \mu_n \frac{\mathrm{SINR}_{\mathrm{opt}} |\hat{g}_{n,1}|^2}{\|\hat{\mathbf{g}}_n\|^4}\right)\hat{\mathbf{g}}_n + \frac{\mu_n \mathrm{SINR}_{\mathrm{opt}}}{\|\hat{\mathbf{g}}_n\|^2}\breve{\mathbf{g}}_n, \quad (3.38)$$

where $\mathrm{SINR}_{\mathrm{opt}}$ stands for the non-zero eigenvalue, i.e., the optimal result, and

$$\breve{\mathbf{g}}_n \in \mathbb{R}^{N_r} \triangleq [\hat{g}_{n,1}\ 0\ \cdots\ 0]^T.$$

Having a close look on (3.38), it is immediately noticed that the convergence behaviour of the first element $\hat{g}_{n,1}$ and the other elements are different, and given respectively as

$$\hat{g}_{n+1,1} = \hat{g}_{n,1}\left(1 + \frac{\mu_n \mathrm{SINR}_{\mathrm{opt}}}{\|\hat{\mathbf{g}}_n\|^2}\left(1 - \frac{|\hat{g}_{n,1}|^2}{\|\hat{\mathbf{g}}_n\|^2}\right)\right), \quad (3.39)$$

$$\hat{g}_{n+1,k} = \hat{g}_{n,k}\left(1 - \frac{\mu_n \mathrm{SINR}_{\mathrm{opt}}}{\|\hat{\mathbf{g}}_n\|^2} \cdot \frac{|\hat{g}_{n,1}|^2}{\|\hat{\mathbf{g}}_n\|^2}\right), \quad (3.40)$$

for $k = 2, \ldots, N_r$.

Assuming a nonzero positive starting value for $\hat{g}_{0,1}$ and noticing that $|\hat{g}_{n,1}|^2/\|\hat{\mathbf{g}}_n\|^2 \leq 1$ for any n, convergence of $\hat{g}_1$ can be immediately concluded from the fact that at each iteration the error with respect to the optimal result, i.e., $\hat{g}_1^\star = 1$, reduces by a factor of

$$\frac{\mu_n \mathrm{SINR}_{\mathrm{opt}}}{\|\hat{\mathbf{g}}_n\|^2}\left(1 - \frac{|\hat{g}_{n,1}|^2}{\|\hat{\mathbf{g}}_n\|^2}\right) \geq 0,$$

which attains equality only when $\hat{g}_{n,1} = 1$, i.e., the optimal result is obtained. Likewise, following a similar reasoning and assuming nonzero positive starting value for any $\hat{g}_k$, $k \in \{2, \ldots, N\}$, the error with respect to the optimal result, i.e., $\hat{g}_k^\star = 0$, reduces by a factor of

$$\frac{\mu_n \mathrm{SINR}_{\mathrm{opt}}}{\|\hat{\mathbf{g}}_n\|^4}|\hat{g}_{n,1}|^2 \geq 0,$$

which attains equality only when $\hat{g}_{n,k} = 0$, i.e., the optimal result is obtained. From both (3.39) and (3.40), we immediately observe that choosing $\hat{g}_{0,1} = 0$ would lead the algorithm not to converge, which emphasizes the importance of an appropriate starting gain vector.

The Overall Algorithm After adapting the gradient algorithm to our coherent relaying scenario and correspondingly proposing distributed solutions to each step of it, we summarize the final overall SINR maximization algorithm with **Algorithm 3**.

Algorithm 3 Distributed Gradient Algorithm for SINR Maximization

initiate $\epsilon > 0, \iota > 1$, feasible starting vector $\mathbf{g}_0$.
repeat "Outer search" - the nth iteration
  Step 1. Estimate/measure $\left\{ \begin{array}{l} \mathbf{h}_{\text{srd},i}, \text{SINR}_i, \|\mathbf{M}^{\frac{1}{2}}\mathbf{g}_n\|^2, \|\mathbf{M}^{\frac{1}{2}}\nabla_{\mathbf{g},n}\text{SINR}_i\|^2, \\ \text{Re}\{(\nabla_{\mathbf{g},n}\text{SINR}_i)^H\mathbf{M}^{\frac{1}{2}}\mathbf{g}_n\}, \|\nabla_{\mathbf{g},n}\text{SINR}_i)\|^2. \end{array} \right\}$
  Step 2. Feedback the information obtained in Step 1 to the relays.
  Step 3. Each relay computes β_n and the related element of the gradient.
  Step 4. Update $\phi_n := \nabla_{\mathbf{g},n}\text{SINR}_i + \beta_n\nabla_{g,n-1}\text{SINR}_i$.
  Step 5. "Inner search" - the optimal step-size
    - Employ either decentralized or centralized protocol.
  Step 6. Update $\mathbf{g}_{n+1} := \mathbf{g}_n + \mu_n^\star\phi_n$.
until $\text{SINR}_{i,n+1} - \text{SINR}_{i,n} \leq \epsilon$.

3.3.2. Multiple Source-Destination Pairs

In the case of multiple S-D pairs, the relays can employ the local gradient calculation scheme in Section 3.3.1.1 to determine the gradient of the SINR for each destination. However, the important decision to make in the multiuser case is how to employ the gradients, i.e., what is the objective function? As it is of common practical interest while designing multiuser systems, in this thesis we choose the sum rate of the N S-D pairs as the primary objective function to be maximized. Besides, we briefly investigate the *max-min* fairness as an another cost function for the multiuser setup, and present concisely the corresponding distributed algorithm.

3.3.2.1. Sum Rate Maximization

We recall the sum rate expression for N S-D pairs

$$\text{R}_{\text{sum}} = \sum_{i=1}^{N} \log_2(1 + \text{SINR}_i).$$

The corresponding sum rate maximization problem is readily given in (3.3). A conjugate gradient based algorithm combined with the barrier method has been proposed to solve it assuming that global CSI knowledge is available (see **Algorithm 1**). Note that simplifications made on the SINR expression and the power constraint in this limited feedback section, which are due to the assumed single-antenna relays, do not change the structure of the problem or the proposed algorithm, but just affect the input data, i.e., the size of channel matrices.

The gradient of the sum rate with respect to the relay gain vector $\mathbf{g}$ is

$$\nabla_{\mathbf{g}}\mathrm{R}_{\mathsf{sum}} = \frac{1}{\log_e(2)} \cdot \sum_{i=1}^{N} \frac{1}{1+\mathrm{SINR}_i} \nabla_{\mathbf{g}}\mathrm{SINR}_i. \tag{3.41}$$

As each relay can compute its elements of the gradients $\nabla_{\mathbf{g}}\mathrm{SINR}_i$, and has acquired knowledge of all SINR_i in this process, it can as well compute its contribution to $\nabla_{\mathbf{g}}\mathrm{R}_{\mathsf{sum}}$.

An algorithm similar to the SINR maximization is used to maximize $\mathrm{R}_{\mathsf{sum}}$ by incorporating (3.41) in the update equation. As explained for the single S-D pair case in Section 3.3.1.1, the destination nodes use orthogonal training sequences to estimate the respective rows of the equivalent channel matrix $\mathbf{H}_{\mathsf{srd}}$ and the receive SINR_i, and feed them all back to the relays (N estimate and feedback cycles). Consequently, each relay has the full knowledge of the equivalent channel $\mathbf{H}_{\mathsf{srd}}$.

On the other hand, in order to find the optimal step-size μ_n during the sum rate maximization, the following unconstrained optimization is performed over μ_n:

$$\underset{\mu_n \geq 0}{\mathbf{arg\,max}} \sum_{i=1}^{N} \log_2\left(1 + \frac{(\mathbf{g}_n + \mu_n \cdot \nabla_{\mathbf{g},n}\mathrm{R}_{\mathsf{sum}})^H \breve{\mathbf{A}}_{\Gamma,i} (\mathbf{g}_n + \mu_n \cdot \nabla_{\mathbf{g},n}\mathrm{R}_{\mathsf{sum}})}{(\mathbf{g}_n + \mu_n \cdot \nabla_{\mathbf{g},n}\mathrm{R}_{\mathsf{sum}})^H \breve{\mathbf{B}}_{\Gamma,i} (\mathbf{g}_n + \mu_n \cdot \nabla_{\mathbf{g},n}\mathrm{R}_{\mathsf{sum}})}\right). \tag{3.42}$$

Both protocols for step-size calculation are applicable with a few modifications and some additional *estimate and feedback* cycles due to the increased number of cross-terms of the signal, interference and noise.

Since both approaches need a scaling to keep the instantaneous transmit power fixed during the step-size search, $\|\mathbf{M}^{\frac{1}{2}}\mathbf{g}_n\|^2$, $\|\mathbf{M}^{\frac{1}{2}}\nabla_{\mathbf{g},n}\mathrm{R}_{\mathsf{sum}}\|^2$, $\mathrm{Re}\{(\nabla_{\mathbf{g},n}\mathrm{R}_{\mathsf{sum}})^H\mathbf{M}\mathbf{g}_n\}$ should be measured at (any) ith destination and fed back to the relays. To this end, all relays make three concurrent transmission, where the kth relay transmits the symbol values

$$|g_{n,k}|^2 \cdot (\mathbf{h}_{\mathsf{s}k}^H\mathbf{h}_{\mathsf{s}k} + \sigma_{n_{\mathsf{d}}}^2)/f_{i,k},$$
$$|\nabla_{\mathbf{g},n}\mathrm{R}_{\mathsf{sum},k}|^2 \cdot (\mathbf{h}_{\mathsf{s}k}^H\mathbf{h}_{\mathsf{s}k} + \sigma_{n_{\mathsf{d}}}^2)/f_{i,k},$$
$$\mathrm{Re}\{\nabla_{\mathbf{g},n}\mathrm{R}_{\mathsf{sum},k} \cdot \mathbf{g}_{n,k}^*\} \cdot (\mathbf{h}_{\mathsf{s}k}^H\mathbf{h}_{\mathsf{s}k} + \sigma_{n_{\mathsf{d}}}^2)/f_{i,k},$$

respectively. Then, the corresponding received signal amplitudes are estimated at the chosen ith destination, and fed back to the relays. Further details regarding the extension of the protocols to the multiuser setup are given as follows.

The Decentralized Protocol The protocol follows the same principles explained in Section 3.3.1.2 with an exception that a dedicated destination can not here optimally decide when to stop the inner loop. In other words, since each destination can only measure its own SINR (hence, rate), but not the others', the decision regarding the sum rate should be given jointly in between the destinations. Such a joint decision-making may need several inter-channel uses, and hence, impairs the spectral efficiency. Henceforth, we propose that as soon as one of the destinations detects that its SINR is reduced with respect to the previous step-size choice, it feeds back a stop bit, and the inner loop is terminated. Nevertheless, such a suboptimal approach would not necessarily yield the optimal step-size, as the sum rate may not decrease while individual rate of some links reduces.

The Centralized Protocol For the centralized approach, any one destination out of N (say the ith) is chosen as the master that computes the optimal $\mu_n^\star$. Hence, $\mathbf{h}_{\mathsf{srd},j}$ and $\mathrm{SINR}_j, \forall j \in \{1,\dots,N\}, j \neq i$, are also needed to be fed back to this master destination. Expanding the ith summand in (3.42), we obtain $\mathbf{g}_n^H \breve{\mathbf{A}}_{\Gamma,i}\mathbf{g}_n$, $\mathbf{g}_n^H \breve{\mathbf{B}}_{\Gamma,i}\mathbf{g}_n$, $(\nabla_{\mathbf{g},n}\mathrm{R}_{\mathsf{sum}})^H \breve{\mathbf{A}}_{\Gamma,i}(\nabla_{\mathbf{g},n}\mathrm{R}_{\mathsf{sum}})$, $(\nabla_{\mathbf{g},n}\mathrm{R}_{\mathsf{sum}})^H \breve{\mathbf{B}}_{\Gamma,i}(\nabla_{\mathbf{g},n}\mathrm{R}_{\mathsf{sum}})$, $\mathrm{Re}\{\mathbf{g}_n^H \breve{\mathbf{A}}_{\Gamma,i}(\nabla_{\mathbf{g},n}\mathrm{R}_{\mathsf{sum}})\}$, and $\mathrm{Re}\{\mathbf{g}_n^H \breve{\mathbf{B}}_{\Gamma,i}(\nabla_{\mathbf{g},n}\mathrm{R}_{\mathsf{sum}})\}$, which should be known by the master destination $\forall i \in \{1,\dots,N\}$. For the sake of brevity, we drop further details, and note that the master destination can collect these terms following very similar but increased number of *measure* and *feedback* cycles explained in Section 3.3.1.2.

3.3.2.2. Maximizing the Minimum SINR

The max-min (fairness) beamforming problem has been introduced in Section 3.2.2, and an SDP based solution has been proposed by assuming that global CSI knowledge is perfectly available at each relay. In this section, we aim to adapt the distributed gradient approach to max-min beamforming for coherent relaying.

Unfortunately, we can not directly apply the gradient approach to (3.10), since the gradient, i.e., derivative, of $\min_{1\leq i\leq N} \mathrm{SINR}_i$ with respect to $\mathbf{g}$ may not exist for some $\mathbf{g}$. However, benefiting from the SINR maximization approach proposed in Section 3.3.1, we can still solve the max-min problem but with relatively large number of iterations with respect to sum rate maximization. The consequent *max-min* algorithm follows similar steps as **Algorithm 3** with the exception that at the end of each outer iteration, the link with minimum SINR is checked, and the corresponding S-D pair's SINR is maximized in the next outer iteration, i.e., the gradient with respect to that link's SINR is computed. In other words, the resultant algorithm alternates between optimizing different links per outer iteration, and hence, may result in rather large number of iterations. The algorithm terminates when maximizing the minimum $\mathrm{SINR}_{i,n}$ at the nth iteration can not be further increased without leading a decrement in the other SINRs, which results in a worse performance than $\mathrm{SINR}_{i,n}$.

3.3.3. Key Issues for Algorithm Implementation

In this subsection, we study the adjustment of the implementation parameters of the gradient algorithm in order to improve its performance in terms of convergence speed. We propose and elaborate on further distributed schemes for starting vector computation, step-size searches, etc.

3.3.3.1. Starting Vector Computation

It is well known that the starting vector has considerable impact on the convergence speed of gradient searches. Moreover, for our case of distributed gradient method, we note that the initial vector should be as well calculated *locally* at the relays only with the available local CSI per relay. In the following, we propose heuristic starting vector calculation methods for both setups of single and multiple S-D pairs, which will be later shown in Section 3.3.4 to perform very efficiently and to result in fast convergence properties.

Single S-D Pair For the case of a single S-D pair in the presence of multiple interferers, we propose to use the following local gain factor for the kth relay

$$g_{0,k} = \frac{h_{k,i}^* \cdot f_{i,k}^*}{\sigma_{n_\mathrm{r}}^2 |f_{i,k}|^2 + (\sigma_{n_\mathrm{d}}^2 / P_\mathrm{r}) \mathbf{h}_{\mathrm{s}k}^H \mathbf{h}_{\mathrm{s}k}}, \tag{3.43}$$

where i denotes the index of the S-D pair in the focus of interest. Up to a constant factor, this gain vector is optimal in the absence of interference, since it implements a maximum ratio combining of the desired signal contributions at the destination. The derivation of (3.43) can be found in the Appendix A.5. As the relay noise $\sigma_{n_\mathrm{r}}^2$ gets more dominant in the denominator, $g_{0,k} \sim h_{k,i}^* / f_{i,k}$, which implements an MRC of the relay received signals at the destination. Whereas, if destination noise $\sigma_{n_\mathrm{d}}^2$ gets larger with respect to $\sigma_{n_\mathrm{r}}^2$, then $g_{0,k} \sim h_{k,i}^* \cdot f_{i,k}^*$, which approaches the MRC of the destination signal contributions. As it will be demonstrated with computer simulations, when the proposed starting gain factors are used, the algorithm converges in an extremely few, e.g., 2-3, iterations for small number of interferers.

Multiple S-D Pairs We cannot apply the initial vector from the previous subsection for multiple S-D pairs, because it is optimized to favor one specific S-D pair and neglects the others. We propose the following heuristic two-stage algorithm where the resultant gain vector serves as an initial vector for the sum rate maximization.

- `Stage 1:` We set the relay noise variance to be very high, i.e., a relatively low average SINR at the relays. In this case, the destination noises become irrelevant and the distributed gradient algorithm converges in a few outer iterations. Note that maximizing $\mathrm{R}_{\mathsf{sum}}$ is equivalent to maximizing $\prod_{i=1}^{N}(1+\mathrm{SINR}_i)$ due to monotonicity of the logarithm function. Hence, dropping the summand "1", we use the gradient of the product of SINR_is in this stage:

  $$\nabla_1 = \sum_{i=1}^{N} \nabla_{\mathbf{g}}\mathrm{SINR}_i \cdot \frac{\prod_{j=1}^{N}\mathrm{SINR}_j}{\mathrm{SINR}_i}.$$

  Whereas, throughout the inner iterations we use the minimum SINR_i, i.e., $\min_i \mathrm{SINR}_i$, as the stopping criterion instead of product of SINRs. The advantage of this choice is that when the decentralized step-size search is used, the destinations do not need to exchange SINR_i to decide whether the product is reduced or not as proposed in Section 3.3.2.1 for sum rate maximization. Moreover, although it is not completely equivalent, as the minimum SINR_i decreases, it is most likely that product also diminishes.

- `Stage 2:` We use the resultant gain vector from `Stage 1` as the starting vector for this stage and set the noise variances to the actual values. We again employ the gradient of the product SINR with a difference that the weakest S-D pair is emphasized by squaring the denominator:

  $$\nabla_2 = \sum_{i=1}^{N} \nabla_{\mathbf{g}}\mathrm{SINR}_i \cdot \frac{\prod_{j=1}^{N}\mathrm{SINR}_j}{(\mathrm{SINR}_i)^2}.$$

  We have observed that convergence speed has considerably improved with the square term in the denominator. The stopping criterion for the line search is again $\min_i \mathrm{SINR}_i$.

The resultant vector from the `Stage 2` is used as the initial gain vector for the actual sum rate maximization algorithm. Note that in order to find a good starting vector, we do not need to iterate `Stage 1` and 2 until they converge. A few number of iterations for both stages may also be enough to obtain quite an efficient initial gain vector. Further discussions on the performance of this heuristic approach will be given in Section 3.3.4 in details.

3.3.3.2. Step-Size Computation

The Decentralized Protocol Since efficient resource utilization is a crucial aspect in coherent relaying, for the decentralized step-size protocol it is important to reach the optimal step-size with as few transmissions as possible from relays to the destination. Assuming that global CSI is available, the commonly used step-size selection methods are constant step-size, diminishing step-size, line minimization, Armijo rule, and Goldstein rule [22]. Due to the distributed structure of our step-size calculation, the commonly used second derivative based step-size calculations for conjugate directions can not be employed in our relay gain allocation [112]. Instead, in Section 3.3.1.2, we propose a very basic search where starting from an initial value μ_{min}, we iterate until μ_{max} is reached or the cost function does not improve anymore, and we use an appropriate increment δ at each inner iteration. Choosing this increment δ very small results in a more accurate step-size calculation but an increased number of inner iterations. On the other hand, for a rather larger increments, the accuracy of the inner iterations is disturbed, which may lead to an increase in the number of outer iterations. Such a scheme resembles the line minimization [22], which performs efficiently but is rather slow in convergence. For the sake of comparison and reducing the forward channel uses for the relays to the destination, we present here two more distributed step-size protocols:

1) A variant of the algorithm presented in Section 3.3.1.2,

2) Almijo rule adapted to a distributed scenario.

Note that for the sake of simplicity, the focus of optimization for the following algorithms is the receive SINR, but they can be trivially extended to the multiuser case.

• **Spiral Search:** The advantage of the original increment algorithm in Section 3.3.1.2 is that it employs only one bit stop-feedback. However, allowing the number of feedbacks to increase, it is possible to design protocols which result in a significant reduction in the forward channel uses. In the spiral search, the update equation for the

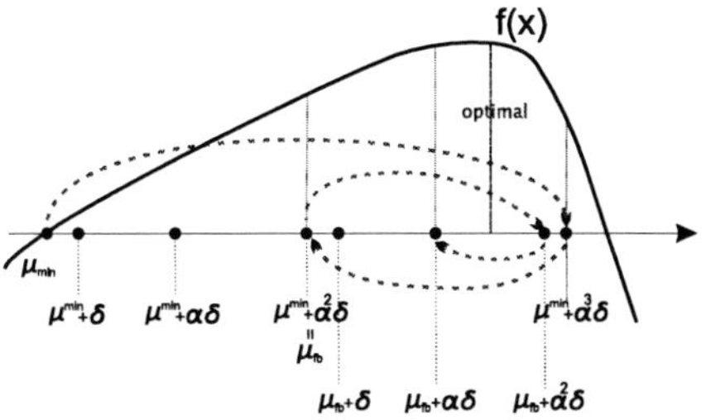

Figure 3.8.: The illustration for the spiral search algorithm. In the first step, it iterates until $\mu^{\min} + \alpha^3\delta$, which is beyond the optimal. In the second step, it re-starts from the previous decision, i.e., $\mu_{\text{fb}} := \mu^{\min} + \alpha^2\delta$, and iterates until $\mu_{\text{fb}} + \alpha^2\delta = \mu^{\min} + 2\alpha^2\delta$, which is again beyond the optimal. In the third step, it re-starts from the previous decision, i.e., $\mu_{\text{fb}} + \alpha\delta = \mu^{\min} + \alpha^2\delta + \alpha\delta$, and iterates with this fashion until a δ-optimal convergence is obtained.

qth step cycle modifies to $\mu_{n,q} \triangleq \mu_{\text{fb}} + \alpha^{(q-1)}\delta$, where $\alpha > 1$ accelerates the search and μ_{fb} is an update for accuracy. Initially, we set $\mu_{\text{fb}} := \mu_n^{\min}$. As before, starting from μ_{fb} and iterating with a monotonically increasing step-size increment $\alpha^{(q-1)}\delta$, the network performs multiple measure cycles. As soon as the destination detects that SINR is reduced with respect to the previous choice, it feeds back a stop bit. In contrary with the original algorithm, instead of quiting here, we update the accuracy parameter $\mu_{\text{fb}} := \mu_{n,\tilde{q}}$, where $\tilde{q}$ is the index of the last cycle before the stop-feedback, set $q := 1$, and then restart the algorithm. With each stop-feedback, we improve the initial point to start the search, and hence, the overall trajectory of the algorithms follows a spiral structure (see Figure 3.8). The algorithm terminates when δ-accuracy is reached at the step-size.

- **Armijo Rule:** The step-sizes $\kappa^t\delta$, $t = 0, 1, \ldots$, are tried successively until the SINR obtained through $\mathbf{g}_n$ is larger than the SINR obtained through $\mathbf{g}_n + \kappa^t\delta\nabla_{\mathbf{g},n}\text{SINR}_i$ with a difference of at least $\varsigma\kappa^t\delta\|\nabla_{\mathbf{g},n}\text{SINR}_i\|^2$ for $t = t_n$. The $\varsigma \in (0,1)$ represents the tolerance parameter and $\kappa \in (0,1)$ is the amount by which the step-size is reduced in every iteration. The Armijo rule can be implemented in a distributed manner by employing an *measure* cycle for each of these successive step-size choices, and as soon as the destination detects that the difference between the successive SINR estimations is at least as much as the value specified above, it feeds a stop bit back.

The Centralized Protocol In the centralized protocol, the optimal step-size is computed by a single node (any destination), which has collected all information needed for solving (3.33) optimally. The necessary computations are done within the digital processor of this node. In other words, resources are not utilized (in terms of channel uses) for inner iterations, but only for gathering the necessary information. Hence, it is a redundant issue in terms of resource management to consider which step-size method would be used for the centralized protocol, because it is deterministically known and fixed. For instance, considering the scenario of a single S-D pair in the presence of multiple interference sources, we use the source-to-relay and relay-to-destination channels once for estimating the equivalent channel with the gradient as the gain vector, e.g., for SINR_i^{∇} and $\mathbf{h}_{\mathsf{srd},i}^{\nabla}$. Recall that we have assumed one channel-use for perfect estimation. Then, we need two more channel uses for the further measurement cycles related with (3.35), e.g., for $\mathrm{Re}\left\{\sigma_{n_r}^2 \mathbf{g}_n^H \mathbf{\Gamma}_i^H \mathbf{\Gamma}_i (\nabla_{\mathbf{g},n} \mathrm{SINR}_i)\right\}$ and $\mathrm{Re}\left\{\frac{\sigma_{n_d}^2}{P_r} \cdot \mathbf{g}_n^H \mathbf{M} (\nabla_{\mathbf{g},n} \mathrm{SINR}_i)\right\}$. Summing up, independent from system parameters and the number of outer iterations, we only require four channel uses for collecting the aforementioned necessary information for each outer iteration. This calculation can be trivially extended to the multiuser case. Nevertheless, it is immediately noticed that the channel-use overhead is much less when compared with the acquisition phase of the decentralized protocol.

3.3.3.3. Relay Power Scaling

The instantaneous relay power constraint is embedded in the SINR expressions as presented in (3.24). However, as explained in 3.3.1, in order to measure the SINRs at the destinations, we are scaling the relay gain vector for each inner/outer iteration such that the power constraint is not violated. It is worth emphasizing that such a scaling affects neither the optimality nor the convergence of the gradient algorithm, but only the number of outer iterations since the optimal step-size computation is disturbed by the scaling. In essence, at the expense of increased outer iterations, the scaling per inner iteration can be dropped. This, in return, reduces the number of channel uses required to gain the terms needed for the scaling at each relay. Hence, a channel use trade-off between inner and outer iterations is observed.

As a second approach to employ the power constraint, one may use the penalty/barrier methods [26] as we did in Section 3.2.1.1, which converts the power constrained problem to an unconstrained one by penalizing the cost function when the power constraint is violated.

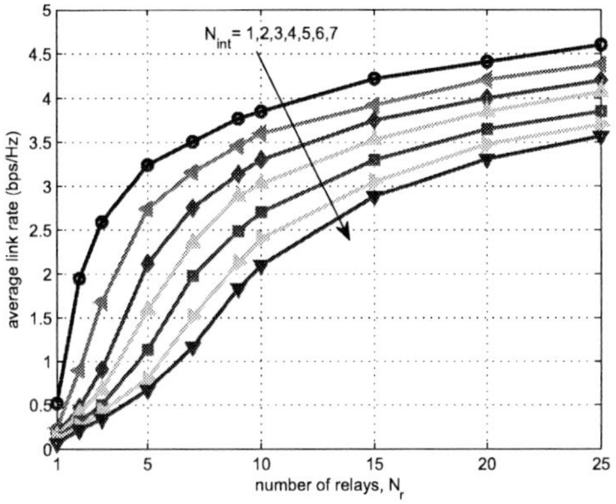

Figure 3.9.: Single S-D pair in the presence of N_{int} interference sources: Average achievable link rates versus number of relays N_{r} with optimal relay gain allocation (SNR $= 20$ dB).

3.3.4. Performance Results

Simulation Setup In this section we present Monte Carlo simulation results, where we aim at providing insight into the speed of convergence of the considered conjugate gradient method, and at elaborating on the implementation issues of the algorithm. As our focus of interest is rather on the convergence of the algorithm, instead of the previously employed $P_{\mathrm{r}} = NP_{\mathrm{s}}$ power constraint, we simply fix $P_{\mathrm{s}} = P_{\mathsf{r}} = 1$. Further, we assume equal relay and destination noise variances, i.e., $\sigma^2 \triangleq \sigma^2_{n_{\mathsf{r}}} = \sigma^2_{n_{\mathsf{d}}}$. Recall that the average SNR is defined as $\mathrm{SNR} = P_{\mathrm{s}}/\sigma^2$. All CSI needed for updating the gain vector is assumed to be known perfectly at the respective nodes. The step-size increment is fixed to $\delta = 10^{-3}$, and normalized with the norm of the gradient at each outer iteration.

3.3.4.1. Single S-D Pair under Interference

Figures 3.9-3.11 illustrate the performance of the coherent AF relaying scheme with single S-D pair in the presence of interference sources. For the results in Fig. 3.9, the gain vector is determined by directly solving the generalized Rayleigh quotient problem (3.24). As such Fig. 3.9 provides an upper bound on the performance of any gradient based scheme. Specifically, it depicts the average achievable link rate versus N_{r} for various numbers of interference sources N_{int}. We observe a strictly increasing achievable

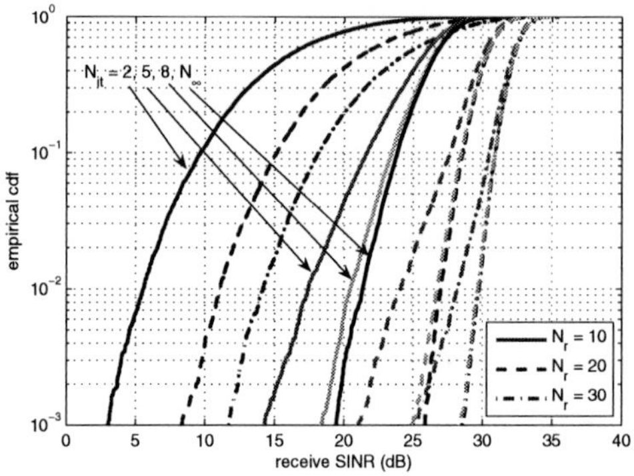

Figure 3.10.: Single S-D pair in the presence of two interference sources: Empirical CDFs of receive SINR for different numbers of relays N_r and outer iterations N_{it} in the gradient algorithm (SNR $= 25$ dB).

rate in N_r. Up to the point $N_r = N_{int}$, this increase is explained by the capability of suppressing more and more interfering streams. Moreover, for $N_r > N_{int}$, the relay nodes are capable of further improving the SINR by the realization of a coherent superposition of the desired signal at the destination node (distributed array gain).

Fig. 3.10 shows the performance of the distributed gradient based gain allocation scheme for $N_{int} = 2$. We study the convergence behaviour of the gradient algorithm from the perspective of the necessary number of outer iterations. Here and in the sequel, while focusing on the number of outer iterations, unless otherwise stated we employ the decentralized protocol with the *spiral* search, but disregard the number of inner iterations during the step-size search. Fig. 3.10 depicts the empirical CDF of the receive SINR in three sets of curves, which corresponds to $N_r = 10, 20$, and 30 relays. The sets are identified by line style (solid, dashed, dashed-dotted). For each set, the CDF after $N_{it} = 2, 5, 8$ and N_∞ outer iterations is shown, where N_∞ corresponds to the optimal result obtained by the generalized Rayleigh quotient problem (3.24). In all cases $N_{it} = 8$ iterations (green) suffice to achieve almost optimal performance (N_∞). For a given number of iterations, the performance improves dramatically with N_r. After $N_{it} = 5$ iterations, for example the $N_r = 10$ configuration (solid) achieves an outage SINR of 17 dB at 1% outage probability, whereas with $N_r = 20$ relays the outage SINR improves by 7 dB to 24 dB. Adding 10 more relays ($N_r = 30$) further improves the performance by 4 dB.

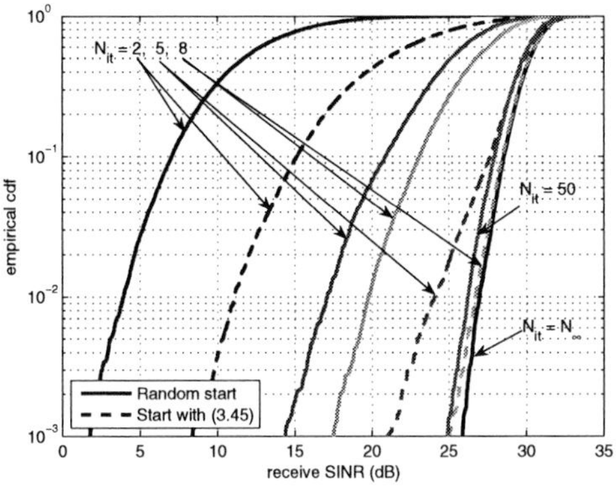

Figure 3.11.: Single S-D pair in the presence of two interference sources: Empirical CDFs of receive SINR for different starting gain allocation choices and outer iterations N_{it} in the gradient algorithm (SNR $= 25$ dB).

Throughout the simulations in Fig. 3.10, we employ the local gain factors proposed in (3.43) for initializing the gradient algorithm. In Fig. 3.11, we investigate the performance of this starting gain factors in (3.43) with respect to a random initialization. We identify the random start and the (3.43) with solid and dashed line styles, respectively. For any given number of iterations, the designed initialization provides almost 7 dB of outage SINR improvement at 1% outage probability. Moreover, not even $N_{\text{it}} = 50$ iterations are enough for the random initialization to reach the performance of the 8th iteration (dashed-green) of the initialization design in (3.43). Although not depicted in the figure, we observe a tendency that when we initialize the gradient algorithm with (3.43), the necessary number of iterations to reach the optimal result varies directly proportional to the number of interferers N_{int}. This is expected since the allocation (3.43) is optimal in the absence of interference, and hence, requires more iterations to combat the increased number of interference sources.

3.3.4.2. Multiple Source-Destination Pairs

In this subsection we discuss the performance of the distributed gradient based gain allocation for multiple S-D pairs. Objective function is the sum rate of the N links. In contrast to the single S-D pair case here no analytical solution is available. Fig. 3.12 illustrates the convergence behaviour (outer iterations) of the conjugate gradient algo-

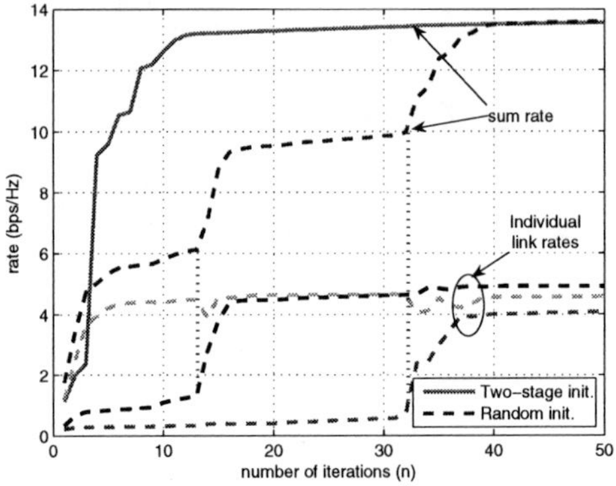

Figure 3.12.: Three S-D pairs: Sample trajectory of achievable rates at each outer iteration for a typical channel realization with $N_r = 30$.

rithm for a typical channel realization, $N = 3$ S-D pairs and $N_r = 30$ relays. The dashed lines refer to a random gain vector initialization. According to the trajectories of the individual link rates after 35 iterations all three links support a rate of about 4.5 bps/Hz each. The convergence behaviour of the three trajectories is dramatically different however. The links (spatial subchannels) open up one after the other with intermediate periods without much improvement. This leads to the plateau structure of the sum rate trajectory. While this basic behaviour can be observed for most channel realizations, the length of the plateaus between the transitions varies substantially. Consequently, for a given number of iterations the number of open spatial subchannels varies considerably. On the other hand, as depicted, the speed of convergence is substantially affected by the initial gain vector. We also plot in Fig. 3.12 the achieved sum rate trajectory for the same channel, but initialized with the heuristic two-stage algorithm proposed in Section 3.3.3.1. Note the dramatic improvement, which arises from the fact that this smart initialization leads several subchannels to open simultaneously. The impact of the two-stage initialization algorithm will be further investigated with the following figures. Here and in the sequel, unless random initialization is performed, the first ten outer iterations of the gradient algorithm is occupied by the two-stage initialization algorithm, where the first stage lasts for three iterations, and the second for seven iterations. Note that these numbers are not optimized, and kept fixed for all channel realizations. As emphasized in Section 3.3.3.1, there are possibly different iteration configurations which may lead better or worse initializations.

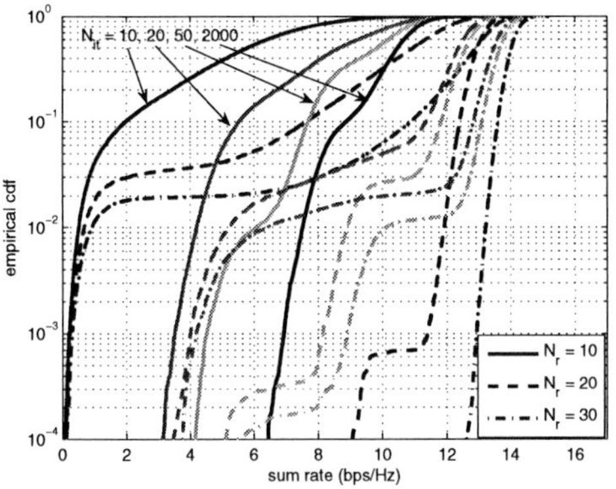

Figure 3.13.: Three S-D pairs: Empirical CDFs of the achievable sum-rate for various numbers of relays N_r and outer iterations N_{it} in the gradient algorithm (SNR $= 25$ dB).

In order to capture this statistical influence, we consider the empirical CDF of the sum rate in Fig. 3.13. The results are based on 100,000 channel realizations, and we employ the heuristic two-stage initialization. Three sets of curves are shown, which refer to $N_r = 10$ (solid), 20 (dashed) and 30 (dashed-dotted) relays. Parameter within each set is the number of iterations N_{it}. A comparison of the curves with the same N_{it} shows that with increasing number of relays, the CDF tends to develop one or several plateaus. We attribute this to the different number of subchannels, which are open after N_{it} iterations. This effect becomes more pronounced as the number of relays grows. Let us consider the $N_r = 30$ case in more detail. After $N_{it} = 10$ iterations either all three subchannels are open, which leads to a sum rate of about 13 bps/Hz or no subchannel is open. The latter happens with a probability of about 1%. After 20 iterations (dashed-dotted) either three or one subchannel (sum rate 4 bps/Hz) are open in our experiment. We conjecture the existence of another plateau, which corresponds to no open subchannels. This case occurs with such a low probability however, that it is not observable in the figure. After 50 iterations (dashed-dotted-green), we observe either three (13 bps/Hz), two (9 bps/Hz) or one (5 bps/Hz) open subchannels. Finally after 2000 iterations, with probability 0.9999 all subchannels are open (no plateau within the depicted probability range).

The conjectured multi-plateau structure of the CDF is nicely supported by curves of $N_{it} = 50$ iterations (green), which exhibit two plateaus each. A comparison between

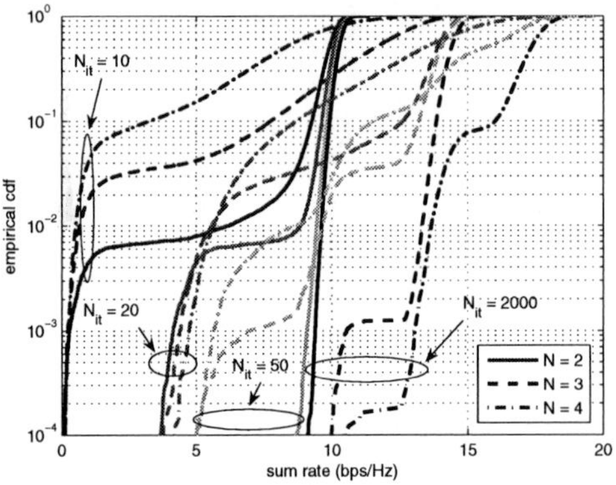

Figure 3.14.: Multiple S-D pairs: Empirical CDFs of the achievable sum-rate for various numbers of S-D pairs N and iterations N_{it} in the gradient algorithm (SNR $= 25$ dB, $N_{\text{r}} = 20$).

these curves reveals that the probability of difficult channel matrices, which require a large number of iterations to open up at least two subchannels, drops with increasing number of relays, e.g., the plateaus between three and two open subchannels occur respectively at decreasing probabilities of 3×10^{-1}, 3×10^{-2}, 1×10^{-2}. This is a result of the increasing diversity in the system. Thus, the gradient scheme benefits from the number of relays. Lastly, we remark that for a random initialization basically the same CDFs would be obtained, if we increase the number of iterations by an order of magnitude.

We study the impact of the number of S-D pairs N in Fig. 3.14 for $N_{\text{r}} = 20$ relays. Different N values are identified by line styles: solid ($N = 2$), dashed ($N = 3$), and dashed-dotted ($N = 4$). Moreover, for the sake of fair comparison, we modify our sum relay transmit power as $P_{\text{r}} = NP_{\text{s}}$. In general, the multi-plateau structure is observed for all N. We observe that $N_{\text{it}} = 20$ outer iterations suffice for all N to open at least one subchannel (4 bps/Hz) with a probability almost 5×10^{-3}. After $N_{\text{it}} = 50$ iterations (green), one more subchannel is opened only for the $N = 2$ case, whereas $N = 3, 4$ need further iterations. Consequently, iterating 2,000 times, we observe that three subchannels are opened for $N = 3, 4$ pairs (around 12-13 bps/Hz) with probability 10^{-3}. In summary, 50 and 2,000 iterations suffice to reach the corresponding optimal sum rate results for $N = 2, 3$ respectively, for at least 99.99% of all channel realizations. However, with $N_{\text{it}} = 2000$ iterations, we can only guarantee 90% of all channel realizations to

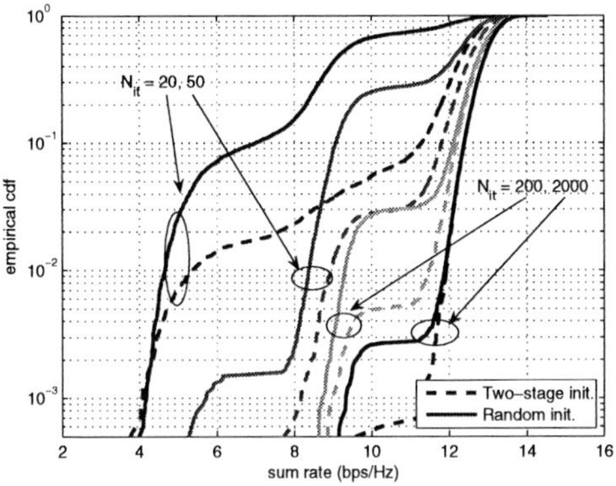

Figure 3.15.: Three S-D pairs: Empirical CDFs of the achievable sum-rate with different initializations of the gradient algorithm ($N_r = 30$, SNR $= 25$ dB).

converge to the optimal result for the $N = 4$ pairs case. As a final remark, we should keep in mind that the number of relays plays a major role in convergence (see Fig. 3.13). Hence, choosing $N_r > 20$ would accelerate convergence for all cases and change the previously stated results.

In Fig. 3.15, we examine the effect of initialization on the convergence behaviour of the outer iterations. Considering both random initialization and the heuristic two-stage initialization algorithm, we plot the CDFs of the sum rate of $N = 3$ S-D pairs for several outer iterations. Note that as mentioned above, when the latter approach is employed, the first ten iterations of the gradient algorithm is used for initialization purpose. At $N_{\text{it}} = 20$ outer iterations, unless there are more than one open subchannels (sum rate is larger than 5 bps/Hz), the heuristic initialization method (dashed) provides at least one order of magnitude outage probability improvement to the random initialization (solid) at any outage rate less than 12 bps/Hz. Considering the $N_{\text{it}} = 50$ case, we notice that the plateau at 3×10^{-1} outage probability (solid) is shifted down to 3×10^{-2} (dashed) with the impact of the two-stage initialization. In other words, the probability that all subchannels have been opened at a given sum rate is increased from 0.7 to 0.97. Similar shifts of plateaus are also observed at $N_{\text{it}} = 200$ (green) or 2000 outer iterations. While concluding, we remark that the number of iterations that each stage of the heuristic initialization algorithm is employed can be further optimized per channel realization to obtain better rate results.

After developing an understanding of the convergence behaviour of the outer itera-

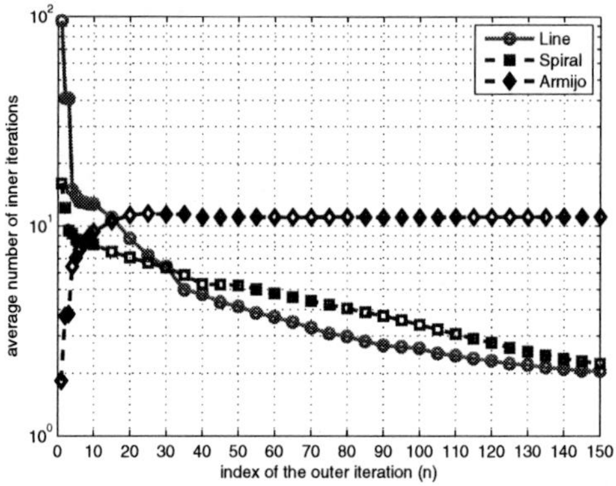

Figure 3.16.: Three S-D pairs: Average number of inner iterations of the line, the spiral, and the Armijo searches for step-size optimization per each outer iteration in the gradient algorithm ($N_r = 30$, SNR $= 25$ dB).

tions, now we turn our attention to the inner iterations for the step-size. The resources allocated to the step-size optimization is more crucial for the decentralized protocol, since it is not deterministic as it is the case for the centralized protocol (see Section 3.3.3.2). Although similar algorithms are performed, the implementation through the centralized protocol is done within the digital processor of the master destination without occupying the forward channel between the relays and destinations.

In Figures 3.16 and 3.17, we compare the performances of different step-size search algorithms, i.e., line search, spiral search, Armijo rule. In Fig. 3.16, we plot the average number of inner iterations per each outer iteration for $N_r = 30$ relays, $N = 3$ S-D pairs and 10,000 channel realizations. The parameters of the Armijo method are set as $\varsigma = 10^{-3}$ and $\kappa = 1/2$. Moreover, we allow at most 100 inner iterations for all methods, since more of it would abandon the practicality of the distributed step-size search due to the excess resource occupation. During the first few outer iterations (acquisition phase), the spiral and Armijo methods provide almost one order of magnitude improvement (in terms of less number of inner iterations) to the original line search algorithm proposed in Section 3.3.1.2. The Armijo rule quits with one to three inner iterations when rather larger step-sizes are need to be taken; whereas it requires much larger number of inner iterations (with respect to the other two methods) for the tracking phase, since the optimal step-sizes are very small for after almost ten outer iterations. The difference between the Armijo rule and the others follows from the fact that the Armijo rule based

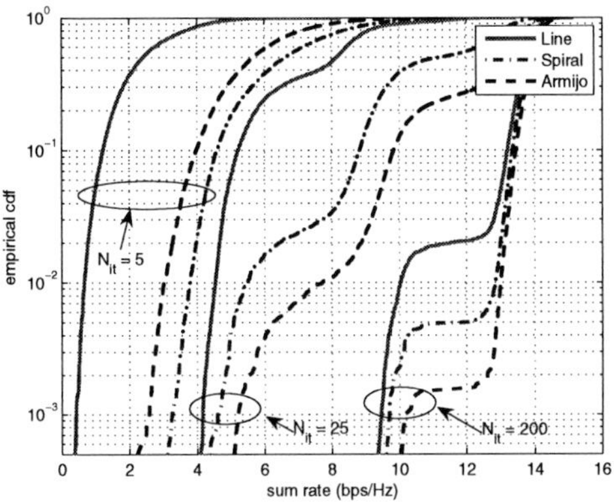

Figure 3.17.: Three S-D pairs: Empirical CDFs of the achievable sum-rate for the line, the spiral, and the Armijo searches for step-size optimization in the gradient algorithm ($N_r = 30$, SNR $= 25$ dB).

search starts from a large step-size and iterates by diminishing it at each inner iteration, whereas its counterparts start with a small step-size and iterates by incrementing it. In summary, for the sake of minimum inner iterations, we recommend to use Armijo rule for the first few outer iterations (acquisition phase) and to shift to any of the other two for further iterations (tracking phase).

Lastly, we study the impact of step-size optimization on the convergence behaviour of the outer iterations in Fig. 3.17. We employ $N = 3$ S-D pairs with $N_r = 30$ relays, and implement the gradient algorithm with the same parameters except that step-sizes are either computed by line search, spiral search or Armijo method. The poor performance of the line search for $N_{\text{it}} = 5$ outer iterations is due to the maximum limit on the inner iterations, i.e., 100 iterations, which leads the method to quit before reaching the optimal step-size. Since these suboptimal step-size decisions trigger increased number of outer iterations by propagation, the line methods performs the worst for any N_{it}. However, we note here that dropping the limit on the maximum number inner iterations, we expect the line and spiral searches to result in the same optimal step-sizes, and hence, identical CDFs for the sum rate. On the other hand, there is a nearly 1 bps/Hz difference between the CDFs of the spiral search and the Armijo methods for $N_{\text{it}} = 5$ favoring the former. This is occurred since the Armijo rule misses the optimal step-size in the acquisition phase while maintaining "enough" improvement on the cost function. However, as we increase the number of outer iterations, Armijo appears to perform better for a given number of iterations, e.g., $N_{\text{it}} = 25$, since now the accuracy of the line search is limited by the δ. To sum up, we observe that the chosen step-size optimization method has a crucial effect on the performance for a given N_{it}. Furthermore, missing the optimal step-size per outer iteration leads reduced rate results, and correspondingly increased number of outer iterations to reach a given rate target.

3.3.5. Comparative CSI Dissemination Analysis

In this section, we aim at answering the question of how much gain the distributed gradient scheme offers in terms of required channel-use for CSI dissemination. We focus on a multiple S-D pairs scenario. Sum rate maximization is performed either through the proposed distributed gradient method or directly through the knowledge of global CSI at each relay. In the following, we distinguish two modes of operation individually: acquisition mode (static channel) and tracking mode (time-varying channel).

3.3.5.1. Acquisition Mode

For the sake of brevity, we assume ideally that one channel-use is enough for error-free channel estimation, dissemination or feedback of a complex coefficient. As both conventional and distributed approaches require that local CSI is available at each relay, first N S-D pairs broadcast training sequences to the relays, which employ $2N$ channel-uses in total.

Attaining global CSI requires each relay to broadcast its local CSI to the others. In order for all other relays to receive this information, i.e. to decode it perfectly, they need to estimate the channel coefficients from themselves to the CSI broadcasting relay. Hence, each relay first occupies one channel-use to transmit a training symbol to the others for inter-relay channel estimation, and $2N$ channel-uses afterwards to broadcast its local CSI, i.e., sum of the sizes of $\mathbf{h}_{\mathsf{s}k}$ and $\mathbf{f}_{k\mathsf{d}}$. That is, the total number of required channel-uses sums up to $C_1 = N_{\mathsf{r}}(2N+1)+2N$. Since $N_{\mathsf{r}} > N(N-1)$ must be satisfied for perfect multiuser orthogonalization, we re-define C_1 as an implicit function of N as: $C_1 = (2N^3 - N^2 + 3N + 1) + N_{\mathsf{r,ex}}(2N+1)$, where $N_{\mathsf{r,ex}} = N_{\mathsf{r}} - N(N-1) - 1$ is the number of excess relays on top of the minimum required.

For the distributed gradient approach, we only account for the channel-uses for the outer iterations and neglect the ones for the step-size search, e.g., we follow the reasoning in Section 3.3.1.2 for the decentralized protocol, and for the centralized protocol, channel

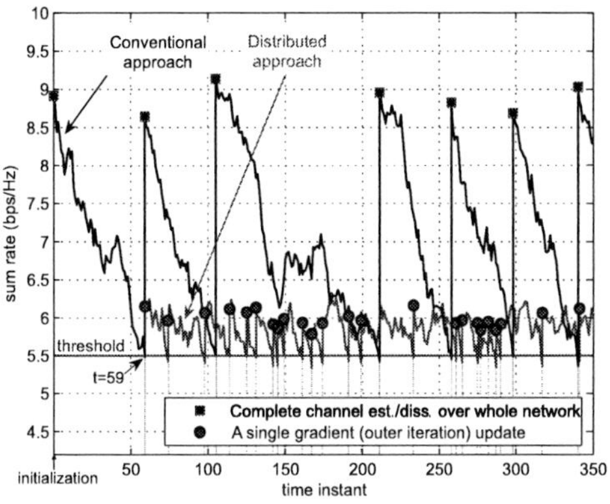

Figure 3.18.: Three source-destination pairs: Sample trajectory of achievable sum rate over time for a typical correlated channel realization campaign with $N_\mathsf{r} = 30$ relays and SNR $= 15$ dB. We hereby compare the frequency of relay gain allocation updates for the conventional and distributed approaches. We set the sum rate threshold and correlation factor respectively as $\bar{\mathrm{R}}_{\mathrm{sum}} = 5.5$bps/Hz and $\rho_{\ell,\ell+1} = 0.999$.

occupation is anyhow independent from N or N_r. In each outer iteration, $\mathbf{H}_{\mathsf{srd}}$ and N SINR values should be estimated and fed back to the relays. That is, we require $2N$ and N^2 channel-uses respectively for estimating and feeding $\mathbf{H}_{\mathsf{srd}}$ back, and N channel-uses for feeding SINRs back. Note that SINRs can be derived from $\mathbf{H}_{\mathsf{srd}}$, and that there is no additional training needed for decoding the feedbacks perfectly, since the relays already have the required relay-to-destination channel knowledge. Hence, the total number of required channel-uses sums up to $C_2 = (N^2 + 3N)N_{\mathrm{it}} + 2N$, where N_{it} denotes the number of outer iterations and $2N$ stands for initial local CSI estimation at each relay.

Note first that C_2 is independent from $N_{\mathsf{r,ex}}$. Recalling the numeric evidence that N_{it} and N_r are inversely proportional, we conclude that C_2 stays at most constant if not decrease, as N_r increases. Moreover, comparing C_1 and C_2, the respective growing rates reveal that CSI overhead of the proposed method grows an order of N slower than that of the conventional approach. However, for a finite N, the comparison is highly dependent on the values of $N_{\mathsf{r,ex}}$ and N_{it}.

3.3.5.2. Tracking Mode

The above comparison applies to optimization without a-priori channel variation knowledge. In practice, channel blocks are correlated in time. Accordingly, using the outdated gain vector from the previous channel block as a starting vector for the new channel block is typically more effective than initialization without this a-priori knowledge. In other words, channel variations can be tracked efficiently to update the relay gains. Consider two consecutive channel realization blocks $(\mathbf{H}_{(\ell)}, \mathbf{F}_{(\ell)})$ and $(\mathbf{H}_{(\ell+1)}, \mathbf{F}_{(\ell+1)})$, which have a correlation factor $\rho_{\ell,\ell+1}$ in between such that

$$\begin{aligned} \mathbf{H}_{(\ell+1)} &= \rho_{\ell,\ell+1}\mathbf{H}_{(\ell)} + \sqrt{1-\rho_{\ell,\ell+1}^2}\,\tilde{\mathbf{H}}_{(\ell)} \\ \mathbf{F}_{(\ell+1)} &= \rho_{\ell,\ell+1}\mathbf{F}_{(\ell)} + \sqrt{1-\rho_{\ell,\ell+1}^2}\,\tilde{\mathbf{F}}_{(\ell)}, \end{aligned} \tag{3.44}$$

where $\tilde{\mathbf{H}}_{(\ell)}, \tilde{\mathbf{F}}_{(\ell)}$ represent random Gaussian channel variations, and their elements are distributed with the same statistics as that of $\mathbf{H}_{(\ell)}, \mathbf{F}_{(\ell)}$. In general, the relay gains should be updated, as well as the channel estimation/dissemination, as soon as the target figure of merit (sum rate here) falls below a predetermined threshold value, say $\bar{\mathrm{R}}_{\mathsf{sum}}$.

Fig. 3.18 shows a sample trajectory of sum rate over time for $N = 3$ and $N_{\mathsf{r}} = 30$. We compute the initial gain vector $\mathbf{g}_{t_0}$ at $t = t_0$ through the conventional approach for a (single) typical channel realization. For each time instant $t_{\ell+1}$, $\ell = 0, 1, \ldots$, we generate a new channel realization block correlated with the previous block at t_ℓ through (3.44). Hence, in principle, the channel blocks of time instants t_{ℓ_1} and t_{ℓ_2} are correlated with $\rho_{\ell_1,\ell_2} = \rho_{\ell,\ell+1}^{(\ell_2-\ell_1)}$ on average. Setting $\bar{\mathrm{R}}_{\mathsf{sum}} = 5.5$bps/Hz and $\rho_{\ell,\ell+1} = 0.999$, we plot the sum rate trajectory for 350 time instants, which corresponds to that the initial and the 350th channel realization blocks are correlated with $0.999^{350} \simeq 0.7$ on average. As seen in the figure, the sum rate threshold is violated at time instance $t = 59$. At this point with the conventional approach, a complete CSI estimation and dissemination process is performed over the whole network to update the relay gains, which results in a sum rate of 8.6 bps/Hz. However, with the proposed distributed approach, only a single outer iteration (one gradient update) is enough to increase the sum rate well above the threshold to about 6.2 bps/Hz. As the time passes on, we observe that the distributed approach needs to adapt the relay gains more often than the conventional one, e.g., for this sample example, conventional and distributed approaches respectively require 6 and 24 updates during $t = 0, \ldots, 350$. Nevertheless, when the number of channel-uses required for updating each approach is compared, the advantage of the

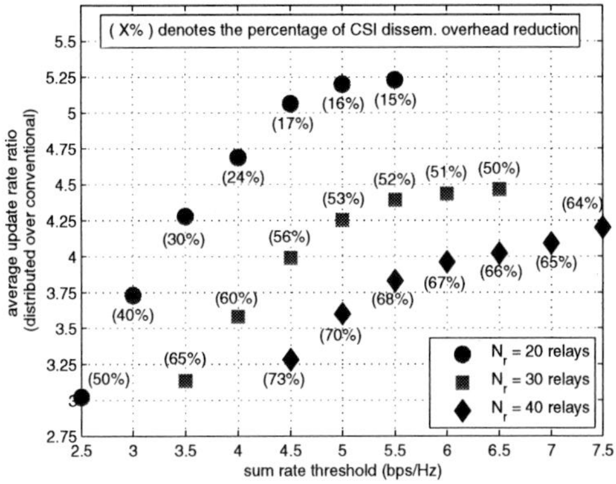

Figure 3.19.: Three source-destination pairs: Average update rate ratio (distributed over conventional) for various number of relays and sum rate thresholds (SNR = 15 dB). While computing improvements percentages, we incorporate average required number of outer iterations for each update in the distributed case, which was surprisingly observed to lie in an interval of $[1, 1.05]$ independently from network parameters.

distributed approach immediately reveals itself. For this example, the conventional approach requires $6C_1 = 1296$ channel-uses in total as opposed to $24(N^2 + 5N) = 576$ channel-uses required by the distributed one (note the difference to C_2 that in the tracking mode distributed approach needs also to update local CSI). That is, the distributed approach offers $100 \cdot (1296 - 576)/1296 \simeq 56\%$ CSI dissemination overhead reduction.

As the previously presented numbers are highly dependent on the specific channel realizations, we present in Fig 3.19 ergodic performance results of update rates for different combinations of N_r and $\bar{R}_{sum}$. To this end, we perform 1000 correlated channel realization campaigns similar to the one presented above. We compute the average required number of updates for each approach over simulated campaigns, and plot in the figure the ratio of these, which we call as *average update rate ratio*, for different $\bar{R}_{sum}$ values. For each parameter set $(N_r, \bar{R}_{sum})$, we calculate the percentage of CSI dissemination overhead reduction attained by the distributed proposal.

Fig. 3.19 depicts that for $N_r = 20$ the distributed approach's update requirement is at most 5.25 times as much as the conventional one (corresponds to $\bar{R}_{sum} = 5.5$ bps/Hz). However, even for such a poor update rate performance, the distributed approach offers an overhead reduction of 15%, which goes up to 50% for $\bar{R}_{sum} = 2.5$ bps/Hz. Adding 10 (or 20) more relays, i.e. $N_r = 30$ (or 40) further improves the overhead reduction to 50−65% (or 64−73%). In general, there is a trend that increasing the threshold value favors the conventional approach. The intuition behind this is as follows: Although a few (usually one) iterations of the distributed approach suffice to increase the sum rate over the threshold, it still falls below the optimal sum rate obtained by the conventional approach. Correspondingly, the trajectory of sum rate with the distributed approach degrades below the threshold faster through time than that of the conventional approach. Hence, more and more gradient updates are required as the threshold increases. Summing up, the distributed approach should definitely be the choice to take in the tracking mode, as it provides remarkably less CSI dissemination overhead independent from network parameters.

3.3.6. Further Application Scenarios for Distributed Gradient

Another contribution of this limited feedback section is to trigger research on adaptation of distributed gradient to some other similar applications where the CSI dissemination overhead appears to be the impractical bottleneck of the system. In the following we first apply the distributed gradient to the optimal power allocation problem in the conventional multi-relay channel. Later, we demonstrate the advantage of employing the iterative distributed gradient when the propagation channels are assumed to be time-varying.

3.3.6.1. Power Allocation for AF Relay Networks

The conventional multi-relay channel, where the unidirectional communication between two nodes is aided by multiple AF relays, is a simplified version of the aforementioned case of single S-D pair in the presence of multiple interferers. Similar scenarios have been considered in previous works [54, 97, 144], where either Lagrangian multiplier method or conic programming has been employed to maximize the receive signal-to-noise ratio (SNR) at the destination with the assumption that global CSI is available during the optimization. Moreover, in [46] a heuristic iterative algorithm has been proposed to perform distributed beamforming by using multiple one-bit feedbacks instead of global CSI. In the following we apply the distributed gradient scheme to the optimal relay

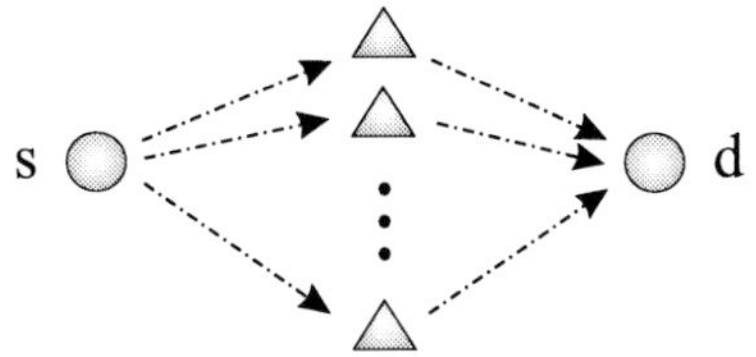

Figure 3.20.: Two-hop network configuration for single S-D pair with half-duplex relays.

power allocation problem for maximizing the receive SNR.

The primary difference with the multiuser setup is that as we have a single S-D pair in the network without any interference, the relay gains can be directly matched to the uplink/downlink channels for efficient coherent combination at the destination. Hence, the *complex valued* relay gain allocation problem boils down to a *real valued* relay power allocation problem.

Dropping the interference terms in (3.21), the received signal model at the destination node modifies to

$$y_{\mathsf{d}} = \mathbf{p}^T(\mathbf{h}_{\mathsf{r}} \odot \mathbf{f}_{\mathsf{r}} \odot \mathbf{g})s + \mathbf{p}^T(\mathbf{g} \odot \mathbf{f}_{\mathsf{r}} \odot \mathbf{n}_{\mathsf{r}}) + n_{\mathsf{d}}, \tag{3.45}$$

where the subscript i is dropped since there is a single S-D pair. The relay power vector

$$\mathbf{p} \in \mathbb{R}_+^{N_{\mathsf{r}}} = [p_0 \cdots p_k \cdots p_{N_{\mathsf{r}}}]^T,$$

whose kth element $p_k \geq 0$ stands for the square root of the kth relay's power. The power vector $\mathbf{p}$ is subject to a sum relay transmit power constraint, i.e., $\sum_{k=1}^{N_{\mathsf{r}}} p_k^2 \leq P_{\mathsf{r}}$. The vector $\mathbf{g} \in \mathbb{C}^{N_{\mathsf{r}}}$ denotes the normalized gain factors given by appropriate matching to uplink and downlink channels per relay [80, 97, 133], and its kth element is written as:

$$g_k = \frac{1}{(P_{\mathsf{s}}|h_k|^2 + \sigma_v^2)} \cdot \frac{h_k^* \cdot f_k^*}{|h_k| \cdot |f_k|}.$$

In order to maximize the received SNR at the destination node, the optimal power allocation vector should be found through the optimization

$$\max_{\mathbf{p}} \ \mathrm{SNR} = \frac{\mathbf{p}^T \boldsymbol{\Upsilon} \mathbf{p}}{\sigma_{n_{\mathsf{r}}}^2 \mathbf{p}^T \boldsymbol{\Xi} \mathbf{p} + \sigma_{n_{\mathsf{d}}}^2} \quad \textbf{subject to } \mathbf{p}^T\mathbf{p} \leq P_{\mathsf{r}}, \mathbf{p} \succeq 0, \tag{3.46}$$

where P_{s} is set to 1, $\boldsymbol{\Upsilon} \triangleq (\mathbf{h}_{\mathsf{r}} \odot \mathbf{f}_{\mathsf{r}} \odot \mathbf{g})(\mathbf{h}_{\mathsf{r}} \odot \mathbf{f}_{\mathsf{r}} \odot \mathbf{g})^H$, $\boldsymbol{\Xi} \triangleq \mathrm{diag}\{|\mathbf{f}_{\mathsf{r}} \odot \mathbf{g}|^2\}$. Assuming global CSI knowledge is available, i.e., $\mathbf{h}_{\mathsf{r}}$, $\mathbf{f}_{\mathsf{r}}$, the optimization in (3.46) can be efficiently solved through gradient based schemes, interior point algorithms [26], Lagrangian multiplier method [54, 144] and/or conic programming [97]. However, as the number of relays increase, the global CSI knowledge assumption loses its practicality. Henceforth, we propose to use distributed gradient based power allocation.

In the following, we first convert the SINR maximization problem to a unconstrained problem by using the logarithmic barrier method. Then, we very briefly elaborate on the achievability of local computation of the related gradient at each relay.

The constrained problem in (3.46) can be modified to an unconstrained problem by using logarithmic barrier method [26] as follows

$$\min_{\mathbf{p}} \quad -\nu \cdot \frac{\mathbf{p}^T \boldsymbol{\Upsilon} \mathbf{p}}{\sigma_{n_\mathsf{r}}^2 \mathbf{p}^T \boldsymbol{\Xi} \mathbf{p} + \sigma_{n_\mathsf{d}}^2} - \ln(P_\mathsf{r} - \mathbf{p}^T \mathbf{p}) - \sum_{k=1}^{N_\mathsf{r}} \ln(p_k), \tag{3.47}$$

where $\nu \in \mathbb{R}_+$ is an accuracy parameter. Similar to **Algorithm 1** of sum rate maximization in Section 3.2.1.1, the unconstrained problem in (3.47) is solved through basic (distributed) gradient method for an initially given ν. The parameter $\nu := \nu\iota$ is successively updated with an $\iota > 1$, and the minimization in (3.47) is repeated for each updated ν until $(N_\mathsf{r} + 1)/\nu < \epsilon$, where $\epsilon > 0$ is sufficiently small.

Denoting the objective function in (3.47) with $f_\mathsf{p}(\mathbf{p})$, the local gradient of $f_\mathsf{p}(\mathbf{p})$ at the kth relay with respect to p_k is written as

$$\nabla_{p_k} f_\mathsf{p}(\mathbf{p}) = \frac{\nu \sigma_{n_\mathsf{r}}^2 p_k \cdot |f_k g_k|^2}{\sigma_{n_\mathsf{r}}^2 \mathbf{p}^T \boldsymbol{\Xi} \mathbf{p} + \sigma_{n_\mathsf{d}}^2} \mathrm{SNR} - \frac{\nu \cdot \boldsymbol{v}_k^T \mathbf{p}}{\sigma_{n_\mathsf{r}}^2 \mathbf{p}^T \boldsymbol{\Xi} \mathbf{p} + \sigma_{n_\mathsf{d}}^2} + \frac{2p_k}{P_\mathsf{r} - \mathbf{p}^T \mathbf{p}} - \frac{1}{p_k},$$

where $\boldsymbol{v}_k \in \mathbb{C}^{N_\mathsf{r}}$ is the transpose of the kth row of $\boldsymbol{\Upsilon}$.

In order to compute $\nabla_{p_k} f_\mathsf{p}(\mathbf{p})$, the kth relay needs to know the information set

$$\mathcal{I}_p = \left\{ \mathrm{SNR},\ h_\mathsf{srd} \triangleq \boldsymbol{v}_k^T \mathbf{p},\ \mathbf{p}^T \mathbf{p},\ h_k,\ f_k,\ \sigma_{n_\mathsf{d}}^2 + \sigma_{n_\mathsf{r}}^2 \mathbf{p}^T \boldsymbol{\Xi} \mathbf{p} \right\},$$

where we assume that $\nu, \iota, P_\mathsf{r}, P_\mathsf{s}$ are globally known over the network. It is readily assumed that each relay has local CSI, i.e., h_k, f_k. Moreover, the relays can be supplied with SNR and h_srd information after an *estimate and feedback* cycle. With these, they can further compute $\sigma_{n_\mathsf{d}}^2 + \sigma_{n_\mathsf{r}}^2 \mathbf{p}^T \boldsymbol{\Xi} \mathbf{p} = |h_\mathsf{srd}|^2/\mathrm{SNR}$. In order to obtain the last missing information $\mathbf{p}^T \mathbf{p}$, and we perform one *measure and feedback* cycle. There, a concurrent transmission is performed from all relays to the destination, where the kth relay transmits the value p_k^2/f_k, and the destination measures the received signal amplitude, and feeds it back to the relays. Summing up, the information set required by the kth relay to compute $\nabla_{p_k} f_\mathsf{p}(\mathbf{p})$ locally, is

$$\mathcal{I}_p = \left\{ h_k, f_k, h_\mathsf{srd}, \mathrm{SNR} \right\},$$

which is similar but correspondingly reduced version of the required information set for multiuser setup.

Shifting our focus to the distributed step size calculation, we can adapt, for instance, the centralized proposal in Section 3.3.1.2 to this specific problem, as well. Skipping the details for the sake of brevity and referring to the Section 3.3.1.2, it can be shown that the destination can gather the necessary information to compute the step size after several *estimate and feedback* cycles. However, we should note here that the estimation of the third summand in (3.47) needs N_r time slots, which may turn out to be impractical for large N_r. Thus, we propose to modify the step size algorithm such that this term is neglected. This neglected term is related with the positiveness of p_k values, and thus, may naturally increase the number of iterations related with ν, which inherently controls the validity of each constraint in (3.46).

We remark finally that, contrary to the multiuser setup, the overall iteration cycle in this scenario is in three-folds: 1) the step-size, 2) the gradient, 3) the accuracy parameter ν. While concluding, we propose that the distributed gradient scheme can be efficiently employed for power allocation problems in order to cope with local CSI dissemination overhead in a rather large AF relay network,.

3.3.6.2. Channel Tracking for Time Varying Channels

In Section 3.3.5, we have shortly touched on channel tracking ability of the distributed gradient approach. In this short subsection, we take a different perspective and focus on how efficient channel tracking is with respect to the case without tracking. In other words, depending on the correlation of two neighbouring blocks, we will show that the necessary number of outer iterations for gain allocation can be reduced by feeding the previous gain vector as an initial vector to the next one. To this end, we perform a simulation campaign, in which we compare the performances of the two cases:

1) *Without tracking:* For each channel block, independent from the correlation in between, we start the algorithm with a random initial relay gain vector.

2) *With tracking:* We start the algorithm with a random initial relay gain vector for the first channel block $(\mathbf{H}_{(1)}, \mathbf{F}_{(1)})$, find the maximal sum rate resultant gain vector allocation, and use this choice as an initial vector while implementing the algorithm for the second block of channel $(\mathbf{H}_{(2)}, \mathbf{F}_{(2)})$.

In Fig. 3.21, we plot the empirical cdf of the necessary number of iterations to achieve 99.99% of the optimal sum rate in the second block of channel for both cases

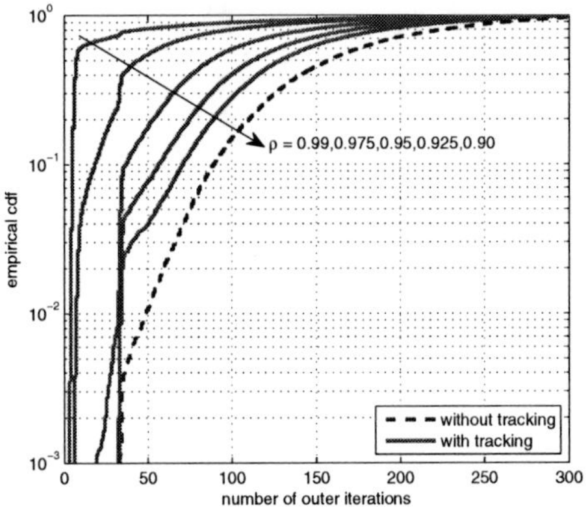

Figure 3.21.: Three S-D pairs: Empirical CDFs of the necessary number of iterations to achieve 99.99% of the optimal sum rate in the second block of channel for both cases of with and without tracking ($N_r = 30$, SNR $= 25$ dB).

of with and without tracking. We repeat the simulations for the set of correlation coefficients $\rho \in \{0.99, 0.975, 0.95, 0.925, 0.9\}$. It is observed that when the correlation coefficient is large enough, the proposed distributed algorithm employing tracking can exploit correlation efficiently, and hence, converges much faster with respect to the case without tracking. Specifically, considering $\rho = 0.95$ and $N_{\text{it}} = 50$, the probability that the algorithm needs less than $N_{\text{it}} = 50$ iterations is 0.01 for the case without tracking; whereas channel tracking improves the same probability more than one order of magnitude to 0.2. Regarding to the extreme case of highly correlated blocks $\rho = 0.99$, this probability increases up to 0.8. On the other hand, decreasing ρ to 0.90, we notice that the improvement with tracking is somewhat limited, i.e., only increased from 0.01 to 0.04. Hence, we conjecture that for correlations less than 0.9, tracking will lose its effect substantially, because the directions of the beams formed for the previous block of channel appears to be misplaced by the increased dominance of the random additions, i.e., $\tilde{\mathbf{H}}_{(1)}$ or $\tilde{\mathbf{F}}_{(1)}$, in the second block.

Summing up, due to the iterative nature of the distributed gradient approach, we can efficiently take advantage of the correlation between succeeding channel blocks, given that correlation is high enough, and hence, the beamforming directions do not need to change significantly.

3.4. Concluding Remarks

Our contribution in this chapter can be summarized in two stages. First, we identify the gains that can be achieved through relay gain optimization, and later, we search for efficient distributed designs to realize these gains in practice.

In the first part, we showed that relay gain optimization is a *must* for efficient utilization of available spatial degrees of freedom. Depending on the needs of the network, one of the proposed beamforming methods can be employed. For example, we should prefer max-min type optimizations for such cases where the network is composed of S-D pairs whose need for information rate is similar, or that the supported data service requires robustness (outage sensitive) over different channel realizations. In case that the supported service, say multimedia, over a certain link requires a minimum fixed rate, then we shall introduce QoS constraints to the optimization. On the other hand, if we are limited by hardware complexity for optimization, we should go over low-complexity ZF solution, which is anyway penalizes the performance only marginally.

Further, we have shown that additional spatial degrees of freedom on top of the amount required for full spatial multiplexing gain, can be converted completely to an effective diversity gain. To exploit this conjecture, we proposed two schemes that achieve the maximal diversity $N_\mathsf{r} - N(N-1)$: relay selection and max-min type optimization.

In the second part, we identified *distributed gradient* based methods as an effective means for substantially reducing the CSI dissemination overhead in coherent AF relaying networks. Considering two typical AF relay network structures, we showed that SINR related gradient can be locally calculated at each relay with only local CSI plus limited feedback. We performed a comprehensive study on the distributedness of each step of the algorithm in order to provide a complete picture of the distributed conjugate gradient algorithm. Moreover, for the favor of the rapid convergence, we addressed efficient starting vector computations.

The main aim of this section was to make the distributed orthogonalization practical for implementation in real-world scenarios, which was previously impeded by the heavy load of local CSI dissemination of N_r^2 distributed antennas. However, with the distributed gradient method, we have reduced the scaling of this CSI dissemination overhead from N_r to N, i.e., linear in N instead of being quadratic. Hence, while the network enjoys high throughput rates arising from large (but finite) numbers of relays, it does not need to pay off, but gains on top in the sense that the convergence is significantly accelerated with increased relay numbers.

Finally, we report that the complexity in terms of the amount of dissemination overhead is scalable within our gradient based proposals independent of being distributed or not. For instance, the gradient algorithm can be terminated at any outer iteration resulting in a sustainable cost function value, and the corresponding relay gains can be employed to attain that value through the network. In other words, we provide a trade-off between performance and complexity.

Chapter 4

Robust Gain Allocation Against Imperfections in Coherent Relaying

The two *sine qua non* assumptions for the efficiency of coherent multiuser relaying are perfect CSI knowledge per relay and globally phase-synchronous relays. This chapter takes a practical look on these two assumptions and assumes that there are imperfections with both. After modeling the corresponding data mismatches within given uncertainty sets, we follow a worst-case framework and design SDP based robust counterparts for max-min beamforming. The robust beamformer designs are shown to achieve substantial worst-case performance improvements with respect to their non-robust counterparts.

4.1. Introduction

While designing the relay gain allocation schemes in the previous chapters, we have assumed that each relay has the required information set for efficient distributed beamforming perfectly in the sense that there are no errors. More specifically, we have considered perfect phase- and frequency-synchronous relays and that any channel state information is perfectly estimated at the corresponding node. However, the practical real-world implementations are not as flawless as we hope them to be. So to say, an efficient system design should take any system imperfection that may harm the predicted performance into account for the sake of reliability (robustness).

State-of-the-Art Robust system design and optimization nowadays have gained an increasing attention in control, communication and network theories [9–15, 25, 42, 43,

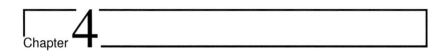

59, 78, 95, 98, 106, 126–130, 148]. Primarily, the aim in robust design is to model the uncertainty within the available data, and then provide an appropriate robust solution. Essentially, this solution should be obtained through a computationally tractable algorithm. The robust optimization approaches can be classified in two main classes. If the problem consists of strict constraints which must be fulfilled for any data set and any realization of the uncertainty within the given sets, then we refer to this as *worst-case* optimization. Alternatively, if the robust problem exploits the stochastic nature of the considered system and enforces probabilistic constraints, then this approach is referred as *probabilistic* optimization.

A general survey of robust convex optimization follows from the references [9–14, 42, 43]. Applying these fundamentals to the general communication theory, an overview has been presented in [78]. More details on robust optimization theory can be found in Appendix A.1.

Robust optimization theory has been extensively applied on the downlink beamforming problem [15, 25, 59, 95, 106, 126–130, 148], where the perfect feedback of downlink channel(s) to the transmitter is highly questionable, and hence, is assumed to be uncertain in a pre-defined set. In [59, 106, 126, 127], point-to-point MIMO beamforming is in the focus of interest, whereas the designs have been extended to the multiuser downlink case in [15,25,95,128–130,148]. Throughout most of these works, researchers have investigated the worst-case performance related designs [15,25,59,95,106,126,128,129,148], by assuming that the CSI errors at the transmitter are defined within spherical/ellipsoidal uncertainty sets. Such kind of worst-case analysis have been shown to be rather conservative in terms of performance. Although it is more complex to model and to approximate in convex formulations (when compared to the former worst-case choice), there are a few attempts to provide probabilistic robust solutions in [127, 128].

The simplest case of point-to-point MISO beamforming has been first addressed in references [126, 127], where either a worst-case analysis [126] or a probabilistically-constrained optimization [127] was performed. Further, for a MIMO setup, an orthogonal space-time block coding based transmitter was employed in [59], where robust power allocation among eigenmodes of the estimated channel was aimed to be designed. Similarly in [106], worst-case robust eigen-beamforming was investigated on the basis of receiver SNR at the output maximal ratio combiner.

Most of the robust precoder optimization studies on multiuser setups have assumed single-antenna receivers for mathematical simplicity [15, 25, 95, 129, 148], with the two exceptions [130, 148] that considered a multiuser MIMO scenario. In one of earliest multiuser reference [15], uncertainty was defined for the covariances of the channel vectors and minimum transmit power was optimized for given uncertain QoS constraints on SINRs. In [95, 128–130], the MSE per receiver was the focus of interest, whereas in [25, 129] QoS constraints on SINR were investigated. Specifically, for instance, both references [25, 129] independently focused on the NP-hard intractable problem of minimizing the transmit power by assuming deterministically bounded (ball) uncertainty sets, and proposed approximate robust counterparts with efficient solutions. Lastly, as an alternative approach, probabilistic requirements are enforced on the MSE based constraints of a multiuser MISO setup in [128].

The reference [98] was the first in cooperative-diversity field that properly addressed a robust distributed beamforming design for the conventional multi-AF-relay channel. There, assuming a worst-case approach, a fixed total sum relay power is distributed to the relays such that the received SINR at the destination is maximized in the presence of uncertainty at the CSI of the relays. On the other hand, reference [57] has also addressed the same scenario with an imperfect knowledge on channel coefficients. However, imperfect knowledge there translated to the knowledge of only second-order statistics of the channel coefficients.

Imperfections at the LO have also decisive affects on the quality of the channel estimation. In essence, if we do not compensate the LO impacts on up- and down-conversion with a perfect synchronization scheme or through a presumed global phase reference, the estimated channel, independent of the optimality of the estimation algorithm, is not the pure propagation channel. It is only an equivalent channel consisting of the real propagation channel and the phase offsets at both transmitter and receiver. In other words, although it can be practical to assume full knowledge on the magnitude of the channels, the phase information about the channels is not a trivial to attain. Addressing this issue, the authors of [94] investigated the optimization and the evaluation of mutual information in MIMO systems assuming that channel's magnitude is perfectly estimated, but the phase estimation is imperfect.

On the other hand, LO based phase and frequency imperfections in a distributed beamforming network have been thoroughly studied in [81], where the authors also proposed a master-slave architecture for coordinating synchronization. Similar proposals for carrier phase synchronization can also be found in [19, 120]. In [18], the impact of noisy carrier phase synchronization on coherent AF relaying has been investigated. Moreover, recently in [20], efficient channel estimation protocols have been proposed with which global carrier phase synchronization becomes redundant. However, to the best of our knowledge, none of the proposals in the literature has yet considered to incorporate LO phase imperfection with the design of distributed beamforming.

Contribution of this chapter We have perceived the phase synchronization between the LOs of the relays and the estimation of CSI as the two main sources of imperfections that threat the performance and the practicality of coherent AF relaying. Regarding any other, for instance, the frequency synchronization of the LOs, which is closely related with phase-synchronization, we assume that it can be simply assumed to be perfect, i.e., $f_{\mathsf{x}} = f_{\mathsf{c}}\ \forall \mathsf{x}$, with the motivation that the frequency offsets are rather stable when compared with LO phase offsets, and can be trivially compensated using well-known techniques proposed in the literature [6,8,27]. For further details on some other possible imperfections that are out of the scope of this thesis, such as frequency offsets, IQ imbalance, quantization noise, amplifier nonlinearities, we refer the interested reader to [16].

As it is suggested by the general received signal model (2.5) at the terminals and the consequent simplification from (2.6) to (2.7), it has been shown in [17, 20] that relay phase synchronization can be in fact avoided with appropriate channel estimation protocols. That is, during the processes of reception in the first hop and forwarding in the second hop, local relay phase offsets cancel each other, i.e., $\mathbf{\Phi}_{\mathsf{r}}\mathbf{G}\mathbf{\Phi}_{\mathsf{r}}^{H} = \mathbf{G}\mathbf{\Phi}_{\mathsf{r}}\mathbf{\Phi}_{\mathsf{r}}^{H} = \mathbf{G}$. Albeit this simplification advantage, there is still a random phase (noise) uncertainty $\mathbf{\Theta}_{\mathsf{r}}$ at the relays, characterized by the time instant differences between reception and forwarding. Being independent from any possible global phase reference, this phase noise hampers the coherence and causes the anticipated SINR at the destinations to be uncertain (random). In Section 4.2, we address this phase offset randomness. We first quantify its impact on received SINRs at the destination terminals. Afterwards, we bound the phase noise uncertainty within pre-determined sets and design a robust relay gain decision which incorporates this random imperfection.

On the other hand, channel estimation related errors come into the imperfection picture in two different aspects: the estimation protocol, i.e., where to estimate which channel, and the accuracy of the corresponding estimation. The former one is crucial in terms of the aforementioned *inherent* phase-synchronization at the relays, and also affects the uncertainty at the phase offsets while deciding on the relay gains. These issues will be covered in Section 4.2 in the context of phase imperfections. Whereas, the latter type of imperfection is commonly triggered by noisy channel estimation, quantization errors, phase offset uncertainty or by being outdated. Towards developing an understanding on these, we devote the Section 4.3 to the case of imperfect channel state information. Therein, we study on worst-case robust gain allocation designs for max-min beamforming. In general, we follow the robust convex optimization methodology of *A. Ben-Tal and A. Nemirovski* [9–14], where the uncertainty is defined within bounded sets and computationally tractable exact or approximate robust counterparts are derived.

4.2. LO Phase Imperfections at the Relays

In this section, we investigate the system imperfections caused by the non-phase synchronous relays. In particular, we focus on two issues: distributed relay phase synchronization and randomness triggered by the phase noise. In Section 4.2.1, we study the impact of LO phase offsets on the beamforming performance, and later present a simple solution to the global phase synchronization problem through a certain channel estimation protocol. Therein, we expose another critical, but usually neglected, problem of phase noise uncertainty. Addressing this randomness in LO phase offsets in Section 4.2.2, we devise a worst-case robust gain allocation design.

4.2.1. Impact of LO Phase Imperfections

Here and in the remainder of this chapter, we consider single-antenna relays. We recall from Section 2.3.2 that the received signal vector at the destinations is

$$\mathbf{y}_{\mathsf{d}} = \mathbf{\Phi}_{\mathsf{d}}^{H}\mathbf{F}\mathbf{\Theta}_{\mathsf{r}}\mathbf{G}\mathbf{H}\mathbf{\Phi}_{\mathsf{s}}\mathbf{s} + \mathbf{\Phi}_{\mathsf{d}}^{H}\mathbf{F}\mathbf{\Phi}_{\mathsf{r}}\mathbf{\Theta}_{\mathsf{r}}\mathbf{G}\mathbf{n}_{\mathsf{r}} + \mathbf{n}_{\mathsf{d}}, \tag{4.1}$$

where we have incorporated the fact that the local relay phase offsets cancel each other, i.e., $\mathbf{\Phi}_{\mathsf{r}}^{H}\mathbf{\Phi}_{\mathsf{r}} = \mathbf{I}$. However, there is still a random phase uncertainty matrix $\mathbf{\Theta}_{\mathsf{r}}$, which is a function of the processing time at the relays.

Let us at first focus on this random term, and model it appropriately. Suppose that the reception at the relays happened at time instant t_1. Due to the down-conversion process, the received signal at the antenna of the kth relay is multiplied by (see (2.1))

$$e^{-j\left(2\pi f_c t_1 + \bar{\theta}_{\mathsf{r}_k}^{(t_0)} + \theta_{\mathsf{r}_k}^{(t_1 - t_0)}\right)},$$

which is then followed by the multiplication with the corresponding gain factor g_k. Finally, assuming that the forwarding is done at time instant t_2, the relay transmit signal is multiplied with

$$e^{j\left(2\pi f_c t_2 + \bar{\theta}_{r_k}^{(t_0)} + \theta_{r_k}^{(t_2-t_0)}\right)},$$

through the up-conversion process. After all, the equivalent processing applied on the received signal at the kth relay becomes

$$\begin{aligned} g_k \cdot e^{j\left(2\pi f_c t_2 + \bar{\theta}_{r_k}^{(t_0)} + \theta_{r_k}^{(t_2-t_0)}\right)} \cdot e^{-j\left(2\pi f_c t_1 + \bar{\theta}_{r_k}^{(t_0)} + \theta_{r_k}^{(t_1-t_0)}\right)} &= g_k \cdot e^{j\left(2\pi f_c (t_2-t_1)\right)} \cdot e^{j\left(\theta_{r_k}^{(t_2-t_0)} - \theta_{r_k}^{(t_1-t_0)}\right)} \\ &= g_k \cdot e^{j\breve{\theta}_k}, \end{aligned} \tag{4.2}$$

where $\breve{\theta}_k$ denotes the equivalent phase uncertainty term and it is a Gaussian random variable with mean $\zeta \triangleq 2\pi f_c(t_2 - t_1)$ and variance $\sigma_{\breve{\theta}}^2 \triangleq (t_2 + t_1 - 2t_0)\sigma_p^2$. Consequently, we can write the vector of *random phase uncertainty* at the relays as

$$\boldsymbol{\theta} = \left[e^{j\breve{\theta}_1}\ e^{j\breve{\theta}_2}\ \ldots\ e^{j\breve{\theta}_{N_r}}\right]^T.$$

Then, the corresponding phase uncertainty matrix becomes $\boldsymbol{\Theta}_r \triangleq \operatorname{diag}\{\boldsymbol{\theta}\}$.

Further, assuming single-antenna relays and incorporating $\boldsymbol{\theta}$ with the SINR definition in (3.23), the received SINR at the ith destination can be equivalently written as

$$\mathrm{SINR}_i = \frac{\left|(\mathbf{f}_{ri} \odot \mathbf{h}_{ir} \odot \mathbf{g})^T \boldsymbol{\theta}\right|^2}{\sum_{j=1, j\neq i}^{N} \left|(\mathbf{f}_{ri} \odot \mathbf{h}_{jr} \odot \mathbf{g})^T \boldsymbol{\theta}\right|^2 + \sigma_{n_r}^2 \mathbf{g}^H \boldsymbol{\Xi}_i \mathbf{g} + \sigma_{n_d}^2}, \tag{4.3}$$

where $\boldsymbol{\Xi}_i = \operatorname{diag}\{|\mathbf{f}_{ri}|^2\}$ is defined similar to the definition of $\boldsymbol{\Xi}$ in (3.46). Observe from (4.3) that the phase offsets at the source and destination terminals do not affect the received SINR, and the power of amplified noise is independent from any phase offsets within the system. However, we notice primarily that for given channel gains and gain allocation, SINR is a random variable dependent on $\boldsymbol{\theta}$.

At this point, the channel estimation comes into the picture in the sense that the relays need to know $\{\mathbf{h}_{ir}, \mathbf{f}_{ri} \forall i, j\}$ to compute the SINR at each destination, and consequently, to design $\mathbf{g}$. We have already mentioned that the estimation protocol is essential for the sake of the *inherent* relay phase offset cancellation. Although there are several protocols dependent on the direction of the channel estimation [16], i.e., where to estimate which channel, here we present only the case that removes the need for global relay phase synchronization.

Channel Estimation For the sake of generality, we focus on the compound uplink and downlink channels. The estimated channel matrices $\hat{\mathbf{H}}$ and $\hat{\mathbf{F}}$ are different from the real propagation channel $\mathbf{H}$ and $\mathbf{F}$ due to the up- and down-conversion processes. Specifically, they are functions of the time of estimation although the real propagation channel may stay constant.

Assuming that the sources transmit orthogonal training sequences and the channel of source-to-relay link is estimated at the relays at time instant $\hat{t}_1$, the estimated channel matrix becomes

$$\hat{\mathbf{H}}(\hat{t}_1) = \mathbf{\Phi}_{\mathsf{r}}^H(\hat{t}_1)\mathbf{H}\mathbf{\Phi}_{\mathsf{s}}(\hat{t}_1). \tag{4.4}$$

We have slightly modified our notation in (4.4) for the clarity in the following derivations and introduced the time instant to the notations of the channel estimation and phase offsets. Note that the estimations are assumed to be noiseless.

There are two options for estimating the channels of the relay-to-destination links: either the destinations transmit training sequences and the channel estimation is performed at the relays side at time instant $\hat{t}_2$, i.e.,

$$\hat{\mathbf{F}}^{(1)}(\hat{t}_2) = \mathbf{\Phi}_{\mathsf{r}}^H(\hat{t}_2)\mathbf{F}\mathbf{\Phi}_{\mathsf{d}}(\hat{t}_2), \tag{4.5}$$

or the relays transmit orthogonal training sequences and the channel estimation is performed at the destinations at time instant $\hat{t}_2$, i.e.,

$$\hat{\mathbf{F}}^{(2)}(\hat{t}_2) = \mathbf{\Phi}_{\mathsf{d}}^H(\hat{t}_2)\mathbf{F}\mathbf{\Phi}_{\mathsf{r}}(\hat{t}_2). \tag{4.6}$$

Although the propagation channel is assumed to be reciprocal, the effects of up and down converters result different channel estimates for the two options. That is, if (4.4) and (4.5) are combined the effective two-hop channel becomes

$$\hat{\mathbf{F}}^{(1)}(\hat{t}_2)\mathbf{G}\hat{\mathbf{H}}(\hat{t}_1) = \mathbf{\Phi}_{\mathsf{r}}^H(\hat{t}_2)\mathbf{F}\mathbf{\Phi}_{\mathsf{d}}(\hat{t}_2)\mathbf{G}\mathbf{\Phi}_{\mathsf{r}}^H(\hat{t}_1)\mathbf{H}\mathbf{\Phi}_{\mathsf{s}}(\hat{t}_1),$$

whereas if (4.4) and (4.6) are combined, it modifies to

$$\hat{\mathbf{F}}^{(2)}(\hat{t}_2)\mathbf{G}\hat{\mathbf{H}}(\hat{t}_1) = \mathbf{\Phi}_{\mathsf{d}}^H(\hat{t}_2)\mathbf{F}\mathbf{\Phi}_{\mathsf{r}}(\hat{t}_2)\mathbf{G}\mathbf{\Phi}_{\mathsf{r}}^H(\hat{t}_1)\mathbf{H}\mathbf{\Phi}_{\mathsf{s}}(\hat{t}_1) = \mathbf{\Phi}_{\mathsf{d}}^H(\hat{t}_2)\mathbf{F}\mathbf{G}\mathbf{H}\mathbf{\Phi}_{\mathsf{s}}(\hat{t}_1), \tag{4.7}$$

where we neglect the random phase uncertainty matrix for the moment. It is important to note here that employing the latter option releases the relays from the requirement to be coherent in terms of LO phase offsets while calculating $\mathbf{g}$, and let us have the removal of $\mathbf{\Phi}_\mathsf{r}$ in (4.1). However, the convenience of using the latter option brings an additional estimation complexity and the requirement that the estimated channels at the destinations should be fed back to the relays. For further details about channel estimation in coherent relaying, we refer the interested reader to [16, 17, 20]. To sum up, we assume that each relay has the perfect knowledge of the estimations $\hat{\mathbf{H}}(\hat{t}_1)$ and $\hat{\mathbf{F}}^{(2)}(\hat{t}_2)$ so that the global phase synchronization is avoided.

Although the equation (4.7) shows that synchronization becomes redundant, it does not reflect the observed signal model at the destinations exactly. We have to further incorporate the random phase noise impact which comes from two sources. First, as we studied above there is an inherent phase noise term $\mathbf{\Theta}_\mathsf{r}$ due to forwarding process. Second, we have an additional randomness on phases which is now due to different time instants of channel estimation, i.e., $\hat{t}_1 \neq \hat{t}_2$. In the following, we show that these two induced randomnesses can be added up in the pre-defined single $\boldsymbol{\theta}$.

The received signal model (4.1) informs us that, independent of the channel realization, the received signal and SINR are random variables of $\boldsymbol{\theta}$. Hence, from the kth relay's perspective, it employs the aforementioned channel estimation in (4.1) to compute SINRs, but also knows a priori that there is the phase randomness $\mathbf{\Theta}_\mathsf{r}$. That is to write, the equivalent channel (4.7) becomes

$$\begin{aligned}\hat{\mathbf{F}}^{(2)}(\hat{t}_2)\cdot\mathbf{\Theta}_\mathsf{r}\mathbf{G}\cdot\hat{\mathbf{H}}(\hat{t}_1) &= \mathbf{\Phi}_\mathsf{d}^H(\hat{t}_2)\mathbf{F}\mathbf{\Phi}_\mathsf{r}(\hat{t}_2)\cdot\mathbf{\Theta}_\mathsf{r}\mathbf{G}\cdot\mathbf{\Phi}_\mathsf{r}^H(\hat{t}_1)\mathbf{H}\mathbf{\Phi}_\mathsf{s}(\hat{t}_1)\\ &= \mathbf{\Phi}_\mathsf{d}^H(\hat{t}_2)\mathbf{F}\cdot\tilde{\mathbf{\Theta}}_\mathsf{r}\mathbf{\Theta}_\mathsf{r}\mathbf{G}\cdot\mathbf{H}\mathbf{\Phi}_\mathsf{s}(\hat{t}_1)\end{aligned}\tag{4.8}$$

where in the second line we used the equality that

$$\mathbf{\Phi}_\mathsf{r}(\hat{t}_2)\cdot\mathbf{\Theta}_\mathsf{r}\mathbf{G}\cdot\mathbf{\Phi}_\mathsf{r}^H(\hat{t}_1) = \mathbf{\Phi}_\mathsf{r}(\hat{t}_2)\cdot\mathbf{\Phi}_\mathsf{r}^H(\hat{t}_1)\cdot\mathbf{\Theta}_\mathsf{r}\mathbf{G}\cdot \triangleq \tilde{\mathbf{\Theta}}_\mathsf{r}\mathbf{\Theta}_\mathsf{r}\mathbf{G}.$$

The matrix $\tilde{\mathbf{\Theta}}_\mathsf{r}$ denotes the random phase uncertainty triggered by the channel estimation time instant difference, and is defined similar to $\mathbf{\Theta}_\mathsf{r}$ with parameters $\hat{t}_1$ and $\hat{t}_2$. As both $\mathbf{\Theta}_\mathsf{r}$ and $\tilde{\mathbf{\Theta}}_\mathsf{r}$ consists of exponential functions of Gaussian random variables (with similar statistics), these two matrices can be added up. Nevertheless, the statistics of phase uncertainty per relay would change, but the structure of the combined phase uncertainty matrix would have the same structure as $\mathbf{\Theta}_\mathsf{r}$. Hence, for the sake of simplicity in the sequel, we assume that all phase noise related uncertainty is embedded in $\mathbf{\Theta}_\mathsf{r}$ (and hence $\boldsymbol{\theta}$). While concluding the discussion on the uncertainty model, we once again highlight that the phase uncertainty has two effective sources: *processing time at the relays* and *the estimation time instant difference*.

SINR Loss due to Random Phase Uncertainty After characterizing the received SINRs in terms of the random phase uncertainty vector $\boldsymbol{\theta}$, we are now interested in the corresponding impact of $\boldsymbol{\theta}$ on SINRs. We study two performance measures: the average SINR over random phase uncertainty and the worst-case SINR over random phase uncertainty. To this end, we perform a set of Monte Carlo simulations with a setup of $N = 2$ S-D pairs, where we simply perform a numerical search over sufficiently large sets of uncertainty and channel realizations.

We choose the relay gain decisions according to the non-robust Max Min Link Rate through $\mathcal{P}_{\text{mm}}$ in Section 3.2.2.1, where we ignore the presence of $\boldsymbol{\theta}$. It is assumed that $P_{\mathsf{s}} = P_{\mathsf{r}} = 1$, and $\sigma^2 \triangleq \sigma_{n_r}^2 = \sigma_{n_{\mathsf{d}}}^2$. The average SNR, SNR $= P_{\mathsf{s}}/\sigma^2$, is set to 20 dB.

Without loss of generality, we assume that uncertainty is zero-mean and σ_θ^2 is the same for all relays. Hence, each realization of $\boldsymbol{\theta}$ is generated such that its kth element's exponential argument $\breve{\theta}_k$ is distributed according to $\mathcal{N}(0, \sigma_\theta^2)$. However, such a Gaussian realization does not accurately capture the "worst-case" impact of the uncertainty. Because, in theory what ever the variance is, there is always a certain probability larger than zero that $\breve{\theta}_k$ can reach a value close to $+\infty$ or $-\infty$. For our case of phase offset realizations, this translates to that independent of σ_θ^2, $\breve{\theta}_k$ can take any value between $-\pi$ and π. In other words, for a given set of channel realizations and N, N_{r}, there is a single worst-case impact of phase randomness. Since this statement is "too conservative" and it is highly improbable that $\boldsymbol{\theta}$ will take on a value that hits this worst-case impact, our numerical approach renders the *practical* impact of phase randomness rather than the *theoretical* one.

In Fig. 4.1, we show both the average and the worst-case impact of uncertainty for $N = 5$ and $N = 10$ relays, where the minimum of the two link-rate is plotted. The results are averaged over 1000 channel realizations. The effect of uncertainty on the average is limited when compared with that of the worst-case effect, and causes at most an SNR loss of 5-10 dB. Naturally, the average loss due to uncertainty increases for larger N_{r}, since the number of uncertain elements grows linearly with N_{r}. Independent of N_{r}, we realize that the minimum worst-case SINR falls below 0 dB beyond $\sigma_\theta^2 = 0.1$, which is well below any practical-operation value when average SNR of 20 dB is taken into account. The worst-case performance for $N_{\mathsf{r}} = 5$ saturates after $\sigma_\theta^2 = 0.6$ at around -17.5 dB. This value represents the aforementioned ultimate theoretical worst-case value for $N_{\mathsf{r}} = 5$, e.g., a global search over $[\pi, -\pi]$ for each $\breve{\theta}_k$ reveals this result. In general, we observe that the performance of both measures for $N = 10$ is strictly better than $N = 5$, except of the worst-case regime beyond $\sigma_\theta^2 = 0.4$. As increasing the number

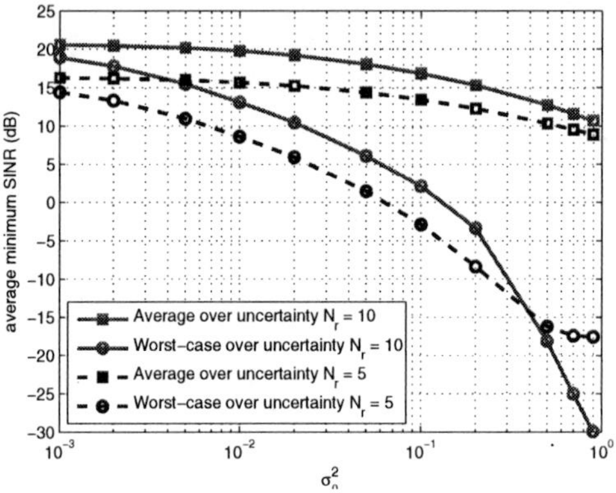

Figure 4.1.: The impact of phase uncertainty on SINR performance of the non-robust coherent AF relaying for $N = 2$ S-D pairs, $N_\mathsf{r} = 5, 10$ relays, and $\mathrm{SNR} = 20$dB.

of relays introduces more uncertainty to the system, the theoretical worst-case impact is much lower for $N = 10$, which results in a cross of the two worst-case curves. This can be interpreted as that the excess spatial (relay) diversity can not mitigate with the uncertainty when the variance of LO phase offsets are larger.

Summing up, we identify the worst-case performance as the most vulnerable regime that non-robust gain allocation schemes suffer. Addressing this, in the following we design a worst-case robust gain allocation scheme against the ruinous effects of random phase uncertainty.

4.2.2. Robust Optimization for Random Phase Uncertainty

In this section, we aim at designing a robust relay gain vector $\mathbf{g}$ against the uncertainty induced by the phase noise. As we observed that the destruction of the uncertainty is concentrated more on the outage-regime, we adapt a worst-case approach, which can be translated as *outage at zero-probability*. On the other hand, we consider a multi-user setup, where we do not assume any priority between S-D links. In other words, robustness of each link should be fairly treated. Summing up, the robust problem we are considering is a worst-case maximization in two aspects:

1. We focus on the worst-case SINR performance of each link with respect to the uncertainty, i.e., the resultant optimal gain vector considers all $\breve{\theta}_k$ realizations

within given bounds.

2. The minimum "worst-case" SINR is maximized over all S-D links.

Hence, the worst-case robust max-min beamforming problem is stated as

$$\max_{\mathbf{g}} \min_{1\le i\le N} \ \mathrm{SINR}_i \ \textbf{subject to} \ \ \mathbf{g}^H\mathbf{M}\mathbf{g} \le P_{\mathrm{r}},$$
$$\breve{\theta}_k \sim \mathcal{N}(\zeta, \sigma_\theta^2), k\in\{1,\ldots,N_r\}, \tag{4.9}$$

where the first constraint represents the imposed sum transmit power constraint on the relays, and it is independent of the phase noise variations.

As presented in Section 2.2, it is assumed practically that the phase uncertainty terms $\breve{\theta}_k$ are Gaussian random variables following a Levy-Wiener process [40,82,86]. However, it should be noted that the phase in general is not continuous due to the effect of modulo 2π operation. In case that the induced variance becomes very large, i.e., the tails of the distribution are broader, the effective phase distribution is ideally modeled by the von Mises distribution. Nevertheless, as the phase noise variance is assumed to be a very small value in practice, e.g. $\sigma_\theta^2 < 0.1$ [82, 86], (and also for the sake of mathematical tractability), the Gaussian assumption has become a commonly accepted rule-of-thumb by the wireless research society.

Hence, we can assure with probability $\mathrm{erf}(\delta/\sqrt{2})$ that all realizations of each $\breve{\theta}_k$ are bounded within $\delta \in \mathbb{Z}_+$ standard deviation $\sigma_\theta > 0$ away from ζ, i.e.,

$$\breve{\theta}_k \in [\zeta + \delta\sigma_\theta, \zeta - \delta\sigma_\theta],$$

where $\mathrm{erf}(x) = (2\sqrt{\pi}) \cdot \int_0^x e^{-t^2} dt$. For example, choosing $\delta = 1, 2, 3, 4$ results in probability of assurance of 0.68269, 0.95450, 0.99730, and 0.99994, respectively. As the effective mean of the phase noise at the relay nodes is dependent only on time instant differences, which are rather deterministic, the mean ζ of the uncertainty can be compensated. Hence, in the sequel we set $\zeta := 0$, and simply assume that $\breve{\theta}_k \sim \mathcal{N}(0, \sigma_\theta^2)$. Consequently, for the sake of generality, we re-define the bounds of $\breve{\theta}_k$, and assume that each phase $\breve{\theta}_k$ lies in an interval of $[0, 2\pi)\backslash[\alpha_k, \beta_k]$, where $[\alpha_k, \beta_k] := [\delta\sigma_\theta, 2\pi - \delta\sigma_\theta]$ and $\delta\sigma_\theta < \pi$.

Incorporating these bounds on uncertainty terms and introducing an auxiliary term $\tau > 0$, the robust problem in (4.9) can be re-formulated as

$$\mathcal{P}_{\mathrm{pr}}: \ \max_{\mathbf{g}, \tau\ge 0} \ \tau \ \textbf{subject to} \ \mathrm{SINR}_i \ge \tau, \ \mathbf{g}^H\mathbf{M}\mathbf{g} \le P_{\mathrm{r}},$$
$$\forall \breve{\theta}_k \in [0, 2\pi]\backslash[\alpha_k, \beta_k], k\in\{1,\ldots,N_r\}, \tag{4.10}$$

where the subscript "pr" denotes the initials of "phase robust". In the following, we will reach the final robust formulation in three successive steps:

1. We first model the SINR constraint as a linear function of phase uncertainties.

2. Then, we incorporate uncertainty bounds with the SINR constraint.

3. Finally, we combine the corresponding derivations in an unified SDP formulation.

Modeling the SINR constraint We re-formulate the quadratic SINR_i inequality constraint such that they can be appropriately combined with the phase uncertainty constraints. The ith SINR constraint is equivalently written as

$$\frac{1}{\tau}\left|(\mathbf{f}_{ri} \odot \mathbf{h}_{ir} \odot \mathbf{g})^T \boldsymbol{\theta}\right|^2 - \sum_{j=1, j\neq i}^{N} \left|(\mathbf{f}_{ri} \odot \mathbf{h}_{jr} \odot \mathbf{g})^T \boldsymbol{\theta}\right|^2 \geq \sigma_{n_r}^2 \mathbf{g}^H \boldsymbol{\Xi}_i \mathbf{g} + \sigma_{n_d}^2.$$

Employing the trace equality of $\mathbf{g}^H \boldsymbol{\Xi}_i \mathbf{g} = \text{Tr}(\boldsymbol{\Xi}_i \mathbf{g}\mathbf{g}^H) = \text{Tr}(\boldsymbol{\Xi}_i \hat{\mathbf{G}})$ on the left hand side, and writing the right hand side as a quadratic function of $\boldsymbol{\theta}$, we obtain

$$\boldsymbol{\theta}^H \Big(\frac{1}{\tau}(\mathbf{f}_{ri} \odot \mathbf{h}_{ir} \odot \mathbf{g})^* (\mathbf{f}_{ri} \odot \mathbf{h}_{ir} \odot \mathbf{g})^T - \sum_{j\neq i} (\mathbf{f}_{ri} \odot \mathbf{h}_{jr} \odot \mathbf{g})^* (\mathbf{f}_{ri} \odot \mathbf{h}_{jr} \odot \mathbf{g})^T\Big) \boldsymbol{\theta}$$
$$\geq \sigma_{n_r}^2 \text{Tr}(\boldsymbol{\Xi}_i \hat{\mathbf{G}}) + \sigma_{n_d}^2. \quad (4.11)$$

Next, we use a similar trace relation on the left hand side of (4.11) by defining $\boldsymbol{\Theta} \triangleq \boldsymbol{\theta}\boldsymbol{\theta}^H$, and hence, (4.11) becomes

$$\text{Tr}\Big(\Big(\frac{1}{\tau}(\mathbf{f}_{ri} \odot \mathbf{h}_{ir} \odot \mathbf{g})^* (\mathbf{f}_{ri} \odot \mathbf{h}_{ir} \odot \mathbf{g})^T - \sum_{j\neq i} (\mathbf{f}_{ri} \odot \mathbf{h}_{jr} \odot \mathbf{g})^* (\mathbf{f}_{ri} \odot \mathbf{h}_{jr} \odot \mathbf{g})^T\Big) \boldsymbol{\Theta}\Big)$$
$$\geq \sigma_{n_r}^2 \text{Tr}(\boldsymbol{\Xi}_i \hat{\mathbf{G}}) + \sigma_{n_d}^2. \quad (4.12)$$

The Hermitian matrix $\boldsymbol{\Theta} \in \mathbb{C}^{N_r \times N_r}$ of phase uncertainty has the following structure:

$$\boldsymbol{\Theta} = \begin{bmatrix} 1 & e^{j(\breve{\theta}_1 - \breve{\theta}_2)} & e^{j(\breve{\theta}_1 - \breve{\theta}_3)} & \dots & e^{j(\breve{\theta}_1 - \breve{\theta}_{N_r})} \\ e^{-j(\breve{\theta}_1 - \breve{\theta}_2)} & 1 & e^{j(\breve{\theta}_2 - \breve{\theta}_3)} & \dots & e^{j(\breve{\theta}_2 - \breve{\theta}_{N_r})} \\ e^{-j(\breve{\theta}_1 - \breve{\theta}_3)} & e^{-j(\breve{\theta}_2 - \breve{\theta}_3)} & 1 & \dots & e^{j(\breve{\theta}_3 - \breve{\theta}_{N_r})} \\ \vdots & \vdots & \vdots & \ddots & \vdots \\ e^{-j(\breve{\theta}_1 - \breve{\theta}_{N_r})} & e^{-j(\breve{\theta}_2 - \breve{\theta}_{N_r})} & e^{-j(\breve{\theta}_3 - \breve{\theta}_{N_r})} & \dots & 1 \end{bmatrix}.$$

We define $\breve{\theta}_{k,\ell} := \breve{\theta}_k - \breve{\theta}_\ell$, where $\breve{\theta}_{k,\ell} \in [0, 2\pi)\backslash[\breve{\alpha}_{k,\ell}, \breve{\beta}_{k,\ell}]$. Due to the assumption that we made while defining the bounds of $\breve{\theta}_k$, we know that $\alpha_k < \beta_k$ and $\beta_k = 2\pi - \alpha_k$ for all k. Consequently, the bounds of $\breve{\theta}_{k,\ell}$ are given by

$$\breve{\alpha}_{k,\ell} = [\alpha_k - \beta_\ell]^{2\pi} \quad \text{and} \quad \breve{\beta}_{k,\ell} = 2\pi - \breve{\alpha}_{k,\ell},$$

where the operator $[x]^{2\pi}$ adds 2π to x when x < 0. However, note that if $\check{\alpha}_{k,\ell} \geq \pi$, then boundaries vanish and we simply have $\check{\theta}_{k,\ell} \in [0, 2\pi)$. Incorporating these definitions and making some algebraic manipulations, the left hand side of the inequality (4.12) can be written as

$$\begin{aligned}\sum_{k=1}^{N_r} \boldsymbol{\Omega}_{i,k,k} + \sum_{k=1}^{N_r} \sum_{\ell=k+1}^{N_r} (\boldsymbol{\Omega}_{i,k,\ell} e^{-j(\check{\theta}_{k,\ell})} + \boldsymbol{\Omega}^*_{i,k,\ell} e^{j(\check{\theta}_{k,\ell})}) \\ = \sum_{k=1}^{N_r} \boldsymbol{\Omega}_{i,k,k} + 2\sum_{k=1}^{N_r} \sum_{\ell=k+1}^{N_r} \mathrm{Re}\{\boldsymbol{\Omega}_{i,k,\ell} e^{-j(\check{\theta}_{k,\ell})}\}, \end{aligned} \tag{4.13}$$

where we defined an auxiliary matrix $\boldsymbol{\Omega}_i$ with the (k, ℓ)th entry

$$\boldsymbol{\Omega}_{i,k,\ell} = \mathrm{Tr}\left(\mathbf{E}_{k,\ell}\left(\left(\frac{1}{\tau} (\mathbf{f}_{ri} \odot \mathbf{h}_{ir})^* (\mathbf{f}_{ri} \odot \mathbf{h}_{ir})^T - \sum_{j \neq i} (\mathbf{f}_{ri} \odot \mathbf{h}_{jr})^* (\mathbf{f}_{ri} \odot \mathbf{h}_{jr})^T \right) \odot \hat{\mathbf{G}}^T \right) \right).$$

The matrix $\mathbf{E}_{k,\ell}$ stands for an all-zero matrix with a single entry at position $[\ell, k]$ that equals to 1.

In order to express (4.13) in an explicit linear form, we stack the phase uncertainty terms and the elements of the auxiliary matrix $\boldsymbol{\Omega}_i$ in vectors $\boldsymbol{\omega}_i$ and $\check{\boldsymbol{\theta}}$, respectively:

$$\begin{aligned} \boldsymbol{\omega}_i &= \left[\omega_{i,1,2} \cdots \omega_{i,1,N_r}\, \omega_{i,2,3} \cdots \omega_{i,2,N_r} \cdots \omega_{i,N_r-1,N_r} \right]^T \in \mathbb{C}^M, \\ \check{\boldsymbol{\theta}} &= \left[e^{j\check{\theta}_{1,2}} \dots e^{j\check{\theta}_{1,N_r}}\, e^{j\check{\theta}_{2,3}} \dots e^{j\check{\theta}_{2,N_r}} \dots e^{j\check{\theta}_{N_r-1,N_r}} \right]^H \in \mathbb{C}^M, \end{aligned}$$

where $\omega_{i,k,\ell} \triangleq \boldsymbol{\Omega}_{i,k,\ell}$, $M = N_r(N_r - 1)/2$. Finally combining all the above derivations, the ith SINR constraint is expressed as a linear function of $\check{\boldsymbol{\theta}}$ as

$$\mathcal{C}^{(i)}_{\mathrm{SINR}}:\ \omega_{i,0} + 2\mathrm{Re}\{\boldsymbol{\omega}_i^T \check{\boldsymbol{\theta}}\} \geq \sigma^2_{n_r} \mathrm{Tr}(\boldsymbol{\Xi}_i \hat{\mathbf{G}}) + \sigma^2_{n_d},$$

where $\omega_{i,0} \triangleq \sum_k \boldsymbol{\Omega}_{i,k,k}$.

Incorporating the uncertainty bounds Our design target is to find a gain allocation vector $\mathbf{g}$ which satisfies $\mathcal{C}^{(i)}_{\mathrm{SINR}}$ for all possible phase variations within the given bounds. Towards this end, we adapt the following *lemma* from the Theorem 3 in [38] to our scenario:

Lemma 4.2.1: [38, 62] Defining $\mathbf{x} = [1\ e^{jx}]^T$, $\mathbf{y} = [y_1\ y_2]^T$, where $x \in [0, 2\pi)$, $y_1 \in \mathbb{R}$, and $y_2 \in \mathbb{C}$, the constraints in the form

$$\begin{cases} \mathrm{Re}\{\mathbf{x}^H \mathbf{y}\} \geq \varpi, & \forall x \in [\alpha, \beta] \\ \mathrm{Re}\{\mathbf{x}^H \mathbf{y}\} \geq \varpi, & \forall x \in [0, 2\pi) \backslash [\alpha, \beta] \end{cases} \tag{4.14}$$

can be equivalently re-formulated to the following linear matrix inequality (LMI) constraints:

$$\begin{cases} \mathbf{y} + j\varphi\mathbf{e} - \varpi\mathbf{e} = \mathbf{p}(\boldsymbol{\Lambda}) + \kappa\mathbf{d}(\alpha,\beta) \\ \mathbf{y} + j\varphi\mathbf{e} - \varpi\mathbf{e} = \mathbf{p}(\boldsymbol{\Lambda}) - \kappa\mathbf{d}(\alpha,\beta) \end{cases}$$

respectively, where $0 \le \alpha < \beta < 2\pi$, $\varpi \in \mathbb{R}, \varphi \in \mathbb{R}$, $\kappa \in \mathbb{C}$, and $\mathbf{e} = [1\ 0]^T$. Moreover, $\mathbf{d}(\alpha,\beta) \in \mathbb{R} \times \mathbb{C}$ and $\mathbf{p}(\boldsymbol{\Lambda}) \in \mathbb{R} \times \mathbb{C}$ are defined as

$$\mathbf{d}(\alpha,\beta) := \begin{bmatrix} \cos\alpha + \cos\beta - \cos(\beta-\alpha) - 1 \\ (1-e^{j\alpha})(e^{j\beta}-1), \end{bmatrix}$$

and $\mathbf{p}(\boldsymbol{\Lambda}) := [\mathrm{Tr}(\boldsymbol{\Lambda})\ 2\mathrm{Tr}(\boldsymbol{\Pi}\boldsymbol{\Lambda})]^T$, where $\boldsymbol{\Lambda} \in \mathbb{C}^2 \succeq 0$ and $\boldsymbol{\Pi}$ is an all-zero matrix except that $\boldsymbol{\Pi}[2,1] = 1$.

As explained in [62], the given linear constraints in (4.14) can be interpreted as that each requires the real part of a certain trigonometric polynomial to be nonnegative over a segment of the unit circle. In order to apply the presented Lemma 4.2.1 directly to our problem, we modify the constraint $\mathcal{C}_{\mathrm{SINR}}^{(i)}$ and decompose it into $M+1$ inequality constraints which are linearly dependent on each other through scalars $\eta_{i,m} \in \mathbb{R}, m = 1,\ldots,M$:

$$\mathcal{C}_{\mathrm{SINR}}^{(i)} := \begin{cases} \omega_{i,0} + 2\mathrm{Re}\{\boldsymbol{\omega}_i[1]\breve{\boldsymbol{\theta}}[1]\} \ge & \eta_{i,1} \\ 2\mathrm{Re}\{\boldsymbol{\omega}_i[2]\breve{\boldsymbol{\theta}}[2]\} \ge & \eta_{i,2} \\ \vdots & \\ 2\mathrm{Re}\{\boldsymbol{\omega}_i[M]\breve{\boldsymbol{\theta}}[M]\} \ge & \eta_{i,M} \\ \sum_m \eta_{i,m} & \ge \sigma_{n_r}^2 \mathrm{Tr}(\boldsymbol{\Xi}_i\hat{\mathbf{G}}) + \sigma_{n_d}^2, \end{cases}$$

where $\boldsymbol{\omega}_i[m]$ and $\breve{\boldsymbol{\theta}}[m]$ represents the mth element of $\boldsymbol{\omega}_i$ and $\breve{\boldsymbol{\theta}}$, respectively.

Next, applying the aforementioned lemma 4.2.1 on each of the first M inequalities of $\mathcal{C}_{\mathrm{SINR}}^{(i)}$, we obtain the following set of constraints:

$$\mathcal{C}_{\mathrm{SINR}}^{(i)} := \begin{cases} \left[\omega_{i,0}\ 2\boldsymbol{\omega}_i[1]\right]^T + j\varphi_{i,1}\mathbf{e} - \eta_{i,1}\mathbf{e} = & \mathbf{p}(\boldsymbol{\Lambda}_{i,1}) - \kappa_{i,1}\mathbf{d}(\breve{\alpha}_{1,2},\breve{\beta}_{1,2}) \\ 2\boldsymbol{\omega}_i[2]\tilde{\mathbf{e}} + j\varphi_{i,2}\mathbf{e} - \eta_{i,2}\mathbf{e} \quad = & \mathbf{p}(\boldsymbol{\Lambda}_{i,2}) - \kappa_{i,2}\mathbf{d}(\breve{\alpha}_{1,3},\breve{\beta}_{1,3}) \\ \vdots & \\ 2\boldsymbol{\omega}_i[M]\tilde{\mathbf{e}} + j\varphi_{i,M}\mathbf{e} - \eta_{i,M}\mathbf{e} & = \mathbf{p}(\boldsymbol{\Lambda}_{i,M}) - \kappa_{i,M}\mathbf{d}(\breve{\alpha}_{N_r-1,N_r},\breve{\beta}_{N_r-1,N_r}) \\ \sum_m \eta_{i,m} \quad \ge & \sigma_{n_r}^2 \mathrm{Tr}(\boldsymbol{\Xi}_i\hat{\mathbf{G}}) + \sigma_{n_d}^2, \end{cases} \tag{4.15}$$

where $\tilde{\mathbf{e}} = [0\ \ 1]^T$, $\varphi_{i,m} \in \mathbb{R}$, $\kappa_{i,m} \in \mathbb{C}$, $\boldsymbol{\Lambda}_{i,m} \succeq 0$, $m = 1,\ldots,M$. Let us now elaborate on the intuition behind $\mathcal{C}_{\mathrm{SINR}}^{(i)}$ to get more insight about our robust design. The SDP formulation (4.15) characterizes the robust counterpart of the SINR constraint such that it incorporates all possible realizations of the phase noise within given bounds. In other words, assume that $\mathcal{C}_{\text{SINR}}^{(i)}$ is feasible for a given relay gain vector $\mathbf{g}$, lower bound τ and corresponding uncertainty parameters. Then, for any realization of phase noise terms within given bounds, the resultant SINR values are *at least* equal to the pre-set lower bound τ. An interesting question that follows the previous discussion is that what is the maximal lower bound τ for given $\mathbf{g}$ and uncertainty bounds. In essence, the answer is readily available through the following simple SDP optimization:

$$\mathcal{P}_{\text{pr-worst}}^{(i)}(\mathbf{g}) : \max_{\tau,\eta_{i,m},\varphi_{i,m},\mathbf{\Lambda}_{i,m},\kappa_{i,m}} \tau \quad \textbf{subject to} \quad \mathcal{C}_{\text{SINR}}^{(i)}$$

$$\mathbf{\Omega}_{i,k,\ell} - \text{Tr}\left(\mathbf{E}_{k,\ell}\left(\left(\frac{1}{\tau}(\mathbf{f}_{\text{r}i} \odot \mathbf{h}_{i\text{r}})^*(\mathbf{f}_{\text{r}i} \odot \mathbf{h}_{i\text{r}})^T - \sum_{j\neq i}(\mathbf{f}_{\text{r}i} \odot \mathbf{h}_{j\text{r}})^*(\mathbf{f}_{\text{r}i} \odot \mathbf{h}_{j\text{r}})^T\right) \odot \hat{\mathbf{G}}^T\right)\right) = 0$$

$$\eta_{i,m}, \varphi_{i,m} \in \mathbb{R},\ \kappa_{i,m} \in \mathbb{C}, \mathbf{\Lambda}_{i,m} \succeq 0, \tau \geq 0, m \in \{1,\dots,M\},\ k,\ell \in \{1,\dots,N_{\text{r}}\},$$

where the second constraint stands for the definitions in (4.13). The problem can be solved iteratively with a bisection method, where at each iteration we check for the feasibility of another τ. Note that $\mathcal{P}_{\text{pr-worst}}^{(i)}$ deals only with the ith link's SINR. In order for a joint optimization of finding the minimum worst-case SINR over the network, we simply need to lower bound all SINRs with the same τ, and maximize it.

The resultant SDP formulation After lastly noting that the power constraint is modeled as $\text{Tr}(\mathbf{M}\hat{\mathbf{G}}) \leq P_{\text{r}}$, we are now ready to combine the derivations above to obtain the final SDP formulation for the worst-case robust max-min beamforming problem:

$$\tilde{\mathcal{P}}_{\text{pr}} : \max_{\hat{\mathbf{G}},\tau,\eta_{i,m},\varphi_{i,m},\mathbf{\Lambda}_{i,m},\kappa_{i,m}} \tau \quad \textbf{subject to} \quad \mathcal{C}_{\text{SINR}}^{(i)}$$

$$\mathbf{\Omega}_{i,k,\ell} - \text{Tr}\left(\mathbf{E}_{k,\ell}\left(\left(\frac{1}{\tau}(\mathbf{f}_{\text{r}i} \odot \mathbf{h}_{i\text{r}})^*(\mathbf{f}_{\text{r}i} \odot \mathbf{h}_{i\text{r}})^T - \sum_{j\neq i}(\mathbf{f}_{\text{r}i} \odot \mathbf{h}_{j\text{r}})^*(\mathbf{f}_{\text{r}i} \odot \mathbf{h}_{j\text{r}})^T\right) \odot \hat{\mathbf{G}}^T\right)\right) = 0$$

$$\text{Tr}(\mathbf{M}\hat{\mathbf{G}}) \leq P_{\text{r}},\ \eta_{i,m}, \varphi_{i,m} \in \mathbb{R},\ \kappa_{i,m} \in \mathbb{C}, \mathbf{\Lambda}_{i,m} \succeq 0, \hat{\mathbf{G}} \succeq 0, \tau \geq 0,$$
$$m \in \{1,\dots,M\},\ i \in \{1,\dots,N\}, k,\ell \in \{1,\dots,N_{\text{r}}\},$$

which is efficiently solved with bisection method in combination with SDP feasibility checks (see Section 3.2.2 for similar implementations). With the above given $\tilde{\mathcal{P}}_{\text{pr}}$, we assure that the phase noises are within given bounds with probability $\text{erf}(\delta/\sqrt{2})$ and the minimum worst-case link SINR is maximized. Note that $\hat{\mathbf{G}}$ has to be a rank-1 matrix by definition; however we have applied SDR on it. Hence, following the discussions on

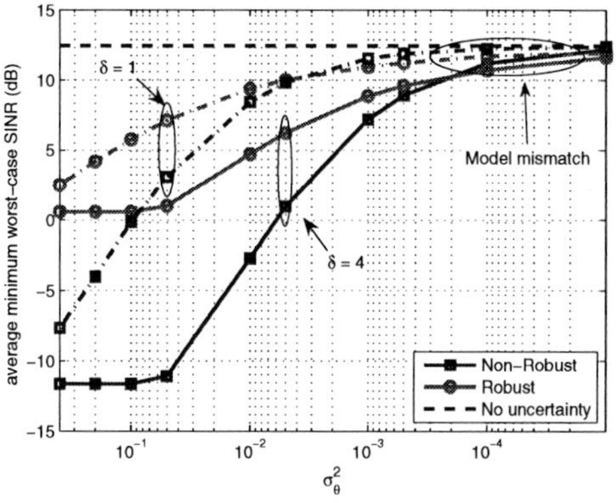

Figure 4.2.: Comparison of average minimum worst-case SINR performances of robust and non-robust designs for $N = 2$ S-D pairs and $N_r = 3$ relays.

the same issue in Section 3.2.2.1, we employ the randomization technique to choose $\mathbf{g}$ out of $\hat{\mathbf{G}}^\star$ [113].

4.2.2.1. Performance Results

We study the performance of the proposed robust solution through Monte Carlo simulations. It is assumed that σ_θ^2 is identical at each relay. Unless otherwise stated, all the assumptions made on the system specifications in the previous non-robust simulation Section 4.2.1 are also valid here.

As a non-robust reference we employ the general max-min beamforming problem $\mathcal{P}_{\mathrm{mm}}$ given in Section 3.2.2.1. The minimum worst-case SINR performance of the resultant relay gain vector can be determined as explained previously with $\mathcal{P}^{(i)}_{\mathrm{pr-worst}}$. However, the corresponding value would reflect an approximate worst-case performance due to the disturbed exactness through the formulation of the phase matrix $\boldsymbol{\Theta}$. Nevertheless, it is natural that our design should sacrifice performance due to the ease in implementation. Therefore, for the sake of fairness, we rather take a brute-force numerical approach to compute the worst-case performance of $\mathcal{P}_{\mathrm{mm}}$ by searching over a sufficiently large test set of phase noises without any bounds on uncertainty, i.e., similar to the one used for Fig. 4.1.

We consider a scenario of $N = 2$ S-D pairs and $N_r = 3$ relays. In Fig. 4.2, we compare the average worst-case SINR performance of both the robust and the non-

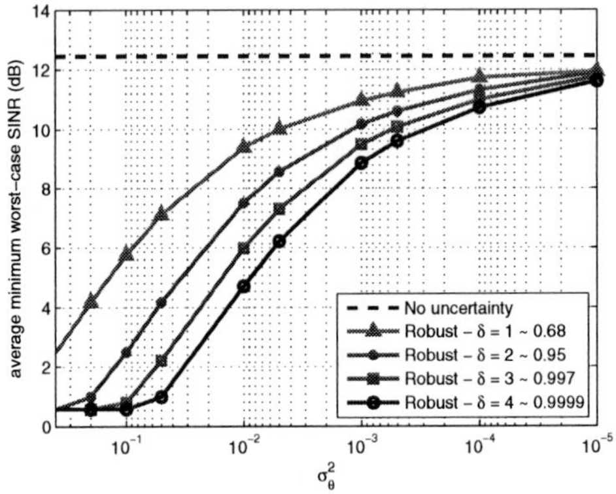

Figure 4.3.: Average minimum worst-case SINR vs. phase uncertainty variance σ_θ^2 for various uncertainty parameters δ.

robust approaches for various phase uncertainty variances. The straight dashed line stands for the average minimum SINR ($\sim$ 12.5 dB) over the network in the absence of phase uncertainty. As expected, both robust and non-robust curves converge to this straight line at the extreme limit of $\sigma_\theta^2 \rightarrow 0$. Independent of δ, we observe that the robust approach proposes a significant worst-case SINR improvement for $\sigma_\theta^2 > 10^{-3}$. Whereas, anyhow for further small values of σ_θ^2, the loss due to uncertainty is marginal. An interesting observation is that for rather large $\sigma_\theta^2 > 10^{-1}$, the uncertainty results in an inoperative worst-case SINR $\ll 0$ dB. Moreover, we see that the worst-case results saturates at around -12 dB, which turns out to be the aforementioned *theoretical* worst-case performance for $N_\text{r} = 3$ relays independent of σ_θ^2. Nevertheless, in general robust optimization recovers these losses and provides gain of up to 10 dB for certain σ_θ^2 values.

Another issue at low-σ_θ^2 regime is that the non-robust approach performs better than its robust counterpart, which is somehow unexpected. The reason to this anomaly is the model mismatch occurred while linearizing $\boldsymbol{\Theta}$ to $\breve{\boldsymbol{\theta}}$. There, we defined M phase terms $\breve{\theta}_{k,l}$ out of N_r phase terms $\breve{\theta}_k$. While modelling these newly defined M terms, for the sake of the simplicity of the model, we assume that they are independently distributed over the defined range, although they are not necessarily so. Despite the aforementioned phase model relaxation, the robust approach performs outstanding at the high-σ_θ^2 regime, where the uncertainty's impact is crucial.

Increasing δ provides a more *outage-secure* performance in the sense that the prob-

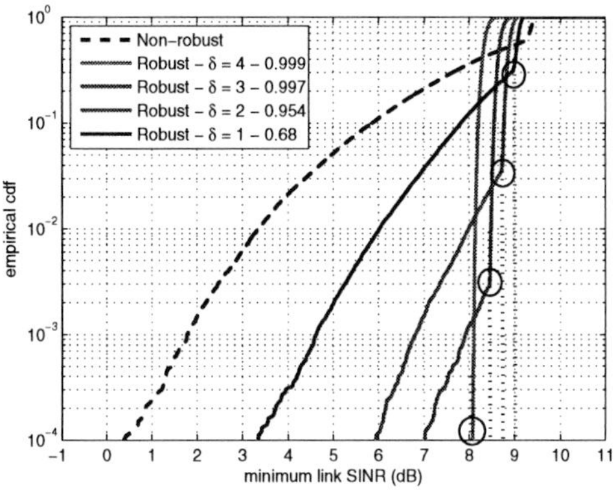

Figure 4.4.: Empirical CDF of minimum SINR (over links) for different uncertainty parameters δ and phase uncertainty variance of $\sigma_\theta^2 = 10^{-2}$.

ability of missing a possible worst-case leading uncertainty is reduced. However, the penalty to pay for this advantage is to be rather *conservative* in performance as the set of uncertainties to cover within the optimization, enlarges. Correspondingly, as depicted in Fig. 4.3, larger δ results in a *worse* worst-case SINR performance. For example, let us focus on a specific comparison of $\delta = 1$ and $\delta = 4$ at $\sigma_\theta^2 = 10^{-2}$. There is around 5 dB difference in the worst-case SINR performances favoring the choice of $\delta = 1$. Although choosing $\delta = 4$ offers a much smaller worst-case SINR (around 4.5 dB), it guarantees this value for almost all uncertainty realizations (with probability 0.9999), which is considerably more reliable when compared with its counterpart's 0.68 probability of assurance. That is, while determining the robust relay gain vector, we can exploit the trade-off between being *outage-secure* and being *conservative.*

We discuss the so called *outage-secureness* in the view of the Fig. 4.4 in details, where we plot the cdf of SINR over a defined range of uncertainty (through δ) for a typical channel realization. The breaking points (circles) of the robust curves represent the worst-case SINR result of the robust optimization for the given δ. Recalling that δ determines the range of uncertainty to take into account through the robust optimization, the tails observed beyond these breaking points are due to this uncontrolled part of the uncertainty. Note that the probability of assurance determined by δ does not directly map to the outage probability of worst-case SINR. But anyhow it has been empirically observed that there is a strong relation in between. For example, $\delta = 3$ implies that 99.7% of all possible uncertainty realizations are considered, and the robust optimization with this setting results in a worst-case SINR of about 8.45 dB. When the corresponding cdf curve (green) is checked in Fig. 4.4, we observe the probability distribution of this SINR value is also approximately $0.003(= 1 - 0.997)$. Note also that although we only depict here the cdf of a single channel realization, the behaviour is very similar for any other channel realizations. Summing up, larger δ causes a smaller robust worst-case SINR value, but assures that this is achieved with a high probability.

4.3. Imperfect Channel State Information

We have already reported that CSI knowledge per terminal is the most critical performance determining factor for coherent relaying. Any insufficiency on it limits the performance in two different aspects. First, the amount CSI per terminal, which scales from no CSI (non-coherent) to global CSI, enforces a corresponding local CSI dissemination overhead to the system. As the number of terminals increases, the relative portion of the time allocated for CSI dissemination with respect to actual data transmission tends to increment quadratically with N_r, and hence, threatens the effective spectral efficiency. Second, assuming an efficient relaying protocol with sufficient CSI knowledge requirement per terminal, the quality of the pre-assumed channel knowledge comes into the picture. More specifically, any mismatch between the actual and estimated CSI hampers the expected (and optimized) performance measures at the destination terminals, e.g., array and/or diversity gain, fairness, etc. This mismatch may arise during the estimation and/or quantization processes, or can even be triggered by not frequently detecting the channel variations, i.e., outdated channel estimation.

The former aspect of the *quantity* of CSI has been already addressed in Section 3.3, and will be further discussed in Chapter 5 by investigating different multiuser relaying protocols. The latter aspect of the *quality* of CSI will be the main topic of this section. We propose different models for the mismatch in the channel estimates and aim at designing worst-case robust gain allocation schemes which are computationally tractable. Note that our focus here is rather on robust gain allocation designs for generic channel uncertainty sets rather than evaluating the impact of noisy channel estimation. We refer the interested reader to [16] for further details on the performance of distributed orthogonalization introduced in Chapter 2 with noisy CSI.

Outline of this Section We start by defining the channel uncertainty model in Section 4.3.1, which in general follows the robust convex optimization methodology studied in [9–14]. In principle, we define a deterministic bounded uncertainty set out of which CSI is assumed to realize. The larger the size of the uncertainty is, the more uncertain the CSI estimate is. Later in the same section, we present a generalized worst-case robust formulation where we assume an independent uncertainty set for each source-to-relay and relay-to-destination channel vectors. As the general problem is intractable to solve, we study two simplified versions in Sections 4.3.2 and 4.3.3, where we respectively assumed that either "only" source-to-relay or relay-to-destination channels are imperfect. Finally, in Section 4.3.4 we present simulation results and discuss the performance of the proposed robust designs.

Note once again that our design problem diverges from its analogous robust proposals for transmit beamforming [15, 25, 59, 95, 106, 126–128], where the power constraint is indifferent to channel knowledge uncertainty. Whereas, our problem involves both worst-case power and worst-case SINR decision rules.

4.3.1. Problem Formulation and Uncertainty Model

We assume that each uplink and downlink channel vector $\mathbf{h}_{ir}$ and $\mathbf{f}_{ir}$ for $i = 1, \ldots, N$, is subject to an independent imperfection source. That is, we define *independent* uncertainty sets for each of these vectors. These sets encompass all possible realizations of the corresponding channel vector. We define the individual uncertainty sets respectively as $\mathcal{U}_{\mathbf{h}_i}$ and $\mathcal{U}_{\mathbf{f}_i}$. Hence, the global CSI knowledge at the relays is considered within an uncertainty set of

$$\mathcal{U} = \mathcal{U}_{\mathbf{h}_1} \times \cdots \times \mathcal{U}_{\mathbf{h}_N} \times \mathcal{U}_{\mathbf{f}_1} \times \cdots \times \mathcal{U}_{\mathbf{f}_N}.$$

Choosing such a global uncertainty set $\mathcal{U}$ is rather conservative since the optimized vector should satisfy all possible realizations of each of the independent uncertainty sets. Consequently, we adapt a worst-case performance approach throughout this section. Similar to [9, 12, 98], we consider spherical uncertainty sets such that

$$\mathcal{U}_{\mathbf{h}_i} = \left\{ \mathbf{h}_{ir} = \mathbf{h}_{ir,0} + \sum_{q=1}^{Q_{\mathbf{h}}} \xi_{\mathbf{h}_i,q} \mathbf{h}_{ir,q} \quad : \quad \|\boldsymbol{\xi}_{\mathbf{h}_i}\| \le \rho_{\mathbf{h}_i} \right\},$$

$$\mathcal{U}_{\mathbf{f}_i} = \left\{ \mathbf{f}_{ri} = \mathbf{f}_{ri,0} + \sum_{q=1}^{Q_{\mathbf{f}}} \xi_{\mathbf{f}_i,q} \mathbf{f}_{ri,q} \quad : \quad \|\boldsymbol{\xi}_{\mathbf{f}_i}\| \le \rho_{\mathbf{f}_i} \right\},$$

where $Q_{\mathbf{h}}, Q_{\mathbf{f}}, \rho_{\mathbf{h}_i}, \rho_{\mathbf{f}_i} \in \mathbb{R}_+$, and $\boldsymbol{\xi}_{\mathbf{h}_i} = [\xi_{\mathbf{h}_i,1} \cdots \xi_{\mathbf{h}_i,Q_{\mathbf{h}}}]^T$, $\boldsymbol{\xi}_{\mathbf{f}_i} = [\xi_{\mathbf{f}_i,1} \cdots \xi_{\mathbf{f}_i,Q_{\mathbf{f}}}]^T$.

In the following considerations, we assume for simplicity that there is no random phase uncertainty at the relays, i.e., $\boldsymbol{\Theta}_{\mathsf{r}} = \tilde{\boldsymbol{\Theta}}_{\mathsf{r}} = \mathbf{I}_{N_{\mathsf{r}}}$. Hence, the received SINR at the ith destination given in (4.3) simplifies to

$$\mathrm{SINR}_i = \frac{\left|(\mathbf{f}_{ri} \odot \mathbf{h}_{ir})^T \mathbf{g}\right|^2}{\sum_{j=1,j\neq i}^{N} \left|(\mathbf{f}_{ri} \odot \mathbf{h}_{jr})^T \mathbf{g}\right|^2 + \sigma_{n_{\mathsf{r}}}^2 \mathbf{g}^H \boldsymbol{\Xi}_i \mathbf{g} + \sigma_{n_{\mathsf{d}}}^2}, \tag{4.16}$$

where $\mathbf{f}_{ri}$ and $\mathbf{h}_{ir}$ are uncertain, and are considered within the corresponding uncertain sets defined previously. We adapt a worst-case max-min robust optimization approach, where we aim at maximizing the minimum "worst-case" SINR over different S-D pairs and all uncertainty sets. In other words, similar to the previous robust optimization $\mathcal{P}_{\text{pr}}$, the design problem is worst-case in two folds.

The robust counterpart of the max-min SINR approach can be written as

$$\mathcal{P}_{\text{csi}}: \quad \max_{\mathbf{g},\tau\geq 0} \ \tau \quad \textbf{subject to} \quad \begin{aligned}&\text{SINR}_i \geq \tau, \ \forall(\mathbf{h}_{ir}, \mathbf{f}_{ri}) \in \mathcal{U}, \\ &\mathbf{g}^H\mathbf{M}\mathbf{g} \leq P_{\text{r}}, \ i \in \{1,\dots,N\}.\end{aligned} \tag{4.17}$$

With $\mathcal{P}_{\text{csi}}$, we seek for the maximal value of the auxiliary variable τ, which serves as a lower bound for all link SINRs. Considering that each SINR is a deterministic but is still an uncertain variable within¡ defined sets, τ gives the minimum minimal SINR over all links. Moreover, as the matrix $\mathbf{M}$ is also affected by the channel uncertainty, the design problem also delivers a worst-case instantaneous power allocation to the relays. In the sequel we refer $\mathcal{P}_{\text{csi}}$ as the *generalized robust counterpart.*

The SINR constraints consist of the products of uplink and downlink channels, which in return result in multiplication of different uncertainty sets. Moreover, although the SINR constraints can be expressed as a quadratic function of $\mathbf{g}$, the left and the right side of the equations are affected by dependent uncertainty sets. All of these are well-known to cause a computationally intractable robust counterpart [9, 10, 14]. Hence, in order to facilitate the solution to the problem, it may be needed to replace the true robust counterpart with an approximate one [14]. Such an approximation would definitely make the robust problem more conservative, and even may cause that there is not any feasible solution for some set of data. Due to the aforementioned difficulties to solve the generalized robust counterpart, we instead investigate simplified scenarios where we can provide computationally tractable solutions to the corresponding true counterparts. Specifically, we consider that only either source-to-relay or relay-to-destination channels are subject to imperfection.

4.3.2. Perfect Uplink but Imperfect Downlink Channel Knowledge

Consider a special scenario where the relays know the uplink channel vectors $\mathbf{h}_{ir}\forall i$ perfectly, but have imperfect knowledge about the downlink channel vectors $\mathbf{f}_{ri}$. Such an imperfection scenario can be a typical example for downlink communications where the sources and the relays are fixed, whereas the destinations are mobile. Hence, the source-relay channel is either constant through time/frequency or can be tracked/estimated efficiently due to very-slow variation. However, as the destinations are mobile, it would be much difficult to estimate the channels correctly or rather to keep the estimations up-to-date.

Hence, the global uncertainty set $\mathcal{U}$ correspondingly reduces to $\mathcal{U}_{\text{down}} = \mathcal{U}_{\mathbf{f}_1} \times \cdots \times \mathcal{U}_{\mathbf{f}_N}$. Substituting the definition of $\mathbf{f}_{ri}$ as given by the uncertainty set $\mathcal{U}_{\mathbf{f}_i}$, the SINR$_i$ definition becomes

$$\text{SINR}_i = \frac{\left|(\mathbf{g} \odot \mathbf{h}_{ir})^T(\mathbf{f}_{ri,0} + \sum_{q=1}^{Q_\mathbf{f}} \xi_{\mathbf{f}_i,q}\mathbf{f}_{ri,q})\right|^2}{\sum_{j\neq i}^{N} \left|(\mathbf{g} \odot \mathbf{h}_{jr})^T(\mathbf{f}_{ri,0} + \sum_{q=1}^{Q_\mathbf{f}} \xi_{\mathbf{f}_i,q}\mathbf{f}_{ri,q})\right|^2 + \sigma_{n_r}^2 \left\|(\mathbf{f}_{ri,0} + \sum_{q=1}^{Q_\mathbf{f}} \xi_{\mathbf{f}_i,q}\mathbf{f}_{ri,q}) \odot \mathbf{g}\right\|^2 + \sigma_{n_d}^2}.$$

Note that the power constraint is not affected by such an uncertain CSI assumption, as it only incorporates with the uplink channels.

In the following, we first derive the robust counterpart for the SINR constraint, and consequently, derive the robust max-min problem through an efficient SDP formulation.

Robust Counterpart of the SINR constraint Our aim now is to model the SINR$_i \geq \tau$ constraint such that the true robust counterpart of this scenario can be solved efficiently. To this end, we equivalently write the SINR constraint as

$$\begin{aligned}\frac{1}{\tau}\left|(\mathbf{g} \odot \mathbf{h}_{ir})^T\mathbf{f}_{ri,0}+\sum_{q=1}^{Q_\mathbf{f}} \xi_{\mathbf{f}_i,q}(\mathbf{g} \odot \mathbf{h}_{ir})^T\mathbf{f}_{ri,q}\right|^2 - \sum_{j=1,j\neq i}^{N} \left|(\mathbf{g} \odot \mathbf{h}_{jr})^T\mathbf{f}_{ri,0}+\sum_{q=1}^{Q_\mathbf{f}} \xi_{\mathbf{f}_i,q}(\mathbf{g} \odot \mathbf{h}_{jr})^T\mathbf{f}_{ri,q}\right|^2 \\ \geq \sigma_{n_r}^2 \left(\mathbf{f}_{ri,0} + \sum_{q=1}^{Q_\mathbf{f}} \xi_{\mathbf{f}_i,q}\mathbf{f}_{ri,q}\right)^T \mathbf{G}^H\mathbf{G}\left(\mathbf{f}_{ri,0} + \sum_{q=1}^{Q_\mathbf{f}} \xi_{\mathbf{f}_i,q}\mathbf{f}_{ri,q}\right)^* + \sigma_{n_d}^2, \qquad (4.18)\end{aligned}$$

where we re-wrote the right hand side of the inequality by using $\mathbf{G} \triangleq \text{diag}(\mathbf{g})$. Defining the following $Q_\mathbf{f} \times N_r$ auxiliary matrices

$$\begin{aligned}\breve{\mathbf{F}}_{ii} &= \Big[(\mathbf{f}_{ri,1} \odot \mathbf{h}_{ir}) \cdots (\mathbf{f}_{ri,Q_\mathbf{f}} \odot \mathbf{h}_{ir})\Big]^T, \\ \breve{\mathbf{F}}_{ij} &= \Big[(\mathbf{f}_{ri,1} \odot \mathbf{h}_{jr}) \cdots (\mathbf{f}_{ri,Q_\mathbf{f}} \odot \mathbf{h}_{jr})\Big]^T, \\ \breve{\mathbf{F}}_{i} &= \Big[\mathbf{f}_{ri,1} \cdots \mathbf{f}_{ri,Q_\mathbf{f}}\Big]^T,\end{aligned}$$

we further modify (4.18) to an interim form which highlights the vector $\boldsymbol{\xi}_{\mathbf{f}_i}$

$$\frac{1}{\tau}\left|(\mathbf{g}\odot\mathbf{h}_{ir})^T\mathbf{f}_{ri,0}+\boldsymbol{\xi}_{\mathbf{f}_i}^T\breve{\mathbf{F}}_{ii}\mathbf{g}\right|^2-\sum_{j=1,j\neq i}^{N}\left|(\mathbf{g}\odot\mathbf{h}_{jr})^T\mathbf{f}_{ri,0}+\boldsymbol{\xi}_{\mathbf{f}_i}^T\breve{\mathbf{F}}_{ij}\mathbf{g}\right|^2$$
$$\geq\sigma_{n_r}^2\left\|\mathbf{f}_{ri,0}^T(\mathbf{G}^H\mathbf{G})^{\frac{1}{2}}+\boldsymbol{\xi}_{\mathbf{f}_i}^T\breve{\mathbf{F}}_i(\mathbf{G}^H\mathbf{G})^{\frac{1}{2}}\right\|^2+\sigma_{n_d}^2. \qquad (4.19)$$

Next, after some algebraic manipulations, we express (4.19) as a regular quadratic function of $\boldsymbol{\xi}_{\mathbf{f}_i}$ as

$$\boldsymbol{\xi}_{\mathbf{f}_i}^T\mathbf{C}_i\boldsymbol{\xi}_{\mathbf{f}_i}^*+2\mathrm{Re}\left\{\boldsymbol{\xi}_{\mathbf{f}_i}^T\mathbf{c}_i\right\}+c_i\geq 0, \qquad (4.20)$$

where

$$\mathbf{C}_i=\frac{1}{\tau}\breve{\mathbf{F}}_{ii}\hat{\mathbf{G}}\breve{\mathbf{F}}_{ii}^H-\sum_{j=1,j\neq i}^{N}\breve{\mathbf{F}}_{ij}\hat{\mathbf{G}}\breve{\mathbf{F}}_{ij}^H-\sigma_{n_r}^2\breve{\mathbf{F}}_i\cdot\mathrm{diag}(\hat{\mathbf{G}})\cdot\breve{\mathbf{F}}_i^H,$$

$$\mathbf{c}_i=\frac{1}{\tau}\breve{\mathbf{F}}_{ii}\hat{\mathbf{G}}(\mathbf{f}_{ri,0}\odot\mathbf{h}_{ir})^*-\sum_{j=1,j\neq i}^{N}\breve{\mathbf{F}}_{ij}\hat{\mathbf{G}}(\mathbf{f}_{ri,0}\odot\mathbf{h}_{jr})^*-\sigma_{n_r}^2\breve{\mathbf{F}}_i\cdot\mathrm{diag}(\hat{\mathbf{G}})\cdot\mathbf{f}_{ri,0}^*,$$

$$c_i=\mathrm{Tr}\left(\left(\frac{1}{\tau}(\mathbf{f}_{ri,0}\odot\mathbf{h}_{ir})^*(\mathbf{f}_{ri,0}\odot\mathbf{h}_{ir})^T-\sum_{j=1,j\neq i}^{N}(\mathbf{f}_{ri,0}\odot\mathbf{h}_{jr})^*(\mathbf{f}_{ri,0}\odot\mathbf{h}_{jr})^T\right)\hat{\mathbf{G}}\right)$$
$$-\sigma_{n_r}^2\mathbf{f}_{ri,0}^T\cdot\mathrm{diag}(\hat{\mathbf{G}})\cdot\mathbf{f}_{ri,0}^*-\sigma_{n_d}^2.$$

While writing the arguments of the quadratic function in (4.20), we have substituted $\hat{\mathbf{G}}$ instead of the product $\mathbf{g}\mathbf{g}^H$, and employed the relation of $\mathrm{diag}(\hat{\mathbf{G}})=\mathbf{G}^H\mathbf{G}$.

Summing up, we have written each SINR constraint as a quadratic function of the uncertainty parameter vector $\boldsymbol{\xi}_{\mathbf{f}_i}$. The length of each of these vectors is constrained by another quadratic function of $\boldsymbol{\xi}_{\mathbf{f}_i}$, i.e., $\|\boldsymbol{\xi}_{\mathbf{f}_i}\|\leq\rho_{\mathbf{f}_i}$. We want any realization of this vector within the defined uncertainty set to also satisfy (4.20). In other words, we enforce to have that the non-negativity of the quadratic function $\rho_{\mathbf{f}_i}-\|\boldsymbol{\xi}_{\mathbf{f}_i}\|\geq 0$ implies the non-negativity of the SINR related constraint. To this end, we use the following lemma called the *S-procedure* (also called *S-lemma*), and mathematically incorporate (4.20) with the size of the uncertainty set.

S-procedure [139] [61]: Let $f_k:\mathbb{C}^m\to\mathbb{R}$, $k=0,...,M$ be defined as

$$f_k(\mathbf{y})=\mathbf{y}^H\mathbf{Q}_k\mathbf{y}+2\mathrm{Re}\{\mathbf{s}_k^H\mathbf{y}\}+w_k,$$

where $\mathbf{Q}_k=\mathbf{Q}_k^H\in\mathbb{C}^{m\times m}$, $\mathbf{s}_k\in\mathbb{C}^m$, $w_k\in\mathbb{R}$. Assume that there exists $\tilde{\mathbf{y}}\in\mathbb{C}^m$ such that $f_k(\tilde{\mathbf{y}})>0$, $\forall=1,...,M$. Then, the following two statements are equivalent:

- $\mathcal{S}_1 : f_0(\mathbf{y}) \geq 0, \forall \mathbf{y} \in \mathbb{C}^m$ such that $f_k(\mathbf{y}) \geq 0, k = 1, ..., M$
- $\mathcal{S}_2$: there exists $\kappa_1, ..., \kappa_M$ such that the following is feasible

$$\begin{bmatrix} w_0 & \mathbf{s}_0^H \\ \mathbf{s}_0 & \mathbf{Q}_0 \end{bmatrix} \succeq \sum_{k=1}^{M} \kappa_k \begin{bmatrix} w_k & \mathbf{s}_k^H \\ \mathbf{s}_k & \mathbf{Q}_k \end{bmatrix}.$$

Consequently, we obtain an LMI of

$$\begin{bmatrix} c_i & \mathbf{c}_i^H \\ \mathbf{c}_i & \mathbf{C}_i \end{bmatrix} - \tilde{\zeta}_i \begin{bmatrix} \rho_{\mathbf{f}_i}^2 & \mathbf{0} \\ \mathbf{0} & -\mathbf{I} \end{bmatrix} \succeq 0, \tag{4.21}$$

where $\tilde{\zeta}_i \in \mathbb{R}_+$. When the left hand side of (4.21) is an active positive semidefinite matrix for given optimization parameters of τ, $\rho_{\mathbf{f}_i}$, $\hat{\mathbf{G}}$, then we can be sure that all possible uncertainty realizations (within given set) leads to an SINR which is at least equal to τ. Such a worst-case oriented robust counterpart, in fact, behaves rather conservative. That is, it may be a very unlikely that a single uncertainty realization takes this "most dangerous" τ value. Nevertheless, it serves as a fully secure lower bound.

Worst-Case SINR Performance Before proceeding to design a robust gain allocation vector which provides the best worst-case performance, we can benefit from (4.21) to compute the worst-case performance of any S-D link for a given gain vector $\mathbf{g}$. The robust counterpart (4.21) of the SINR constraint is called to be active as long as it is feasible for a given $\mathbf{g}$ and τ. That is, τ is an achievable SINR for all possible uncertainty realizations within given uncertainty sets. Hence, a simple search for the maximum τ with which (4.21) is still active, results in the worst-case SINR performance of the ith S-D link for a given gain vector. To summarize, the worst-case SINR performance of the ith S-D link for given $\mathbf{g}$ (assuming that it fulfills the power constraint) is computed through

$$\mathrm{SINR}_i^{\mathrm{worst}}(\mathbf{g}) = \max_{\tau \geq 0} \ \tau \ \textbf{subject to} \ \begin{bmatrix} c_i & \mathbf{c}_i^H \\ \mathbf{c}_i & \mathbf{C}_i \end{bmatrix} - \tilde{\zeta}_i \begin{bmatrix} \rho_{\mathbf{f}_i}^2 & \mathbf{0} \\ \mathbf{0} & -\mathbf{I} \end{bmatrix} \succeq 0$$

which can be solved with bisection algorithm combined with SDP feasibility checks.

SDP for the Robust Max-min Problem Now, we can combine (4.21) with the relay sum transmit power constraint and write the robust max-min problem where the uplink channel is assumed to be perfect. The corresponding semidefinite program is given as

$$\mathcal{P}_{\text{csi}}^{\text{down}}: \max_{\hat{\mathbf{G}},\tau\geq 0} \tau \quad \textbf{subject to} \quad \begin{bmatrix} c_i & \mathbf{c}_i^H \\ \mathbf{c}_i & \mathbf{C}_i \end{bmatrix} - \tilde{\zeta}_i \begin{bmatrix} \rho_{\mathbf{f}_i}^2 & \mathbf{0} \\ \mathbf{0} & -\mathbf{I} \end{bmatrix} \succeq 0,\ i \in \{1,\ldots,N\},$$
$$\operatorname{Tr}(\mathbf{M}\hat{\mathbf{G}}) \leq P_{\mathsf{r}}, \hat{\mathbf{G}} \succeq 0, \tilde{\zeta}_i \geq 0. \tag{4.22}$$

The problem $\mathcal{P}_{\text{csi}}^{\text{down}}$ has been readily written in linear combinations of the semidefinite matrix $\hat{\mathbf{G}}$ and can be efficiently solved with any SDP tool. This conservative robust program guarantees that in the worst-case a max-min SINR value of $\tau^\star$ is supported independent from the actual realization of the uncertainty (within the defined set). In essence, much larger values than $\tau^\star$ can be expected for most of the realizations of the uncertainty. The optimized SINR value can be read as an outage SINR at a probability approaching zero. Although, for practical applications, an optimization based on non-zero outage criteria, i.e., stochastic/probabilistic optimization, would make more sense, such approaches are not tractable for our current scenario (see [127,128] for probabilistic optimization approaches for some other simpler scenarios.)

Finally, note that the $\mathcal{P}_{\text{csi}}^{\text{down}}$ is not an approximation but, on the contrary, it is exactly the true robust counterpart of the max-min problem. However, we have employed Lagrangian relaxation on $\hat{\mathbf{G}}$ by removing the rank-1 constraint on it, and hence provided a lower bound to the true counterpart.

4.3.3. Perfect Downlink but Imperfect Uplink Channel Knowledge

Consider the converse of the previous scenario, and assume that the relays know the downlink channel vectors $\mathbf{f}_{\mathsf{r}i} \forall i$ perfectly, but have imperfect knowledge about the uplink channel vectors $\mathbf{h}_{i\mathsf{r}}$. Substituting the definition of $\mathbf{h}_{i\mathsf{r}}$ as given by the uncertainty set $\mathcal{U}_{\mathbf{h}_i}$, the SINR_i becomes

$$\text{SINR}_i = \frac{\left|(\mathbf{g} \odot \mathbf{f}_{\mathsf{r}i})^T(\mathbf{h}_{i\mathsf{r},0} + \sum_{q=1}^{Q_{\mathbf{h}}} \xi_{\mathbf{h}_i,q}\mathbf{h}_{i\mathsf{r},q})\right|^2}{\sum_{j\neq i}^{N} \left|(\mathbf{g} \odot \mathbf{f}_{\mathsf{r}i})^T(\mathbf{h}_{j\mathsf{r},0} + \sum_{q=1}^{Q_{\mathbf{h}}} \xi_{\mathbf{h}_i,q}\mathbf{h}_{j\mathsf{r},q})\right|^2 + \sigma_{n_{\mathsf{r}}}^2 \mathbf{g}^H \Xi_i \mathbf{g} + \sigma_{n_{\mathsf{d}}}^2},$$

where the term Ξ_i of amplified noise is independent from the assumed CSI uncertainty, because it relates only with the downlink channels.

Contrary to the previous section, the power constraint here is affected by the uncertainty. Recall from (3.22) that the power constraint is given by

$$\mathbf{g}^H\Big(P_{\mathsf{s}} \cdot \operatorname{diag}\Big\{[\mathbf{h}_{\mathsf{s}1}^H\mathbf{h}_{\mathsf{s}1} \ \cdots \ \mathbf{h}_{\mathsf{s}N_{\mathsf{r}}}^H\mathbf{h}_{\mathsf{s}N_{\mathsf{r}}}]\Big\} + \sigma_{n_{\mathsf{r}}}^2\mathbf{I}_{N_{\mathsf{r}}}\Big)\mathbf{g} \ = \ \mathbf{g}^H\mathbf{M}\mathbf{g} \ \leq \ P_{\mathsf{r}}. \tag{4.23}$$

It is immediately noticed that in the SINR constraints we employ the column vectors of the compound uplink matrix $\mathbf{H}$, whereas the power constraint is expressed in terms of the row vectors of $\mathbf{H}$. We have already defined column-wise uncertainty sets for $\mathbf{H}$ and assumed that the channel coefficients from different sources to the relays are subject to independent uncertainties. However, due to the mathematical structure of (4.23), it is not trivial to directly incorporate these sets with the power constraint. Hence, for the sake of tractability, we take a rather conservative approach and on top of column-wise uncertainty, we further assume row-wise uncertainty. That is, each $\mathbf{h}_{\mathsf{s}k}, k = 1, \ldots, N_{\mathsf{r}}$, is defined within independent ellipsoidal set

$$\tilde{\mathcal{U}}_{\mathbf{h}_k} = \left\{ \mathbf{h}_{\mathsf{s}k} = \mathbf{h}_{\mathsf{s}k,0} + \sum_{q=1}^{Q_{\mathbf{h}}} \tilde{\xi}_{\mathbf{h}_k,q} \mathbf{h}_{\mathsf{s}k,q} \;\;:\;\; \|\tilde{\boldsymbol{\xi}}_{\mathbf{h}_k}\| \leq \tilde{\rho}_{\mathbf{h}_k} \right\}, \tag{4.24}$$

which is different and independent than the ones for column vectors of $\mathbf{H}$. Unfortunately, such an assumption overlooks the uncertainty dependencies between any row and column vector pair of $\mathbf{H}$.

Robust Counterpart for the SINR Constraint Following very similar steps to (4.18)-(4.19), we equivalently write the SINR constraint with only uplink uncertainty as

$$\frac{1}{\tau}\left|(\mathbf{g} \odot \mathbf{f}_{ri})^T \mathbf{h}_{ir,0} + \boldsymbol{\xi}_{\mathbf{h}_i}^T \breve{\mathbf{H}}_{ii}\mathbf{g}\right|^2 - \sum_{j \neq i}^{N} \left|(\mathbf{g} \odot \mathbf{f}_{ri})^T \mathbf{h}_{jr,0} + \boldsymbol{\xi}_{\mathbf{h}_j}^T \breve{\mathbf{H}}_{ij}\mathbf{g}\right|^2 \geq \sigma_{n_r}^2 \mathbf{g}^H \Xi_i \mathbf{g} + \sigma_{n_d}^2, \tag{4.25}$$

where we defined the auxiliary matrix

$$\breve{\mathbf{H}}_{ij} = \left[(\mathbf{h}_{jr,1} \odot \mathbf{f}_{ri}) \cdots (\mathbf{h}_{jr,Q_{\mathbf{f}}} \odot \mathbf{f}_{ri}) \right]^T, \; i,j \in \{1, \ldots, N\}.$$

Notice that each magnitude square term in the left hand side of (4.25) is affected by an independent uncertainty set. We introduce N real scalars $\varrho_{ij} \geq 0$ to (4.25) and decompose (4.25) into $N+1$ linearly associated inequalities. We further express each magnitude square term as an explicit quadratic function of the corresponding $\boldsymbol{\xi}_{\mathbf{h}_i}$. Hence, we obtain the following set of inequalities

$$\mathcal{C}_{\mathrm{SINR}_i}^{\mathrm{up}} \;:\; \begin{cases} \begin{cases} \boldsymbol{\xi}_{\mathbf{h}_i}^T \tilde{\mathbf{C}}_{ij} \boldsymbol{\xi}_{\mathbf{h}_i}^* + 2\mathrm{Re}\left\{ \boldsymbol{\xi}_{\mathbf{h}_i}^T \tilde{\mathbf{c}}_{ij} \right\} + \tilde{c}_{ij} \geq \tau \varrho_{ij}, \; j = i, \\ \boldsymbol{\xi}_{\mathbf{h}_i}^T \tilde{\mathbf{C}}_{ij} \boldsymbol{\xi}_{\mathbf{h}_i}^* + 2\mathrm{Re}\left\{ \boldsymbol{\xi}_{\mathbf{h}_i}^T \tilde{\mathbf{c}}_{ij} \right\} + \tilde{c}_{ij} \leq \varrho_{ij}, \;\; j \neq i, \end{cases} \\ \varrho_{ii} - \sum_{j=1, j\neq i}^{N} \varrho_{ij} \; \geq \; \sigma_{n_r}^2 \mathrm{Tr}(\Xi_i \hat{\mathbf{G}}) + \sigma_{n_d}^2, \end{cases} \tag{4.26}$$

for $i, j = 1, \ldots, N$, where

$$\tilde{\mathbf{C}}_{ij} = \breve{\mathbf{H}}_{ij}\hat{\mathbf{G}}\breve{\mathbf{H}}_{ij}^H,$$
$$\tilde{\mathbf{c}}_{ij} = \breve{\mathbf{H}}_{ij}\hat{\mathbf{G}}(\mathbf{h}_{jr,0} \odot \mathbf{f}_{ri})^*,$$
$$c_{ij} = \mathrm{Tr}\Big(\big((\mathbf{f}_{ri,0} \odot \mathbf{h}_{jr})^*(\mathbf{f}_{ri,0} \odot \mathbf{h}_{jr})^T\big)\hat{\mathbf{G}}\Big).$$

We apply the *S-lemma* and incorporate each of the first N quadratic inequality in (4.26) with the corresponding uncertainty size related quadratic function $\|\boldsymbol{\xi}_{\mathbf{h}_i}\| \leq \rho_{\mathbf{h}_i}$. Hence, the SDP formulation of the robust counterpart for the SINR constraint becomes

$$\mathcal{C}^{\mathrm{up}}_{\mathrm{SINR}_i} : \begin{cases} \begin{bmatrix} \tilde{c}_{ij} - \tau\varrho_{ij} & \tilde{\mathbf{c}}_{ij}^H \\ \tilde{\mathbf{c}}_{ij} & \tilde{\mathbf{C}}_{ij} \end{bmatrix} - \tilde{\zeta}_{ij} \begin{bmatrix} \rho^2_{\mathbf{h}_i} & \mathbf{0} \\ \mathbf{0} & -\mathbf{I} \end{bmatrix} \succeq 0, \ j = i, \\ \begin{bmatrix} -\tilde{c}_{ij} + \varrho_{ij} & -\tilde{\mathbf{c}}_{ij}^H \\ -\tilde{\mathbf{c}}_{ij} & -\tilde{\mathbf{C}}_{ij} \end{bmatrix} - \tilde{\zeta}_{ij} \begin{bmatrix} \rho^2_{\mathbf{h}_i} & \mathbf{0} \\ \mathbf{0} & -\mathbf{I} \end{bmatrix} \succeq 0, \ j \neq i, \\ \varrho_{ii} - \sum_{j=1, j\neq i}^{N} \varrho_{ij} \ \geq \ \sigma^2_{n_r}\mathrm{Tr}(\boldsymbol{\Xi}_i\hat{\mathbf{G}}) + \sigma^2_{n_d}, \end{cases} \tag{4.27}$$

where $\tilde{\zeta}_{ij} \in \mathbb{R}_+$.

Robust Counterpart for the Power Constraint: Substituting the definition of $\mathbf{h}_{sk}$ from (4.24) into (4.23) and making some algebraic manipulations, we express (4.23) as

$$\sum_{k=1}^{N_r} |g_k|^2 \Big(\mathbf{h}_{sk,0} + \sum_{q=1}^{Q_\mathbf{h}} \tilde{\xi}_{\mathbf{h}_k,q}\mathbf{h}_{sk,q}\Big)^T \Big(\mathbf{h}_{sk,0} + \sum_{q=1}^{Q_\mathbf{h}} \tilde{\xi}_{\mathbf{h}_k,q}\mathbf{h}_{sk,q}\Big)^* \leq \frac{P_r}{P_s} - \frac{\sigma^2_{n_r}}{P_s}\mathrm{Tr}(\hat{\mathbf{G}}), \tag{4.28}$$

where $|g_k|^2 = \hat{\mathbf{G}}[k,k]$. We re-write (4.28) as a sum of quadratic functions of $\tilde{\boldsymbol{\xi}}_{\mathbf{h}_k}$ as

$$\sum_{k=1}^{N_r} |g_k|^2 \Big(\tilde{\boldsymbol{\xi}}^T_{\mathbf{h}_k}\breve{\mathbf{H}}_k\breve{\mathbf{H}}_k^H\tilde{\boldsymbol{\xi}}^*_{\mathbf{h}_k} + 2\mathrm{Re}\Big\{\tilde{\boldsymbol{\xi}}^T_{\mathbf{h}_k}\breve{\mathbf{H}}_k\mathbf{h}^*_{sk,0}\Big\} + \mathbf{h}^H_{sk,0}\mathbf{h}_{sk,0}\Big) \leq \frac{P_r}{P_s} - \frac{\sigma^2_{n_r}}{P_s}\mathrm{Tr}(\hat{\mathbf{G}}), \tag{4.29}$$

where $\breve{\mathbf{H}}_k = [\mathbf{h}_{sk,1} \ \cdots \ \mathbf{h}_{sk,Q_\mathbf{h}}]^T$. Next, we decompose (4.29) into $N_r + 1$ inequalities:

$$\mathcal{C}_{\mathrm{power}} : \begin{cases} |g_k|^2 \Big(\tilde{\boldsymbol{\xi}}^T_{\mathbf{h}_k}\breve{\mathbf{H}}_k\breve{\mathbf{H}}_k^H\tilde{\boldsymbol{\xi}}^*_{\mathbf{h}_k} + 2\mathrm{Re}\Big\{\tilde{\boldsymbol{\xi}}^T_{\mathbf{h}_k}\breve{\mathbf{H}}_k\mathbf{h}^*_{sk,0}\Big\} + \mathbf{h}^H_{sk,0}\mathbf{h}_{sk,0}\Big) \leq \varsigma_k, k = 1, \ldots, N_r, \\ \sum_{k=1}^{N_r} \varsigma_k \ \leq \frac{P_r}{P_s} - \frac{\sigma^2_{n_r}}{P_s}\mathrm{Tr}(\hat{\mathbf{G}}), \end{cases}$$

where $\varsigma_k \geq 0$. As a final step, we apply the *S-lemma* to incorporate each quadratic inequalities in $\mathcal{C}_{\mathrm{power}}$ with its corresponding uncertainty size related function $\|\tilde{\boldsymbol{\xi}}_{\mathbf{h}_k}\| \leq \tilde{\rho}_{\mathbf{h}_k}$.

Consequently, we obtain the following SDP representation for the robust counterpart power of the power constraint:

$$\mathcal{C}_{\text{power}}: \begin{cases} \begin{bmatrix} -\mathbf{h}_{\text{s}k,0}^H \mathbf{h}_{\text{s}k,0} \cdot |g_k|^2 + \varsigma_k & -(\check{\mathbf{H}}_k \mathbf{h}_{\text{s}k,0}^*)^H \cdot |g_k|^2 \\ -\check{\mathbf{H}}_k \mathbf{h}_{\text{s}k,0}^* \cdot |g_k|^2 & -\check{\mathbf{H}}_k \check{\mathbf{H}}_k^H \cdot |g_k|^2 \end{bmatrix} - \check{\zeta}_k \begin{bmatrix} \tilde{\rho}_{\mathbf{h}_k}^2 & \mathbf{0} \\ \mathbf{0} & -\mathbf{I} \end{bmatrix} \succeq 0, \; k = 1, \ldots, N_{\mathsf{r}} \\ \sum_{k=1}^{N_{\mathsf{r}}} \varsigma_k \leq \frac{P_{\mathsf{r}}}{P_{\mathsf{s}}} - \frac{\sigma_{n_{\mathsf{r}}}^2}{P_{\mathsf{s}}} \operatorname{Tr}(\hat{\mathbf{G}}) \end{cases}$$

where $\check{\zeta}_k \geq 0$.

As a further side-contribution, we can benefit from the robust counterpart $\mathcal{C}_{\text{power}}$ in order to compute the worst-case power consumption for a given gain allocation vector $\mathbf{g}$ and channel uncertainty sets. In other words, in an uncertain channel knowledge scenario, we can determine the transmit power margin that we should tolerate on top of the pre-assumed value with the given $\mathbf{g}$. The SDP formulation of the corresponding optimization problem is simply

$$\mathcal{P}_{\text{wc}-\text{power}} : \min_{P_{\mathsf{r}} \geq 0} \; P_{\mathsf{r}} \quad \textbf{subject to} \quad \mathcal{C}_{\text{power}}, \check{\zeta}_k \geq 0, \varsigma_k \geq 0, k = 1, \ldots, N_{\mathsf{r}}. \tag{4.30}$$

SDP for the Robust Max-min Problem Now that we have derived the robust counterparts for all constraints, we can combine them all and write the resultant approximate robust counterpart for the max-min problem as

$$\begin{aligned} \mathcal{P}_{\text{csi}}^{\text{up}} : \quad \max_{\hat{\mathbf{G}}, \tau \geq 0} \; \tau \quad \textbf{subject to} \quad & \mathcal{C}_{\text{SINR}_i}^{\text{up}}, \mathcal{C}_{\text{power}}, \hat{\mathbf{G}} \succeq 0, \\ & \check{\zeta}_k \geq 0, \varsigma_k \geq 0, k = 1, \ldots, N_{\mathsf{r}}, \\ & \tilde{\zeta}_{ij} \geq 0, \varrho_{ij} \geq 0, i, j = 1, \ldots, N. \end{aligned} \tag{4.31}$$

As emphasized above, $\mathcal{P}_{\text{csi}}^{\text{up}}$ is an approximate solution to the robust max-min problem due to the relaxed definitions of the uncertainty regions and assumed constraint-wise uncertainties. It is important to note here that even with a simplified scenario of only uplink channel uncertainty, it is not trivial to come up with a true robust counterpart.

4.3.4. Performance Results

Simulation Setup In this section we illustrate the performance of the proposed robust designs for relay gain vector allocation in the presence of various sizes of CSI uncertainty sets. We used the MATLAB based semidefinite tool YALMIP [77] to solve the designed semidefinite problems. Unless otherwise stated, $P \triangleq P_{\mathsf{s}} = P_{\mathsf{r}} = 1$ and $\sigma^2 \triangleq \sigma_{n_{\mathsf{r}}}^2 = \sigma_{n_{\mathsf{d}}}^2$. We set the average SNR to $\text{SNR} = P/\sigma^2 = 20\text{dB}$.

In the following we assume that the uncertainty sets are chosen such that all of the corresponding $\boldsymbol{\xi}$, e.g., $\boldsymbol{\xi}_{\mathbf{h}_i}, \boldsymbol{\xi}_{\mathbf{f}_i}, \tilde{\boldsymbol{\xi}}_{\mathbf{f}_k}$, is one-dimensional, and hence, $Q_{\mathbf{h}} = Q_{\mathbf{f}} = 1$. Moreover, for the sake of simplicity, we assume equal sizes of uncertainty for all employed channel vectors, i.e., $\rho = \rho_{\mathbf{h}_i} = \rho_{\mathbf{f}_i} = \tilde{\rho}_{\mathbf{f}_k}$. We model each element of the available channel estimate (nominal data) and the uncertainty perturbation term as an i.i.d. Rayleigh fading coefficient with zero-mean and unit-variance. However, we scale the norm of the perturbation so that it is as big as that of the nominal data. For the sake of clarity, let us give a specific example for the uncertainty set $\mathcal{U}_{\mathbf{h}_i}$, whose definition can be adapted from Section 4.3.1 to the aforementioned assumptions as

$$\mathcal{U}_{\mathbf{h}_i} = \left\{ \mathbf{h}_{ir} = \mathbf{h}_{ir,0} + \xi_{\mathbf{h}_i,1}\mathbf{h}_{ir,1} \quad : \quad |\xi_{\mathbf{h}_i,1}| \leq \rho_{\mathbf{h}_i} \right\}.$$

Here, $\mathbf{h}_{ir,0}$ is the nominal data and $\mathbf{h}_{ir,1}$ is the uncertainty perturbation. With the enforced constraint of $\|\mathbf{h}_{ir,0}\| = \|\mathbf{h}_{ir,1}\|$, we make sure that $\rho_{\mathbf{h}_i} = 1$ corresponds to an uncertainty that is of the same size (in terms of the norm) as the assumed estimated channel. Further note that $\rho_{\mathbf{h}_i} = 0$ indicates that the available channel knowledge is perfect.

As a non-robust benchmark, we consider the general max-min beamforming design $\mathcal{P}_{\text{mm}}$ (without ZF constraint) proposed in Section 3.2.2. Such a relay gain choice neglects the presence of CSI uncertainty. We denote the corresponding non-robust gain vector with $\mathbf{g}_{\text{non-robust}}$. Since the worst-case performance of this gain choice is dependent on the assumed uncertainty model, and hence, differs for each above considered two cases, we present further details on the worst-case performance computation in the corresponding section of simulations.

Perfect Uplink - Imperfect Downlink In this first simplified case, we assume that the uplink channel knowledge is perfect. The *exact* robust counterpart for the max-min problem is given by the $\mathcal{P}_{\text{csi}}^{\text{down}}$ in (4.22). Hence, the optimal value of $\mathcal{P}_{\text{csi}}^{\text{down}}$ for given channel and uncertainty parameters gives the robust minimum worst-case SINR.

On the other hand, the worst-case performance of the non-robust gain allocation choice $\mathbf{g}_{\text{non-robust}}$ is computed through the robust counterpart of SINR (4.21) for given gain vector, i.e.,

$$\mathcal{P}_{\text{csi}}^{\text{down}}(\mathbf{g}_{\text{non-robust}}): \quad \max_{\tau \geq 0} \ \tau \quad \textbf{subject to} \quad \begin{bmatrix} c_i & \mathbf{c}_i^H \\ \mathbf{c}_i & \mathbf{C}_i \end{bmatrix} - \tilde{\zeta}_i \begin{bmatrix} \rho_{\mathbf{f}_i}^2 & \mathbf{0} \\ \mathbf{0} & -\mathbf{I} \end{bmatrix} \succeq 0,$$
$$\tilde{\zeta}_i \geq 0, \ i \in \{1, \ldots, N\}.$$

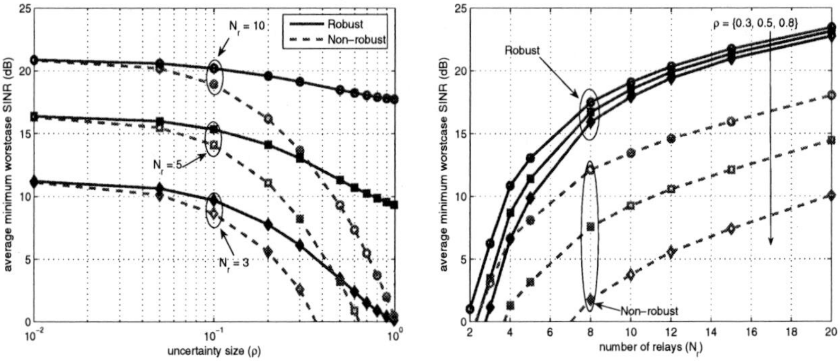

(a) (b)

Figure 4.5.: (a)Average minimum worst-case SINR vs. uncertainty size ρ for $N = 2$ S-D pairs, (b) Average minimum worst-case SINR vs. number of relays N_r for $N = 2$ S-D pairs.

We compare the average minimum worst-case SINR performances of robust and non-robust designs in Fig. 4.5. The average is taken over 1,000 channel realizations and the minimum is performed over N link-based worst-case SINR values. We first focus on the impact of uncertainty size on the worst-case performances. In Fig. 4.5(a) we consider three sets of curves, which corresponds to $N_r = 3, 5, 10$ relays. Note first that an uncertainty size that is smaller than $\rho = 5 \times 10^{-2}$, results only in marginal SINR losses, and hence can be tolerated by the system. Independent from number of relays, we observe that the robust design improves its SINR advantage with respect to the non-robust counterpart, as the size of the uncertainty enlarges. For instance, if $N_r = 5$, the robust design offers a gain of 1 dB at $\rho = 0.1$, which is later improved to almost 10 dB as $\rho \to 1$. Especially for $N_r = 5, 10$, the worst-case performance of the non-robust design degrades very fast and converges to 0 dB just after $\rho = 0.6$. Whereas, the robust design can still provide operational SINR values of about 10 dB. It is interesting to note that as N_r increases, the robust design performs such that it provides higher SNR values for a fixed ρ value. It is primarily due to the increased diversity in the system, which is completely exploited to maximize the minimum worst-case SINR value.

Let us now elaborate on the worst-case performance of the robust design at the extreme limit of $\rho = 1$. It is natural to observe that the lowest SINR values are attained at this regime, but it may be unexpected that the performance of the robust design is quite robust for $\rho = 1$. In fact this is basically dependent on the choice of uncertainty perturbations. Motivated by the common noisy channel estimation considerations, we preferred the perturbation to be independent from the nominal data. However, for example in [98], the nominal and the perturbation are chosen the same. That is, consider for example the channel vector $\mathbf{f}_{ri}$ and say that we choose the perturbation channel vector $\mathbf{f}_{ri,1}$ to be equal to the nominal channel vector $\mathbf{f}_{ri,0}$ (see [98] for such a choice). Then at $\rho = 1$, a trivial choice of $\xi_{\mathbf{f}_i,1}$ is to set it to -1, which results in nothing but nulling the channel. In such a case, robust optimization may lose its control on the SINR. However, as above presented we have preferred an approach, where the perturbation matrix is chosen randomly and its norm is scaled accordingly to the norm of the nominal data. In other words, it is not possible that $\xi_{\mathbf{f}_i,1}$ can project the perturbation channel to the opposite direction of the nominal channel. Although we do not present here, we have observed through simulations that for such an equal nominal and perturbation choice, our robust design still outperforms the non-robust counterpart, but provide less SINR advantage when compared with the results in Fig. 4.5.

We illustrate the worst-case performance from a different perspective in Fig. 4.5(b), where we plot the average minimum worst-case performances versus number of relays N_r. Similar to (a), we observe that for each additional relay in the network, the effect of larger ρ, decreases for the robust design, but increases for the non-robust design.

Perfect Downlink - Imperfect Uplink We now modify our assumption such that the downlink channel knowledge is perfect, but that of uplink is imperfect. In contrary with the precious case, we could not derive an *exact* robust counterpart, but only an *approximate* through the assumption of independent uncertainty sets.

As above, we can again compute the worst-case performance of the non-robust gain allocation choice $\mathbf{g}_{\text{non-robust}}$ through the robust counterpart of SINR derived in (4.27). However, note that this worst-case value does not correspond to the actual worst-case performance, but to only an approximation of it. Because, while deriving (4.27) we have relaxed the uncertainty sets for the sake of mathematical convenience. Similarly, the robust minimum worst-case SINR computed with the approximate robust counterpart $\mathcal{P}_{\text{csi}}^{\text{up}}$ in (4.31), is not the actual value that the robust design attains.

Henceforth, we choose an empirical approach to determine the actual worst-case value of the non-robust design for the given robust and non-robust gain vectors. That is, for each given nominal and perturbation vector, we try sufficiently enough ξ values (10^6

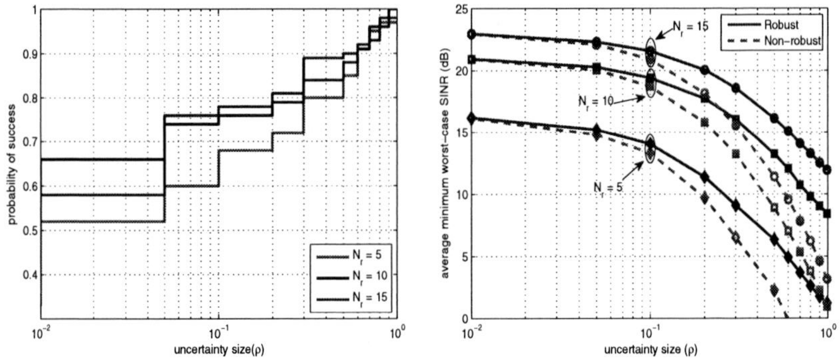

(a) (b)

Figure 4.6.: (a) Probability of success to obtain a feasible robust minimum worst-case SINR that is larger than that of the non-robust design, (b) Average minimum worst-case SINR vs. the uncertainty size ρ for $N = 2$ S-D pairs

trials), and determine the minimum worst-case SINR by checking them all. We should take the power constraint into account here, which is also affected by the channel uncertainty in contrary with the previous case. That is, while comparing the performances of the two design, we should assure through an appropriate scaling that, both leads to an equal instantaneous relay sum transmit power per nominal and perturbation channel realization.

Although the uncertainty model originally does not enforce a distribution for ξ, we choose it to be complex Gaussian. Hence, if the empirically computed minimal worst-case value is not exact, it is at least a good upper bound. In order to verify the accuracy of this empirical worst-case search, we adapt it to the previous case of perfect uplink and imperfect downlink, where we have an exact robust counterpart. The obtained results have revealed that both theoretical and empirical results match one-to-one.

Since the robust design is not exact, for some nominal and perturbation channel vector realizations, it may as well result in a minimum worst-case SINR that is smaller than that of the non-robust design, meaning that it could not succeed to produce a feasible robust solution. In Fig. 4.6(a), we plot this probability of success over the uncertainty size ρ for different number of relay sets. We observe in general that probability of success improves as $\rho \to 1$. For instance, on average 50% of the realizations succeeds to improve the minimum worst-case SINR when $\rho < 5 \times 10^{-2}$, whereas, the success rate grows up to 99% as ρ gets larger. These observations are in fact good news which depict that in a regime of maximal disturbance due to uncertainty, e.g., $\rho > 0.5$, our approximate robust counterpart solution works pretty satisfactorily. Note that for any unsuccessful case, we can stick up with the non-robust design in order not to sacrifice more from performance. To summarize, for all cases, we "at least" improve the worst-case performance with our robust design.

The next question to seek an answer for is how the robust and non-robust designs' minimum worst-case SINRs compare for these successful realizations. To this end, we plot average minimum worst-case SINR (of only the successful realizations) versus the uncertainty size in Fig. 4.6(b). The observations for Fig. 4.5 are also carried over here, where the robust design provides substantial improvement and this gain is emphasized more with larger number of relays.

A natural question is that how the respective impact of uplink and downlink channel imperfections on the minimum worst-case SINR, compare. Basically, while the uplink channels incorporate solely with the SINR, the downlink channels are additionally effective on the power constraint. We can compare the minimum worst-case SINRs for different types of imperfections through Figures 4.5(a) and 4.6(b). However, for the sake of fairness, we should keep in mind the derivation of the corresponding designs. Recall that the design for uplink imperfection is not an exact but an approximate solution due to assumed independencies between uncertainty sets. We focus specifically on the $N_\mathsf{r} = 10$ relays case: downlink imperfection with solid-blue-circle and dashed-red-circle curves in Fig. 4.5 (a) and uplink imperfection with solid-blue-square and dashed-red-square curves in Fig. 4.6 (b). We observe that the performance of the non-robust approach against ρ behaves very similar for both sets of curves and approaches to 0dB monotonically as $\rho \to 1$. On the other hand, while comparing the performances of the robust designs, we focus on the worst-case SINR loss between the regimes $\rho \to 0$ and $\rho \to 1$. We immediately realize that the robust design for imperfect downlink behaves quite resistant against ρ and loses only 5 dB. Whereas, the design for imperfect uplink needs to sacrifice almost 15 dB. Although these findings suggest to be interpreted as that *uplink CSI imperfection is more crucial than that of downlink*, this statement may not be that immediate due to the aforementioned design issues.

4.4. Concluding Remarks

Broadly speaking, the efficiency of a wireless communication system can be quantified with two features: performance and practicality. That is to say, a system with extraordinary transmission performances may be worth nothing if it is impractical to implement it in real-world scenarios. Addressing this issue in this chapter, we considered the negative impact of system imperfections on the distributed orthogonalization performance of coherent AF relaying, and the counter-measures that should be taken. We focused on two typical and crucial imperfection scenarios and followed a robust framework for the design of relay gain vector in the presence of these uncertainties.

First, we studied the LO phase offset synchronization between the relays, which is compulsory for coherent addition at the destination nodes. As it has been reported in [19] that phase synchronization can be avoided through specific channel estimation procedures, we rather focused on the inevitable problem of phase noise uncertainty at AF relays, which causes destination SINRs to be random variables of time. We bounded the phase uncertainty per relay within a pre-determined range dependent on the desired probability of assurance, and proposed an efficient worst-case robust relay gain allocation scheme based on SDP. It has been numerically shown that robust approach provides significant worst-case SINR improvement, while offering a trade-off between being conservative and outage-secure for setting the optimization parameters.

In the second part, we shifted our focus to channel knowledge uncertainty at the relays. In essence, phase noise uncertainty and the quality of channel estimation are directly related with each other. Even if the channel varies very slowly and we may have a perfect channel estimation algorithm with sufficiently enough estimation phase duration, there is the aforementioned phase randomness that change the corresponding channel estimates. In other words, the impacts of different RF chains make the compensation of phase knowledge of estimates very difficult in a real world scenario. This results in a channel uncertainty scenario addressed in [94], where perfect magnitude but imperfect phase knowledge of channel estimates were considered. Although not addressed explicitly in this thesis, such a scenario can also be adapted here, and be shown that the solution of the corresponding robust problem boils down to be very similar to the robust design $\mathcal{P}_{\text{pr}}$ against phase noise.

In this chapter, we have treated the uncertainties arising from magnitude and phase jointly, and defined independent uncertainty sets per given channel vectors associated with the relays. We followed a worst-case max-min philosophy, where the best worst-case SINR performance is attained over all S-D links and uncertainty sets. The original generalized formulation has been shown to be intractable, which leaded us to consider simplified and/or approximate solutions for the robust counterpart. First, we considered that the relays has perfect uplink but imperfect downlink channel knowledge. We derived the exact robust counterpart of the corresponding worst-case max-min problem, where the power constraint was independent from the uncertainty due to perfectness in the second hop. Secondly, we studied the opposite scenario where downlink is perfect but uplink is imperfect. An approximate robust counterpart has been derived through the assumption of side-wise independent uncertainty sets. All of the above mentioned robust designs were solved through efficient SDP techniques. Noting the substantial gains achieved through robust optimization, we conclude that uncertainty awareness is essential in providing reliable multiuser communications services in practical systems.

Chapter 5

Distributed Multiuser Networks with Multiple-Relay-Clusters

In Chapter 3, we have reported that an excess number of relays provides additional gains to the network, in addition to distributed spatial multiplexing, such as sum/outage rate improvement, diversity and fairness. However, these benefits come with a cost of increased local CSI dissemination overhead. In this chapter, we aim at designing multiuser AF relaying protocols, which achieve similar gains of network-wide global CSI requiring solutions, but employing the excess relays in a way that we have a bounded dissemination overhead independent from the number of relays. To this end, we propose two clustered protocols, where the relays are partitioned into multiple sets. We first consider homogeneous relay clusters, where each cluster individually manages the multiuser interference independent of the other clusters. Second, we introduce a hierarchy in between different clusters and consider heterogenous clusters in terms of the targeted gains to achieve and the required amount of CSI knowledge for gain allocation. Finally, we report that the multiuser network can enjoy the benefits of an excess number of relays in the expense of only very limited CSI dissemination overhead.

5.1. Introduction

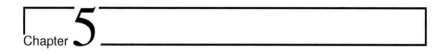

In the previous chapters, we have developed a common understanding on working principles of coherent relaying, and studied extensively the achievable gains of distributed beamforming. Throughout these investigations, we have identified global dissemination of local CSI of each relay as a major overhead coming with coherent multiuser relaying. In spite of the heavy load of global CSI requirement, we also showed that the distributed AF network can exploit MIMO gains more and more as the number of relays increases within the network. Specifically, in addition to the primary target of distributed spectral efficiency, multiple S-D pairs can benefit from supplementary gains such as sum rate, outage improvement and fairness in between the pairs. Moreover, distributed effective diversity gain has been shown to be achievable at each destination through efficient relay selection or max-min type optimizations.

In this chapter, we shift our focus to rather larger AF relay networks, where the aforementioned local CSI dissemination is very likely to result in spectral efficiency loss. Hence, there is a certain need for proper multiuser relaying protocols which can be employed in such large networks and can provide the above mentioned gains without hampering the practicality of the system. Addressing the same drawback of coherent relaying in practical implementations, we have proposed distributed calculation of relay gains in Section 3.3. However, in this chapter we take an alternative approach and try to exploit clustering the relays in smaller independent groups, which may emphasize the proximity of the relays to each other in practice. To this end, we propose two novel clustered relaying protocols, where we confine the information exchange locally in each of these (rather) small clusters. Consequently, we hope for reducing the information exchange within the network. However, not surprisingly, such an approach would destruct the *global* coherence within the network. Hence, smart and cautious measures should be taken while designing the protocols in order to recover the related performance loss as much as possible. In summary, throughout this chapter we aim at providing and exploiting a trade-off between the overhead (in terms of spectral use) and the system performance.

Contribution of this Chapter The first protocol considers that the whole relay network is grouped in *totally* independent C clusters. Each of these is independently serving the N S-D pairs as if there are no other clusters in the network. We assume global CSI knowledge within each cluster and a per-cluster power constraint. Hence, there is no information exchange between the clusters. Such a clustering is referred to be homogeneous as all clusters are responsible for interference mitigation and require similar channel knowledge while doing so. Nevertheless, the fragmented structure of the network results in loss of coherency between relays in different clusters, and this destructs the achieved diversity and array gains at the destinations. In order to recover the loss in diversity, which could have been obtained through max-min beamforming through a conventional globally coherent network (see Section 3.2.2), we propose to apply phase rotations in between clusters and attain distributed diversity gain at the destinations. Moreover, in the context of this section, we provide an analysis on local CSI dissemination in terms of traffic overhead and transmit energy consumption.

As a second approach, we introduce hierarchy in the relaying protocol such that a group of relays have more information about the network than some others. We propose that the total relay set is partitioned into two main hierarchical levels, i.e. clusters, which have different needs to CSI for calculating the relay gain coefficients. In the lower level of hierarchy, each relay employs only *local* CSI to compute its gain factor, and hence, is cost-free in terms of CSI dissemination overhead. However, while these relays provide array gain at the destinations, they also cause multiuser interference. In the higher level of hierarchy, the relays are equipped with further channel information to have effective control on the signal forwarding of the relays in the lower hierarchical level and correspondingly to result in distributed orthogonalization of multiple pairs. Although the number of relays in the former cluster is totally irrelevant for distributed multiplexing, we need a minimum number of relays in the latter cluster for interference cancellation. We study sufficient conditions for achieving full spatial multiplexing, required channel knowledge per hierarchical level, corresponding relay gain allocations, and power allocation strategies among hierarchical levels. We finally identify *relay selection* for different hierarchical levels as an efficient means to recover outage (diversity) and sum rate gains. It is shown that with drastically reduced CSI dissemination overhead, the hierarchical protocol combined with relay selection approaches to the performance of conventional multiuser relaying requiring global CSI knowledge over the whole network.

5.2. Multiple Homogeneous Relay Clusters

Following the general multiuser relaying setup, we consider a wireless network with $2N+N_{\mathsf{r}}$ single antenna nodes. The relays are grouped in C clusters, where the cth cluster is denoted by $\mathscr{C}_c, c \in \{1,\ldots,C\}$ and is composed of $N_{\mathsf{r},c}$ relays, i.e., $\sum_{c=1}^{C} N_{\mathsf{r},c} = N_{\mathsf{r}}$. Each cluster is assumed to be unaware and independent of the others. It is also assumed that there are no direct links between sources and destinations. The $N_{\mathsf{r},c} \times N$ complex channel matrix between the N sources and the cth cluster is denoted by $\mathbf{H}^{(c)}$, and likewise, $\mathbf{F}^{(c)}$ denotes the $N \times N_{\mathsf{r},c}$ complex channel matrix between the cth cluster and the N destinations. The conventional compound channels $\mathbf{H}$ and $\mathbf{F}$, defined in Section 2.2, can be simply obtained by stacking the corresponding cluster based channel matrices one after the other in a single matrix. Moreover, in the sequel of this chapter,

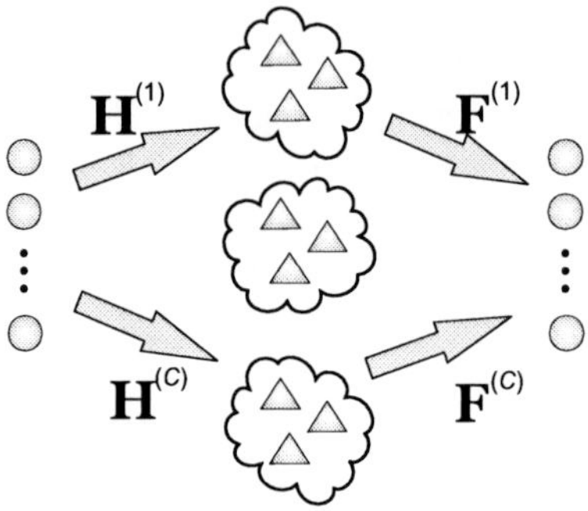

Figure 5.1.: Two-hop network configuration with multiple (half-duplex) relay clusters.

we assume for the sake of simplicity that there is global phase reference throughout the network, and hence, we drop the phase offset related terms from our derivations. Fig. 5.1 summarizes the clustered network configuration.

Outline of this Section The reminder of this section is organized as follows. We present the clustered relaying protocol in Section 5.2.1 with details of transmit and receive signal models. We elaborate on relay gain allocation per cluster in Section 5.2.2. Section 5.2.3 analyzes the CSI dissemination overhead in terms of required channel use and transmit energy consumption. Moreover, there we also investigate the impact of cluster formations on the average information rate performances. Finally, in subsection 5.2.4 we study the performance of the proposed system through computer simulations.

5.2.1. Clustered Relaying Protocol

The traffic pattern follows the conventional two-hop half-duplex scheme presented in Section 2.3. However, in the following we re-formulate the received signal models such that they reflect the transmission and the reception of L consecutive source symbols.

In the first hop, each source transmits L symbols consecutively, and hence, the $N_{\mathsf{r},c} \times L$ received signal matrix at the cth cluster is written as

$$\mathbf{Y}_{\mathsf{r},c} = \mathbf{H}^{(c)}\mathbf{S} + \mathbf{N}_{\mathsf{r},c}, \tag{5.1}$$

where $\mathbf{Y}_{\mathsf{r},c} = [\mathbf{y}_{\mathsf{r},c}^{(1)} \cdots \mathbf{y}_{\mathsf{r},c}^{(L)}]$, $\mathbf{y}_{\mathsf{r},c}^{(\ell)}$ represents the received signal vector of the cth cluster in the ℓth symbol interval, $\mathbf{S} = [\mathbf{s}^{(1)} \cdots \mathbf{s}^{(L)}]$, and $\mathbf{s}^{(\ell)}$ represents the transmit signal vector in the ℓth symbol interval. Moreover, $\mathbf{N}_{\mathsf{r},c} \in \mathbb{C}^{N_{\mathsf{r},c} \times L}$ is the corresponding relay noise matrix with i.i.d. $\mathcal{CN}(0, \sigma_{n_\mathsf{r}}^2)$ entries.

Upon receiving $\mathbf{Y}_{\mathsf{r},c}$, the relays in $\mathscr{C}_c$ multiply the received signals with the corresponding gain factor matrix $\mathbf{G}^{(c)} \triangleq \mathrm{diag}\{\mathbf{g}^{(c)}\}$, and obtain the resultant $\mathbf{R}_c = \mathbf{G}^{(c)}\mathbf{Y}_{\mathsf{r},c}$ to be forwarded to the destinations through the downlink channel $\mathbf{F}^{(c)}$ in the L consecutive symbol periods. As we have learned from the previous chapters, the design of the relay gain factors are based on the knowledge of uplink and downlink channels. We assume here that each relay in the cth cluster knows only $\mathbf{H}^{(c)}$ and $\mathbf{F}^{(c)}$ perfectly, but has no information about the other clusters' CSI. We refer this knowledge interchangeably as either *local cluster-CSI knowledge* or *global CSI knowledge within cluster.*

With the proposed clustered relaying protocol, the global "network-wide" CSI knowledge requirement of the conventional coherent multiuser relaying in Chapter 2 and Section 3.2 is relaxed to be cluster oriented. The corresponding significant reduction in dissemination overhead will be investigated thoroughly in Section 5.2.3. On the other hand, note that we can still apply the distributed gradient approach proposed in Section 3.3 locally per cluster, and further relax the local cluster-CSI knowledge.

Recall that we have been assuming a slowly varying channel environment for our coherent relaying system. Hence, $\mathbf{H}^{(c)}$ and $\mathbf{F}^{(c)}$ stay constant for at least $2L$ symbols transmission duration. If $\mathbf{R}_c$ is transmitted to the destinations without any further signal processing, the compound channel coefficients from all clusters would add up to a single coefficient which stays constant during L symbol periods. Consequently, this prevents any temporal diversity at the destinations. In [50], time-variant phase offsets are introduced at the relays to create an artificial time-variant channel, which is then utilized to achieve diversity by an outer code. It is later shown in [2] theoretically that such an approach results in a full-diversity order of the number of employed relays. Here, we employ this phase rotation concept between the clusters instead of the relays. The spatial diversity is transformed to temporal diversity by time variant and cluster specific phase variations, i.e., the same phase variation is used for all relays in the same cluster. Thus, before second-hop transmission, each relay multiplies the corresponding element of $\mathbf{R}_c$ with time (ℓ) and cluster (c) specific phase offset $\psi_{c,\ell}$, such that they forward the signal $\mathbf{\Psi}_c \odot \mathbf{R}_c$, where

$$\mathbf{\Psi}_c \in \mathbb{C}^{N_{\mathsf{r},c} \times L} = \left[[\psi_{c,1} \cdots \psi_{c,L}]^T \ldots [\psi_{c,1} \cdots \psi_{c,L}]^T \right]^T.$$

Besides, we assume that each cluster transmits with a fixed sum power $P_{\mathsf{r},c}$ per symbol, where $\sum_{c=1}^{C} P_{\mathsf{r},c} = P_\mathsf{r}$. For the sake of fairness and avoiding any transmit power dependence between the clusters, which could be the case for optimizing the cluster-power allocation, we choose $P_{\mathsf{r},c} = P_{\mathsf{r}}/C$. Such an equal power allocation per cluster is also critical for achieving full distributed diversity through phase rotations at the clusters.

In the second hop, the destinations receive a superposition of all signals from C clusters. Hence, the received signal block over L symbols duration is represented as

$$\mathbf{Y}_{\mathsf{d}} = \sum_{c=1}^{C} \left(\mathbf{F}^{(c)} \left(\mathbf{\Psi}_c \odot (\mathbf{G}^{(c)}\mathbf{H}^{(c)}\mathbf{S}) \right) + \mathbf{F}^{(c)} \left(\mathbf{\Psi}_c \odot (\mathbf{G}^{(c)}\mathbf{N}_{\mathsf{r},c}) \right) \right) + \mathbf{N}_{\mathsf{d}}, \qquad (5.2)$$

where $\mathbf{Y}_{\mathsf{d}} = [\mathbf{y}_{\mathsf{d}}^{(1)} \cdots \mathbf{y}_{\mathsf{d}}^{(L)}]$, and $\mathbf{N}_{\mathsf{d}}$ is the corresponding destination noise matrix with i.i.d. $\mathcal{CN}(0, \sigma_{n_{\mathsf{d}}}^2)$ entries. In order to gain more insight, let us write the received signal vector at the ℓth symbol interval, i.e., the ℓth column of $\mathbf{Y}_{\mathsf{d}}$:

$$\begin{aligned} \mathbf{y}_{\mathsf{d}}^{(\ell)} &= \sum_{c=1}^{C} \psi_{c,\ell} \cdot \left(\mathbf{F}^{(c)}\mathbf{G}^{(c)}\mathbf{H}^{(c)}\mathbf{s}^{(\ell)} + \mathbf{F}^{(c)}\mathbf{G}^{(c)}\mathbf{n}_{\mathsf{r},c}^{(\ell)} \right) + \mathbf{n}_{\mathsf{d}}^{(\ell)} \\ &= \sum_{c=1}^{C} \psi_{c,\ell} \cdot \mathbf{H}_{\mathsf{srd}}^{(c)}\mathbf{s}^{(\ell)} + \sum_{c=1}^{C} \psi_{c,\ell} \cdot \mathbf{F}^{(c)}\mathbf{G}^{(c)}\mathbf{n}_{\mathsf{r},c}^{(\ell)} + \mathbf{n}_{\mathsf{d}}^{(\ell)}, \end{aligned} \qquad (5.3)$$

where $\mathbf{H}_{\mathsf{srd}}^{(c)}$ is defined as the equivalent channel through the cth cluster, and $\mathbf{n}_{\mathsf{r},c}^{(\ell)}$ is the corresponding noise vector of the cth cluster in the ℓth symbol interval. The first summation in (5.3) represents the equivalent channel observed at the destinations, which is a non-coherent superposition of source signals through C different clusters. In the absence of time (ℓ) and cluster (c) specific phase rotation term $\psi_{c,\ell}$, the superimposed equivalent channel $\sum_{c=1}^{C} \mathbf{H}_{\mathsf{srd}}^{(c)}$ would stay constant during all L symbols and hence, not provide temporal diversity.

The next question to answer is how to choose the compound phase rotation matrix

$$\mathbf{\Psi} \in \mathbb{C}^{C \times L} = \left[[\psi_{1,1} \cdots \psi_{1,L}]^T \cdots [\psi_{C,1} \cdots \psi_{C,L}]^T \right]^T,$$

which is adapted from $\mathbf{\Psi}_c$'s. First, in order to achieve full diversity, we need to choose L to be at least equal to C, i.e., we aim at attaining a diversity order that is equal to the number of clusters in the network. It is proposed in [50] that an FFT matrix or a diagonal matrix can be a practical choice to be used as $\mathbf{\Psi}$. The interested reader is referred to [2] for further details on the optimization and decision criteria for $\mathbf{\Psi}$.

Choosing $\mathbf{\Psi}$ to be a diagonal matrix of size C implies that, in each symbol duration of the second hop, one single relay cluster is transmitting, whereas the others are silent.

Hence, $\mathbf{y}_{\mathsf{d}}^{(\ell)}$ in (5.3) becomes

$$\mathbf{y}_{\mathsf{d}}^{(\ell)} = \mathbf{H}_{\mathsf{srd}}^{(\ell)}\mathbf{s}^{(\ell)} + \mathbf{F}^{(\ell)}\mathbf{G}^{(\ell)}\mathbf{n}_{\mathsf{r},\ell}^{(\ell)} + \mathbf{n}_{\mathsf{d}}^{(\ell)}, \tag{5.4}$$

where, here and in the sequel, the indices ℓ and c are used interchangeably for such choice of $\boldsymbol{\Psi}$. Assigning each symbol duration to one single cluster provides the advantage of using full relay sum transmit power P_{r} at each cluster instead of P_{r}/C. However, as noted in [50], such an approach of switching for each symbol to another cluster has the disadvantage that the on-off switching sets high requirements at the linearity of the relays power amplifier.

The instantaneous information rate between the ith S-D pair (normalized to phase-rotation block length $L = C$) through this artificially created time-selective channel is given by

$$\mathrm{R}_i = \frac{1}{2L}\sum_{\ell=1}^{L}\mathrm{R}_{i,\ell} = \frac{1}{2L}\sum_{\ell=1}^{L}\log_2(1+\mathrm{SINR}_{i,\ell}), \tag{5.5}$$

where $\mathrm{R}_{i,\ell}$ and $\mathrm{SINR}_{i,\ell}$ denote the achievable rate and the received SINR for the ith link at the ℓth symbol interval. More specifically, when cluster switching is assumed, $\mathrm{SINR}_{i,\ell}$ reflects the distributed beamforming effect of the ℓth cluster at the ith destination.

5.2.2. Relay Gain Allocation per Cluster

In the proposed cluster-based MU relaying system, each cluster is independent and unaware of the others. Hence, depending on the number of relays $N_{\mathsf{r},c}$ and S-D links N, the cluster $\mathscr{C}_c$ can locally choose to use one of the beamforming approaches proposed in Section 3.2. However, due to the superimposed signals from different clusters, the presumed gains (SINR or rate) with the chosen beamforming may not be achievable at the destinations. In other words, without any further information flow in between clusters, none of the clusters can have a direct deterministic control on R_i in general. In essence, independent from the applied phase rotations, this fact is the performance expense of reducing the CSI dissemination overhead through clustering.

However, in the case of cluster switching, i.e., $\boldsymbol{\Psi} = \mathbf{I}_C$, as each symbol duration is allocated to a single cluster, i.e., no superposed cluster signals, the ℓth cluster can determine $\mathrm{R}_{i,\ell}$ by choosing its gain factors correspondingly. But, as in general, it does not still have a direct information on the resultant R_i.

Let us now focus closer on the cluster switching scheme. Each cluster per se tries to optimize its gain vector $\mathbf{g}^{(c)}$ such that the sum rate or minimum link rate is maximized for the symbol that it is in charge to forward. If all clusters choose to employ the same beamforming approach, then we can claim that the optimality achieved with the chosen beamforming for the corresponding $\mathrm{R}_{i,\ell}$ would also be achieved for the overall R_i. For example, say that max-min beamforming with ZF condition, which is presented in Section 3.2.2.2, is chosen to be employed in all clusters for relay gain allocation. Correspondingly, each $\mathrm{R}_{i,\ell}$ is determined by the corresponding cluster such that it is the optimal max-min rate under ZF condition for the ℓth symbol of the ith link. Hence, assuming that there is no information flow between the clusters, the resultant R_i (averaged over all $\mathrm{R}_{i,\ell}$ in (5.5)) is also the optimal max-min information rate that the ith link can attain with ZF condition under given channel conditions. In summary, choosing $\mathbf{\Psi} = \mathbf{I}_C$ and using the same beamforming at each cluster, we can assure optimal sum or max-min R_i under given power and channel conditions without any further information exchange.

Remark 5.2.1: For any given (full diversity leading) phase rotation matrix $\mathbf{\Psi}$, it is possible to jointly optimize the cluster gain vectors $\mathbf{g}^{(c)} \forall c \in \{1, \ldots, C\}$ and directly determine the resultant R_i in (5.5). However, such an approach would need further CSI dissemination in between clusters, which in return would destroy the independence of clusters.

5.2.3. Analysis of CSI Dissemination Overhead

After having introduced the clustered relaying protocol in the previous sections, here we analyze the reduction in CSI dissemination overhead from several different aspects. We assume that each relay in the network perfectly estimates its backward and forward channels (local CSI), which is denoted by $\mathbf{h}_{\mathsf{s}k}$ and $\mathbf{f}_{k\mathsf{d}}$ for the kth relay, $k \in \{1, \ldots, N_\mathsf{r}\}$. Then, each relay, say in the cth cluster, shares its own local CSI with the other $N_{\mathsf{r},c} - 1$ relays in the same cluster.

We simply assume here that the local CSIs are disseminated through a secondary short-range wireless system with only unicast (single or multiple hops) or multicast ability. In a single-hop unicast scheme, each node needs to establish point-to-point links with all the other nodes, which results in a high channel use overhead and transmit energy consumption. Instead, the relays can also establish multiple short-distance hops to disseminate their local CSI (multi-hop unicast). In other words, the relays help each other by forwarding the received local CSI of its neighbours to others. Moreover, if the relays are able to multicast data, they can establish point-to-multipoint links, which in

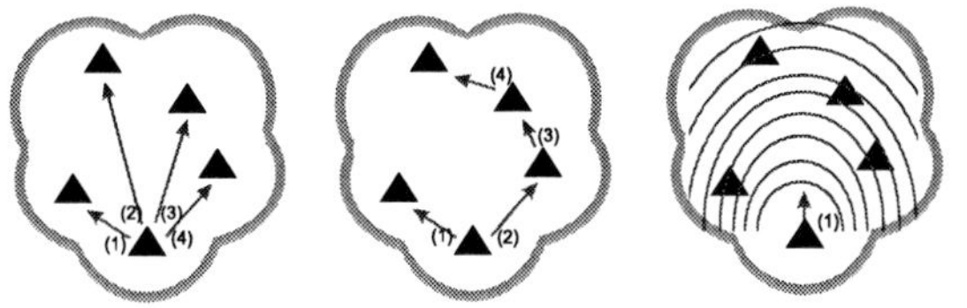

Figure 5.2.: The illustrations of different local CSI dissemination schemes: For the considered 5 node scenario in the figure, unicast schemes employ 4 channel-use, whereas the multicast scheme requires only 1.

turn reduces the amount of transmissions with respect to the unicast schemes. See Fig. 5.2 for the interpretation of different dissemination schemes.

In the following, we investigate these three dissemination schemes in the view of two overhead measures: the number of channel uses and the transmit energy consumption. Throughout this study, special attention will be paid to cluster formations with different number of clusters C and correspondingly with various sets of relays per cluster. Moreover, in the last part, we focus on the effects of cluster formations on the overall information rate performance of the clustered multiuser relaying network.

5.2.3.1. Channel Use for Local CSI Dissemination

Unicast Assuming a secondary short-range wireless system with only unicast ability (single or multi-hop), the amount of channel uses that is needed to disseminate all CSI within the cth cluster is defined as a function of $N_{\mathsf{r},c}$ and written as

$$f_{\text{cu}}^{\text{u}}(N_{\mathsf{r},c}) = 2NN_{\mathsf{r},c}(N_{\mathsf{r},c} - 1), \tag{5.6}$$

where the subscript "cu" denotes "channel use" and the superscript "u" denotes "unicast". The function f_{cu}^{u} in (5.6) follows from that local CSI of each relay is compromised of $2N$ channel coefficients, i.e., N uplink and N downlink, and that $(N_{\mathsf{r},c}-1)$ channel uses are needed to disseminate each channel coefficient independent from number of unicast hops. Note that we assume ideally that one channel use is enough to disseminate one complex channel coefficient perfectly without any error.

The cluster relay configuration $\mathcal{R} = \{N_{\mathsf{r},1}, \ldots, N_{\mathsf{r},C}\}$ defines the set of the number of relays in each cluster, and its cardinality is equal to C. In the view of definitions above, we present the following lemma which states the optimal cluster configuration that minimizes f_{cu}^{u} for given N_{r} and C.

Lemma 5.2.1: Given N_r and C, the most uniform distribution of relays to clusters is the optimum cluster-relay configuration (OCC) that minimizes f_{cu}^{u} independent of N. Thus, if $\mathrm{mod}(N_r, C) = 0$, we have C clusters each with N_r/C relays; otherwise there are $\mathrm{mod}(N_r, C)$ clusters each with $\lceil N_r/C \rceil$ relays, and $C - \mathrm{mod}(N_r, C)$ clusters each with $\lfloor N_r/C \rfloor$ relays.

Proof: See Appendix A.6.

After stating the optimal cluster formation in order to minimize the channel use for dissemination, next we elaborate on the optimal number of clusters C with the following lemma.

Lemma 5.2.2: Given N_r and the design constraint that relays are distributed to clusters as uniform as possible (according to the Theorem 5.2.1), a network with $C > 0$ clusters needs more channel use for CSI dissemination than any network with $C + \hat{v}$ clusters, where $\hat{v} \in \mathbb{Z}_+$.

Proof: See Appendix A.7.

We assume that the minimum allowable cluster size $N_{\min}$ is determined by the network designer. For instance, $N_{\min}$ can be $N(N-1)+1$ which is the minimum number of relays needed for zero-forcing per cluster. Finally, we summarize this section with the following self-explaining theorem.

Theorem 5.2.1: The minimum channel use for CSI dissemination is attained by the most uniform distribution of relays to $C_{\max}$ clusters, where $C_{\max} = \lfloor N_r/N_{\min} \rfloor$.

Proof: The proof immediately follows from the lemmas 5.2.1 and 5.2.2. □

Multicast Assuming the same conditions as in the unicast case, the amount of channel uses that is needed to disseminate all CSI within the cth cluster with multicast dissemination, is simply written as

$$f_{cu}^{m}(N_{r,c}) = 2NN_{r,c},$$

where the superscript "m" denotes "multicast". The f_{cu}^{m} changes linearly with $N_{r,c}$, and hence, the total amount of channel use is not affected by the size of different clusters.

5.2.3.2. Energy Consumption of Local CSI Dissemination

Let E_c be the sensitivity level of any node, which is the minimum amount of receive energy that any node needs to decode the received symbols correctly. The transmit energy reduces proportionally with $\mathbb{d}^{-\alpha_{\text{pl}}}$, where $\mathbb{d}$ is the distance between the transmitter and the receiver, and $2 \leq \alpha_{\text{pl}} \leq 4$ is the path loss exponent. Thus, the energy consumption for one unit of exchange is $E_{\text{c}}\mathbb{d}^{\alpha_{\text{pl}}}$.

Single-hop Unicast Consider the cth cluster with $N_{\text{r},c}$ relays, which are arbitrarily located and use single-hop unicast communication in between. Then, the total energy consumption of the cth cluster for CSI dissemination is

$$E_{\text{csi}}^{\text{su}}(N_{\text{r},c}) = 2NE_{\text{c}} \sum_{k=1}^{N_{\text{r},c}} \sum_{q=1,q\neq k}^{N_{\text{r},c}} \mathbb{d}_{k,q}^{\alpha_{\text{pl}}} \tag{5.7}$$

where the superscript "su" denotes "single-hop unicast" and $\mathbb{d}_{k,q}$ is the distance between the kth and the qth relay.

Multi-hop Unicast Although it has been presented in the previous section that multiple hops for unicating does not bring any advantage to the total amount of channel use for dissemination, establishing multiple short distance point-to-point links reduces the transmit energy consumption drastically with respect to the to single-hop unicast scheme. We define $\mathscr{R}_k$ as the lowest-energy-optimal route, which contains the set of distances for the hops that the kth relay employs while disseminating its local CSI. Then, the total energy consumption in the cth cluster with multi-hop unicast scheme is given by

$$E_{\text{csi}}^{\text{mu}}(N_{\text{r},c}) = 2NE_{\text{c}} \sum_{k=1}^{N_{\text{r},c}} \sum_{\mathbb{d}\in\mathscr{R}_k} \mathbb{d}^{\alpha_{\text{pl}}}, \tag{5.8}$$

where the superscript "mu" denotes "multi-hop unicast". Since the derivation of this optimal route $\mathscr{R}_k$ is out of the scope of this thesis, we assume that it is readily available and known.

Multicast The total transmit energy consumption of a $N_{\text{r},c}$-relay cluster using multicast communication for CSI dissemination is

$$E_{\text{csi}}^{\text{m}}(N_{\text{r},c}) = 2NE_{\text{c}} \sum_{k=1}^{N_{\text{r},c}} d_{k,\max}^{\alpha_{\text{pl}}}, \tag{5.9}$$

where $d_{k,\max} = \max_q (d_{k,q})$, $\forall\, q \neq k,\ k,q = 1,\ldots,N_{\text{r},c}$.

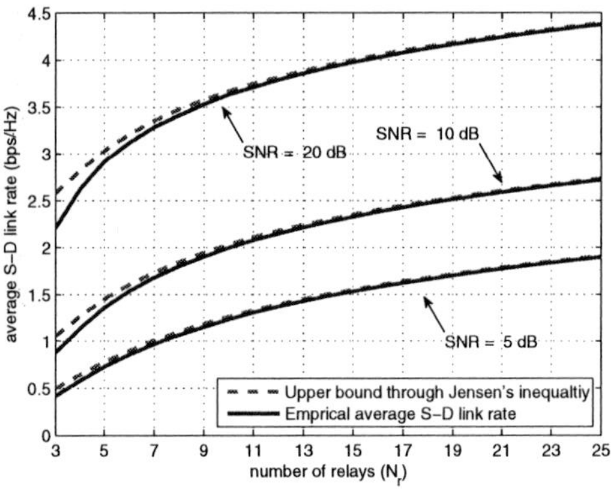

Figure 5.3.: Comparison of average link rates with the corresponding upper bounds for increasing number of relays N_r in a single cluster ($N = 2$ S-D pairs.).

5.2.3.3. Performance of Optimum Cluster Configuration

The proposed cluster based MU system unavoidably suffers from an array gain loss, when compared with MU relaying with global CSI. This is due to the fact that the coherent combination of all relays in the system has been disturbed and reduced to coherent combination of the relays in a single cluster. Thus, the less clusters there are in the system, i.e., more relays per cluster, the higher is the array gain and so is the average rate correspondingly. Besides, the interesting question here is the impact of the optimal cluster configuration proposed in lemma 5.2.1 on the average rate performance of the links.

We claim that the OCC also maximizes the Jensen upper bound of the average link rates, given that all relays are at the same distance to all sources/destinations, cluster powers $P_{r,c}$ are equal for all $c \in \{1, \ldots, C\}$, and N, P_s, $\sigma_{n_r}^2$, and $\sigma_{n_d}^2$ are kept fixed. This claim is proven given in two stages. We have to first prove that the Jensen upper bound of the average link rates grows at most linearly with number of relays. Next, based on this statement, we show that the most uniform distribution of the relays to the clusters maximizes this upper bound.

Using Jensen's inequality, the channel averaged information rate of the ith link is upper bounded by (see (5.5))

$$\mathbb{E}_{\mathbf{H}^{(c)},\mathbf{F}^{(c)}}\left\{\frac{1}{2C}\sum_{c=1}^{C}\log_2\left(1+\{\mathrm{SNR}_{i,c}\}\right)\right\} \leq \frac{1}{2C}\sum_{c=1}^{C}\log_2\left(1+\mathbb{E}_{\mathbf{H}^{(c)},\mathbf{F}^{(c)}}\{\mathrm{SNR}_{i,c}\}\right),$$

where cluster switching is assumed for simplicity. Note that the channel terms in the subscript of the expectation notation denote that the expectation is taken over the channel realizations. We are interested in the behaviour of

$$f_{\mathsf{rate}}(N_{\mathsf{r},c}) = \log_2\left(1+\mathbb{E}_{\mathbf{H}^{(c)},\mathbf{F}^{(c)}}\{\mathrm{SNR}_{i,c}\}\right)$$

as the number of relays $N_{\mathsf{r},c}$ of the cth cluster increases. To this end, we claim that increasing the size of (any) cluster would definitely increase the corresponding f_{rate}, however, the corresponding relative increase would diminish with each additional relay. In other words, we propose that

$$f_{\mathsf{rate}}(M_1) - f_{\mathsf{rate}}(M_2) \leq f_{\mathsf{rate}}(M_2) - f_{\mathsf{rate}}(M_3) \tag{5.10}$$

holds for any set of values

$$\{M_1, M_2, M_3 \in \mathbb{N}\,|M_1 > M_2 > M_3,\, M_1 - M_2 = M_2 - M_3 = 1\}.$$

The assumption in (5.10) is equivalent to

$$\frac{1+\mathbb{E}_{\mathbf{H}^{(c)},\mathbf{F}^{(c)}}\{\mathrm{SNR}_{i,c}\}|_{N_{\mathsf{r},c}=M_1}}{1+\mathbb{E}_{\mathbf{H}^{(c)},\mathbf{F}^{(c)}}\{\mathrm{SNR}_{i,c}\}|_{N_{\mathsf{r},c}=M_2}} \leq \frac{1+\mathbb{E}_{\mathbf{H}^{(c)},\mathbf{F}^{(c)}}\{\mathrm{SNR}_{i,c}\}|_{N_{\mathsf{r},c}=M_2}}{1+\mathbb{E}_{\mathbf{H}^{(c)},\mathbf{F}^{(c)}}\{\mathrm{SNR}_{i,c}\}|_{N_{\mathsf{r},c}=M_3}}. \tag{5.11}$$

In [85], the authors showed that the capacity of a AF relay network can at most increase *linearly* with the number of relays, otherwise it is *sublinear*. Thus, we can deduce that in the most extreme case the linear behaviour of f_{rate} sufficiently means that (5.11) is satisfied with equality. As the rate flattens out with increasing $N_{\mathsf{r},c}$ [85], the inequality in (5.11) will be strictly smaller. In Fig. 5.3, the behaviour of f_{rate} and the strictness of the bound are validated numerically for several SNR values. Based on this numerical validation, the rest of the proof follows similar steps with that of the proof of lemma 5.2.1, and hence presented in Appendix A.8.

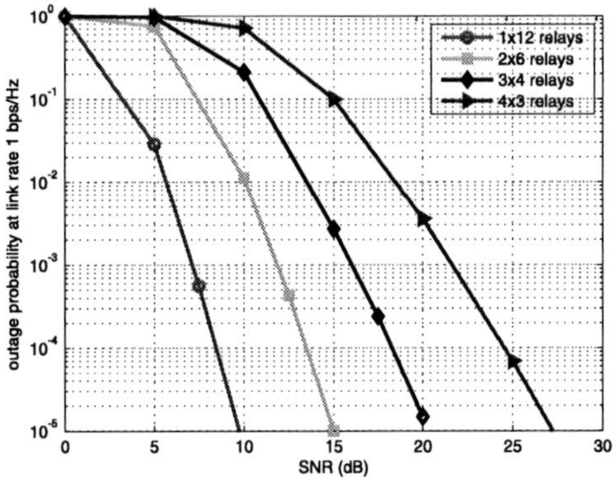

Figure 5.4.: Outage probability at 1 bps/Hz vs. average SNR for different cluster configurations with $N_r = 12$ total relays and $N = 2$ S-D pairs.

5.2.4. Performance Results

Simulation Setup In this section, we present computer simulation results, where we used YALMIP [77] to solve the designed semidefinite problems. Each relay is assumed to have perfect cluster-wide global CSI knowledge. The elements of the channel matrices $\mathbf{H}^{(c)}$ and $\mathbf{F}^{(c)}$ are i.i.d. Rayleigh fading coefficients with zero mean and unit-variance. Furthermore, a local phase reference is assumed to be available within each cluster. We assume $P_\mathsf{s} = P_\mathsf{r} = 1$ and equal relay/destination noise variances $\sigma^2_{n_\mathsf{r}} = \sigma^2_{n_\mathsf{d}}$. Unless otherwise stated, the average SNR is fixed to 20dB. We use cluster switching through all following simulations.

Simulation Results We first study the outage probability performance of the clustered protocol. With the enforced cluster and time specific phase offsets, we expect an effective diversity order of at least as much as the number of clusters in the network. However, in case that there are excess relays per cluster and we choose the relay gain factors through a max-min beamforming approach within each cluster, we may even observe higher gains at the destinations (refer to the optimality discussion in Section 5.2.2). In Fig. 5.4, we show outage probability performance of various different cluster configurations with $N_r = 12$ total relays and $N = 2$ S-D pairs. We assume that each cluster independently employs a max-min zero-forcing gain allocation. We consider four cluster configurations: 4 clusters each with 3 relays, 3 clusters each with 4 relays, 2 clusters each with 6 relays,

1 cluster with 12 relays. The last case of $C = 1$ is equivalent to multiuser relaying with network-wide global CSI knowledge.

Let us first focus to the case of $C = 4$ clusters, where each cluster has as much relays as minimum relay configuration for zero-forcing, i.e., $N(N-1)+1 = 3$. As the minimum relay configuration results in a diversity order of 1, the only source of diversity is through cluster-specific phase rotations. As conjectured, we observe in the Fig. 5.4 that the first configuration of $C = 4$ clusters, achieves an effective diversity order of 4. Next, consider the second cluster configuration of $C = 3$, where we expect third order of diversity to be exploited through cluster phase rotations. However, as now each cluster has excess relays, there is also a possible outage improvement due to max-min optimization. Consequently, we read from the slope of the corresponding figure that in fact there occurs more than 4 order of diversity. As long as we decrease the number clusters and correspondingly increase the number of relays per cluster, we observe a continuous outage improvement (additional diversity). Hence, we identify a trade-off between the two sources of outage improvement, i.e., phase rotations and max-min optimization. Note as well that besides the effective diversity gain, we also obtain an array gain improvement as C decreases. This is due to the improved coherence between clusters.

In Fig. 5.5, we focus on average and outage rate performances of the proposed clustered multiuser relaying network, where we again assume that each cluster uses a max-min ZF gain allocation. As references, we plot the performances of multiuser relaying with network-wide global CSI (with a max-min ZF gain allocation) and asymptotic ZF relaying with only local CSI per relay (see Section 2.4.2). We label each scheme respectively as C-MUR, MUR and AZF. We consider equal-sized clusters each with 9 relays. For any number of clusters C, MUR and AZF use $9C$ relays in total.

Fig. 5.5(a) compares the average S-D link rates. It is immediately realized that C-MUR acts as a balancing proposal between global CSI and local CSI requiring schemes in terms of supplied link rates. The average rate of the C-MUR scheme is indifferent to the additional relay clusters, whereas the two reference schemes improve their rates as C increases. The reason of the insensitive behaviour of C-MUR, is simply because of the employed cluster switching. That is, independent of the number of clusters C, there is always a single cluster of 9 relays that serves the network at each symbol duration. On the other hand, in Fig. 5.5(b), we plot the 1% outage rate performance. In contrary to the average performance, the outage performance of C-MUR improves monotonically with each additional cluster. This improvement arises from the phase

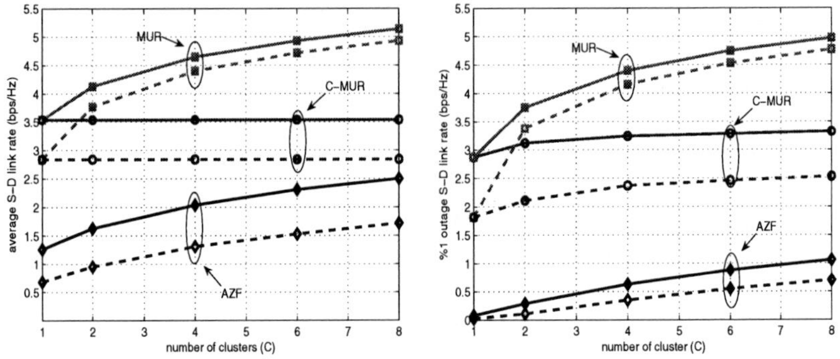

(a) (b)

Figure 5.5.: (a) Average S-D link rates vs. number of clusters C (b) Outage S-D link rates vs. number of clusters C ($N = 2$ (solid lines), $N = 3$ (dashed lines), $N_{\mathsf{r},c} = 9 \forall c$).

diversity, and has already depicted itself through provided effective diversity in Fig. 5.4. Another interesting observation is that AZF performs very poor in outage regime with respect to its counterparts. Nevertheless, C-MUR has an inherent disadvantage of reduced average and outage rates with respect to MUR with global CSI, which is due to the relaxed coherency between the clusters. However, as we will study in the next figure, its advantage evinces itself in terms of drastically reduced local CSI dissemination overhead and transmit energy consumption for CSI dissemination.

Fig. 5.6 depicts this aforementioned trade-off between the information rate performance and the efficiency of resource utilization. First of all, recall that AZF relaying has neither local CSI dissemination overhead nor transmit energy consumption, since there is no need for the relays to disseminate their local CSI. Specifically, Fig. 5.6(a) compares C-MUR and MUR with different dissemination schemes, i.e., multicast and single- and multi-hop unicast, in terms of the transmit energy consumption of CSI dissemination. In order to simplify the analysis in (5.7), (5.8), (5.9), we consider a scenario, where all relays are located on a straight line with equal distances (r) between the neighbouring relays. For instance, such a case simplifies the lowest-energy-optimal route $\mathscr{R}_k$ in multi-hop unicast, and correspondingly boils it down to visit all relays on the way to the destination relay. In general, we notice that the curves for MUR depict much more rapid increase as C increases as opposed to that of C-MUR. The reason is that each

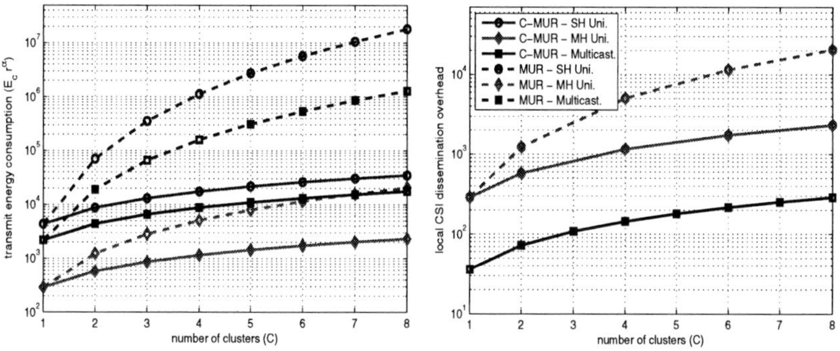

(a) (b)

Figure 5.6.: (a) Transmit energy consumption of CSI dissemination vs. number of clusters C (b) CSI exchange load vs. number of clusters C for unicast and multicast schemes ($N = 2$, $N_{r,c} = 9\forall c$).

additional cluster adds up polynomial load on MUR, whereas this effect is multiplicative for C-MUR. Let us consider the extreme case of $C = 8$ clusters. The C-MUR scheme offers 3, 2, 1 orders of magnitude energy consumption reduction for single-hop unicast, multi-hop unicast, and multicast schemes, respectively, with respect to MUR.

In Fig. 5.6(b), we show the behaviour of local CSI dissemination overhead in terms of required number transmission for increasing number of clusters. As mentioned before in Section 5.2.3.1, single- and multi-hop dissemination schemes result in exactly same overhead for both C-MUR and MUR. However, the CSI exchange overhead of C-MUR with any unicast scheme is only 10 percent of that of MUR for $C = 8$ clusters. In the contrary, both relaying protocols have the same overhead for the multicast dissemination scheme.

Lastly in Fig. 5.7, we present a numerical validation of the proof that the optimum cluster configuration, which aims for equal size of clusters, maximizes the average S-D link rates. The average link rate versus the cluster-relay configuration scenarios is plotted for $C = 2$ and $C = 3$ clusters. The x-axis consists of all possible cluster-relay configurations $\{N_{r,1}, \ldots, N_{r,C}\}$ for given number of relays N_r and clusters C. The maximum average S-D link rates are achieved with the most uniform distribution of relays to clusters which are $\{7, 8\}$, and $\{5, 5, 5\}$ for $C = 2$ and $C = 3$, respectively.

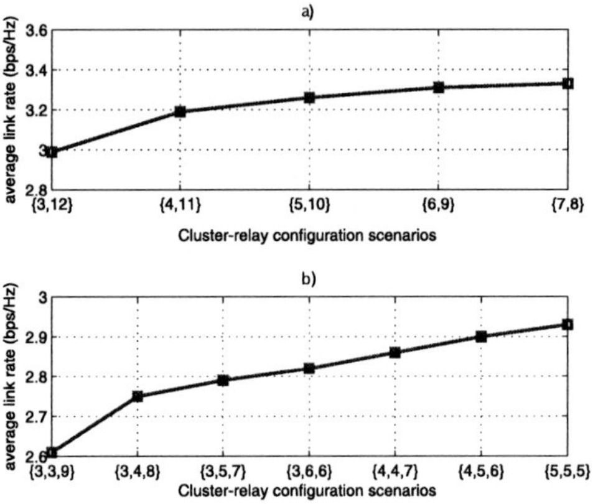

Figure 5.7.: Average rate vs. cluster-relay configuration: a) $C = 2$, b) $C = 3$ for $N_s = 2$, $N_r = 15$.

In summary, we report that relay clustering is an efficient means to drastically reduce the system resources wasted for local CSI dissemination, which is compulsory for coherent relaying. With clustering, we not only improve spectral efficiency (through occupying much less channel-use for dissemination), but also provide longer life-time to the relays by saving from the transmit energy consumption. While attainig these practical gains, we sacrifice from average supportable link rates in between the S-D pairs. However, we recover the lost spatial diversity through phase-rotations per cluster, and provide reliability (effective diversity) to the network.

5.3. A Hierarchical Relaying Protocol with Two Relay Clusters

In the previous section, we considered to group the relays in several independent clusters, and hence, relaxed the CSI dissemination requirement from being global in network-wide to being local in cluster-wide. Despite the tremendous gains on the required channel use and transmit energy for CSI dissemination, there are still two issues that need to be further improved for this clustering approach. First, although the loss in diversity gain is recovered through time and cluster specific phase rotations, there is an inevitable array gain loss due to the non-coherency between clusters. Secondly, we still assume global CSI knowledge within each cluster, where a sufficient number of relays per cluster is needed for efficient management of the interference observed at the destinations.

Addressing these issues in this section, We propose a novel relaying scheme, where hierarchy is introduced into the network such that a group of relays has more channel state information about the network than the rest. That is, relay nodes with different complexities can coexist within the same network. The relays are grouped into two clusters (also called *hierarchical levels* in the sequel) with different CSI requirements and missions to accomplish. In the lower hierarchical level, denoted by $\mathcal{C}_{\text{local}}$, each relay is totally independent of the rest of the network and only needs local CSI for calculating its own gain factor. Hence, even though the relays in $\mathcal{C}_{\text{local}}$ contribute to the array gain at the destinations, they do not have any control on the multiuser interference at the destinations. On the other hand, we provide further channel information in addition to local CSI to the relays in the higher hierarchical level $\mathcal{C}_{\text{global}}$: 1) Local CSI of all other relays within $\mathcal{C}_{\text{global}}$, 2) Very limited feedback about $\mathcal{C}_{\text{local}}$ which is independent of the number of relays in $\mathcal{C}_{\text{local}}$. Contrary to $\mathcal{C}_{\text{local}}$, the relays of $\mathcal{C}_{\text{global}}$ are responsible to choose their gain factors such that the multiuser interference at the destinations is efficiently managed, e.g., multiple communicating pairs are orthogonalized. With such a clustered approach, CSI dissemination overhead scales only with the number of relays in $\mathcal{C}_{\text{global}}$, but not with all relays in the network, i.e., the spatial gains through the relays in $\mathcal{C}_{\text{local}}$ are cost-free in terms of CSI dissemination overhead. Throughout this hierarchical framework, we study power allocation per cluster as well as different distributed beamforming designs, i.e. maximizing minimum link rate or sum rate with and without enforcing an ZF constraint. Moreover, in order to recover the distributed diversity and array gains that had been lost due to clustering, we consider relay selection to form individual clusters by incorporating the instantaneous channel knowledge at each relay. Specifically, we will show that all the degrees of freedom remaining after multiuser interference cancellation, can be fully converted to distributed diversity gain. Aiming at realizing an optimal relay selection in a distributed manner without requiring network-wide global CSI, we finally propose a simple and spectrally efficient relay selection scheme, which operates iteratively while choosing the members of each cluster. To sum up, with drastically reduced CSI dissemination overhead, the proposed hierarchical protocol combined with relay selection will be shown to achieve sum and outage rate performances of conventional coherent multiuser relaying systems requiring *network-wide* global CSI.

Outline of this Section The reminder of this section is organized as follows. The basics of the hierarchial relaying protocol are given in Section 5.3.1. Therein, we state the amount of CSI for each relay per cluster and present the relay gain allocations which result in distributed orthogonalization of multiple S-D destination pairs. We investigate further distributed beamforming approaches in Section 5.3.2, where we consider both max-min and sum rate beamforming schemes with and without ZF constraint. Section 5.3.3 is devoted to the case of imperfect feedback from the $\mathscr{C}_{\text{local}}$ relays, and there we design a worst-case robust gain allocation scheme for $\mathscr{C}_{\text{global}}$. In Section 5.3.4, we focus on relay selection for clusters and elaborate on potential gains to achieve. In the context of this section, we propose a simple and efficient selection scheme that is based on the instantaneous channel conditions and does not require global CSI. Finally, simulation results and corresponding discussions are presented in Section 5.3.5.

5.3.1. Hierarchical Relaying Protocol

We partition the whole relay set $\mathscr{C}_{\text{network}}$ into two main clusters $\mathcal{C}_{\text{local}}$ and $\mathcal{C}_{\text{global}}$, such that $\mathscr{C}_{\text{network}} = \mathcal{C}_{\text{local}} \cup \mathcal{C}_{\text{global}}$. Note that $\mathscr{C}_{\text{network}}$ corresponds to $\mathcal{R}$ in (3.1). The number of relays in $\mathcal{C}_{\text{local}}$ and $\mathcal{C}_{\text{global}}$ are denoted by $N_{\text{r},1}$ and $N_{\text{r},2}$ respectively, such that $N_{\text{r}} = N_{\text{r},1} + N_{\text{r},2}$. The source-to-relay and relay-to-destination channels related to the cluster $\mathcal{C}_{\text{local}}$ are denoted by $\mathbf{H}^{(1)} \in \mathbb{C}^{N_{\text{r},1} \times N}$ and $\mathbf{F}^{(1)} \in \mathbb{C}^{N \times N_{\text{r},1}}$, respectively. Likewise, the corresponding channel matrices of the second cluster $\mathcal{C}_{\text{global}}$ are denoted by $\mathbf{H}^{(2)} \in \mathbb{C}^{N_{\text{r},2} \times N}$ and $\mathbf{F}^{(2)} \in \mathbb{C}^{N \times N_{\text{r},2}}$, respectively (see Fig. 5.8). That is to write, $\mathbf{H} = [(\mathbf{H}^{(1)})^T \ (\mathbf{H}^{(2)})^T]^T$ and $\mathbf{F} = [\mathbf{F}^{(1)} \ \mathbf{F}^{(2)}]$. The relay gain matrices $\mathbf{G}^{(1)} \in \mathbb{C}^{N_{\text{r},1} \times N_{\text{r},1}}$ and $\mathbf{G}^{(2)} \in \mathbb{C}^{N_{\text{r},2} \times N_{\text{r},2}}$ contain the relay gain coefficients of $\mathcal{C}_{\text{local}}$ and $\mathcal{C}_{\text{global}}$ on their

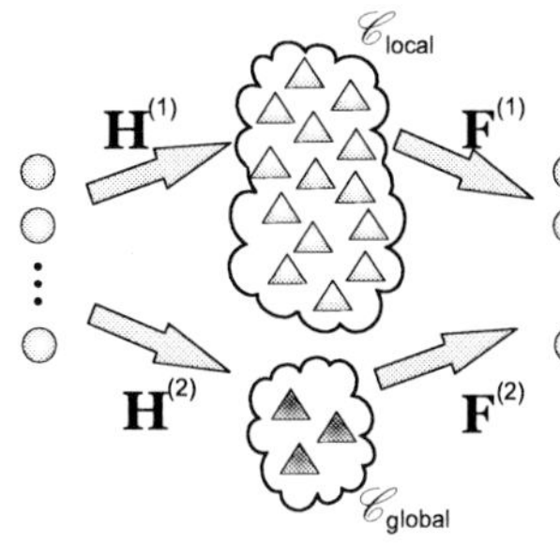

Figure 5.8.: The two-hop network configuration for the hierarchical relaying protocol based on two relay clusters $\mathscr{C}_{\text{global}}$ and $\mathscr{C}_{\text{local}}$.

diagonals, respectively, i.e. $\mathbf{G} = \text{bdiag}\{\mathbf{G}^{(1)}, \mathbf{G}^{(2)}\}$.

The relays are clustered into two hierarchical levels according to the amount of network channel information they have. The relays in $\mathcal{C}_{\text{local}}$ only have local CSI knowledge (lower level of hierarchy). Whereas each relay of $\mathcal{C}_{\text{global}}$ is set to be ranked higher in the hierarchy, and hence, has additional channel information on top of local CSI. These information are

- local CSI of all other relays within $\mathcal{C}_{\text{global}}$, i.e., this refers to global CSI within $\mathcal{C}_{\text{global}}$,
- $N(N-1)$ off-diagonal elements of the equivalent two-hop channel through $\mathcal{C}_{\text{local}}$ at the destinations, i.e.,

$$\mathbf{H}_{\text{srd}}^{(1)} \triangleq \mathbf{F}^{(1)}\mathbf{G}^{(1)}\mathbf{H}^{(1)}.$$

Having access only to local CSI, a member of $\mathcal{C}_{\text{local}}$ can match its received signal from multiple sources appropriately to backward and forward channels [85, 133]. However, with such a channel-matching, it can not manage (or cancel) multiuser interference at the destinations on its own, but only contributes to the received signal power (array gain). In spite of this rather inactive impact on multiuser interference mitigation, each $\mathcal{C}_{\text{local}}$ relay is independent from the other relays in terms of CSI exchange requirement. Hence, the relays in $\mathcal{C}_{\text{local}}$ are cost-free in terms of local CSI dissemination overhead. That is to say, signal contributions of $\mathcal{C}_{\text{local}}$ at the destinations should be supervised such that they do not induce multiuser interference while providing array gain for intended signals. As the name of our proposal implies, the aforementioned active interference management is maintained by $\mathcal{C}_{\mathsf{global}}$ relays which have higher status in the relaying hierarchy and have access to a broader set of CSI knowledge.

In the following, we present a two-phase hierarchical relaying protocol (HRP) through which we explain how to gain the necessary channel information for each relay in the network, and how to choose the relay gain factors for achieving full distributed spatial multiplexing. Before the proposed relaying protocol begins, it is readily assumed that each relay in the network has estimated its own local CSI perfectly. Furthermore, the relays in $\mathcal{C}_{\mathsf{global}}$ have disseminated their local CSI within the cluster. The HRP consists of two phases. In the first phase, which is called *estimation and feedback phase* (EFP), $\mathcal{C}_{\mathsf{global}}$ is silent, whereas $\mathcal{C}_{\mathsf{local}}$ is active and performs a regular two-hop half-duplex relaying routine. The EFP is for estimating $\mathbf{H}_{\mathsf{srd}}^{(1)}$ at the destinations and feeding it back to $\mathcal{C}_{\mathsf{global}}$. In the second phase, which we call the *data transmission phase* (DTP), both clusters are active and simultaneously forward signals from the sources to the destinations. The role of $\mathcal{C}_{\mathsf{global}}$ in the DTP is to manage the multiuser interference at the destinations, and further to optimize its relay gain factors with respect to the chosen figure of merit.

Assuming a low mobility environment, we define a transmission cycle as the duration of the transmission until the channel conditions and relay gain coefficients need to be updated. A transmission cycle lasts for $2(L_1 + L_2)$ symbol intervals, where the first $2L_1$ slots are used for the EFP, whereas the last $2L_2$ slots are occupied for the actual data transmission. For each phase, the sources communicate with relays during the first half of the allocated slots (first hop) and the relays communicate with the destinations in the second half of the allocated slots (second hop). While we are assuming to have constant channel coefficients during $2(L_1 + L_2)$ symbol intervals, we further assume that $L_1 \ll L_2$. That is, the channel uses spent for the estimation and feedback phase do not severely threaten the spectral efficiency of our proposed scheme.

5.3.1.1. Estimation and Feedback Phase

During the first phase, the transmit signal vectors are designed to be orthogonal training sequences, and last for L_1 symbol intervals. While $\mathscr{C}_{\mathsf{global}}$ is silent (not receiving and transmitting) during the EFP, the kth relay, R_k, in $\mathscr{C}_{\mathsf{local}}$ receives an L_1-length block of symbols $\tilde{\mathbf{y}}_{\mathsf{r},k} = \left[y_{\mathsf{r},k}^{(1)} \cdots y_{\mathsf{r},k}^{(L_1)}\right]$ which is expressed as (recall from section 2.3)

$$\tilde{\mathbf{y}}_{\mathsf{r},k} = \mathbf{h}_{\mathsf{s}k}^T \mathbf{S} + \tilde{\mathbf{n}}_{\mathsf{r},k}, \quad \mathsf{r}_k \in \mathcal{C}_{\text{local}},$$

where $\mathbf{S} \in \mathbb{C}^{N\times L_1} = [\mathbf{s}^{(1)} \cdots \mathbf{s}^{(L_1)}]$ is similarly defined as in (5.1), and $\tilde{\mathbf{n}}_{\mathsf{r},k}$ is the corresponding $1 \times L_1$ noise vector at relay r_k. Here and in the sequel, all relay noise elements are assumed to be distributed with $\mathcal{CN}(0, \sigma_{n_\mathsf{r}}^2)$. Before $\tilde{\mathbf{y}}_{\mathsf{r},k}$ is forwarded to the destinations in the next L_1 time slots, each relay multiplies the corresponding element of $\tilde{\mathbf{y}}_{\mathsf{r},k}$ with its complex relay gain factor g_k. This gain factor is based solely on local CSI knowledge per relay and calculated according to *the asymptotic zero-forcing gain allocation* [85, 133] given in Section 2.4.2, i.e.,

$$g_k = \frac{\sqrt{P_{\text{local},k}}}{\sqrt{\sum_{i=1}^{N} |\mathbf{h}_{\mathsf{s}k}^H \mathbf{f}_{k\mathsf{d}}^* \cdot \mathbf{h}_{\mathsf{s}k}[i]|^2 P_\mathsf{s} + |\mathbf{h}_{\mathsf{s}k}^H \mathbf{f}_{k\mathsf{d}}^*|^2 \sigma_{n_\mathsf{r}}^2}} \cdot \mathbf{h}_{\mathsf{s}k}^H \mathbf{f}_{k\mathsf{d}}^*, \quad \text{for } \mathsf{r}_k \in \mathcal{C}_{\text{local}}, \tag{5.12}$$

where the first fractional expression is the scaling to meet the per-node power constraint $P_{\text{local},k}$ and corresponds to ξ in (2.26). We assume that each relay in $\mathscr{C}_{\text{local}}$ transmits with the same instantaneous transmit power, i.e., $P_{\text{local},k} = P_{\text{local}}/N_{\mathsf{r},1} \forall \mathsf{r}_k \in \mathcal{C}_{\text{local}}$. The reason for employing a per-node power constraint in $\mathscr{C}_{\text{local}}$ is to avoid dependence among gain coefficients of other relays in the cluster.

Dropping further details, the L_1-block of destination receive signal vectors $\mathbf{Y}_\mathsf{d} \in \mathbb{C}^{N\times L_1}$ is written as

$$\begin{aligned} \mathbf{Y}_\mathsf{d} &= \mathbf{F}^{(1)}\mathbf{G}^{(1)}\mathbf{H}^{(1)}\mathbf{S} + \mathbf{F}^{(1)}\mathbf{G}^{(1)}\mathbf{N}_{\mathsf{r},1} + \mathbf{N}_\mathsf{d} \\ &= \mathbf{H}_{\mathsf{srd}}^{(1)}\mathbf{S} + \mathbf{F}^{(1)}\mathbf{G}^{(1)}\mathbf{N}_{\mathsf{r},1} + \mathbf{N}_\mathsf{d}. \end{aligned} \tag{5.13}$$

Each destination, say the ith, estimates all N elements of the ith row of $\mathbf{H}_{\mathsf{srd}}^{(1)}$ using the knowledge of the training sequence matrix $\mathbf{S}$ and its column-wise orthogonality. We assume for simplicity that in order to perfectly estimate these N equivalent channel elements, the length of the training sequences L_1 is chosen to be (at least) N, which in return means that the EFP last for $2L_1 = 2N$ symbol intervals.

The last step of EFP is to feedback the necessary information about $\mathbf{H}_{\mathsf{srd}}^{(1)}$ (it can be the full matrix or only off-diagonal elements depending on the chosen beamforming strategy) to the relays of $\mathscr{C}_{\text{global}}$ through a reliable feedback channel. Here and in the sequel, we assume that EFP is error-free and each destination supplies its estimated equivalent channel vector to the $\mathscr{C}_{\text{global}}$ perfectly.

5.3.1.2. Data Transmission Phase

As the first phase is completed and the partial equivalent channel information of $\mathscr{C}_{\text{local}}$ is fed back to the $\mathscr{C}_{\text{global}}$, the actual data transmission phase takes place. In this phase, both clusters aid the communication between S-D pairs following a two-hop relaying routine. Using the global CSI within the cluster and the fed back partial $\mathscr{C}_{\text{local}}$ equivalent CSI knowledge, the activated $\mathscr{C}_{\text{global}}$ relays choose their gain factors such that the multiuser interference between different S/D pairs is managed, e.g., totally removed. In other words, the gain factor of each relay in $\mathscr{C}_{\text{global}}$ is dependent on

- its own local CSI,
- local CSIs of the other relays in $\mathscr{C}_{\text{global}}$,
- and (partially) $\mathbf{H}^{(1)}_{\text{srd}}$.

We impose a sum transmit power of P_{global} on to the $\mathscr{C}_{\text{global}}$, which adds to the total instantaneous power consumption of all relays in the network as $P_{\text{r}} = P_{\text{global}} + P_{\text{local}}$.

Introducing the contribution of the two-hop transmission through $\mathscr{C}_{\text{global}}$ relays into the received signal block $\mathbf{Y}_{\text{d}}$ in (5.13), the received signal block at the destinations during the second phase becomes

$$\begin{aligned}\mathbf{Y}_{\text{d}} &= \underbrace{\mathbf{F}^{(1)}\mathbf{G}^{(1)}\mathbf{H}^{(1)}\mathbf{S} + \mathbf{F}^{(1)}\mathbf{G}^{(1)}\mathbf{N}_{\text{r},1}}_{\text{contribution of local cluster}} + \underbrace{\mathbf{F}^{(2)}\mathbf{G}^{(2)}\mathbf{H}^{(2)}\mathbf{S} + \mathbf{F}^{(2)}\mathbf{G}^{(2)}\mathbf{N}_{\text{r},2}}_{\text{contribution of global cluster}} + \mathbf{N}_{\text{d}} \\ &= \left(\mathbf{H}^{(1)}_{\text{srd}} + \mathbf{H}^{(2)}_{\text{srd}}\right)\mathbf{S} + \mathbf{F}^{(1)}\mathbf{G}^{(1)}\mathbf{N}_{\text{r},1} + \mathbf{F}^{(2)}\mathbf{G}^{(2)}\mathbf{N}_{\text{r},2} + \mathbf{N}_{\text{d}}, \end{aligned} \tag{5.14}$$

where $\mathbf{Y}_{\text{d}} \in \mathbb{C}^{N\times L_2}$, $\mathbf{N}_{\text{r},2} \in \mathbb{C}^{N_{\text{r},2}\times L_2}$ and $\mathbf{H}^{(2)}_{\text{srd}} \in \mathbb{C}^{N\times N} \triangleq \mathbf{F}^{(2)}\mathbf{G}^{(2)}\mathbf{H}^{(2)}$.

Observe in (5.14) that for a given channel realization, $\mathbf{G}^{(2)}$ is the only variable which is not pre-determined and hence, needs to be designed appropriately for the sake of distributed orthogonalization. In other words, using the global CSI within $\mathcal{C}_{\text{global}}$ and (partial) knowledge of $\mathbf{H}^{(1)}_{\text{srd}}$, the activated relays in $\mathcal{C}_{\text{global}}$ choose their gain factors such that the multiuser interference between different S/D pairs is removed, i.e., $\mathbf{H}_{\text{srd}} = \mathbf{H}^{(1)}_{\text{srd}} + \mathbf{H}^{(2)}_{\text{srd}}$ is diagonal. As each of the relays in $\mathcal{C}_{\text{local}}$ is unaware and inactive in the decision of $\mathbf{G}^{(2)}$, at first glance it can be seen doubtful to employ $\mathcal{C}_{\text{local}}$ in the network and to spend a portion of the total transmit power for these relays. However, it will be shown in the following that $\mathcal{C}_{\text{local}}$ can be incorporated constructively to approach the performance of a competitor coherent relaying system, in which each of the N_{r} relays disseminates its local CSI.

We adapt the compound interference matrix definition from Section (2.4.1) to the current clustered scenario and correspondingly define $\mathbf{Z}_{\mathcal{I},2} \in \mathbb{C}^{N(N-1)\times N_{\mathsf{r},2}}$ as the compound interference matrix through the second cluster $\mathscr{C}_{\mathsf{global}}$. It comprises the $N(N-1)$ row vectors $(\mathbf{h}_{jr}^{(2)} \odot \mathbf{f}_{ir}^{(2)})^T$ for all $i, j \in \{1, \ldots, N\}$ and $j \neq i$, where $\mathbf{h}_{jr}^{(2)}$ and $\mathbf{f}_{ir}^{(2)}$ denote the channel vectors associated with the relays only in $\mathscr{C}_{\mathsf{global}}$, i.e. $\mathbf{h}_{jr}^{(2)} = \mathbf{H}^{(2)}\mathbf{e}_j$ and $\mathbf{f}_{ir}^{(2)} = (\mathbf{F}^{(2)})^T\mathbf{e}_i$.

Defining $\mathbf{g}^{(2)} \in \mathbb{C}^{N_{\mathsf{r},2}} \triangleq \mathrm{diag}\{\mathbf{G}^{(2)}\}$, the multiplication $\left(\mathbf{Z}_{\mathcal{I},2} \cdot \mathbf{g}^{(2)}\right)$ delivers a vector containing the equivalent channel coefficients of all multi-user interference terms caused by $\mathscr{C}_{\mathsf{global}}$, the off-diagonal elements of $\mathbf{H}_{\mathsf{srd}}^{(2)}$. Next, we define the vector $\mathbf{z}_{\mathcal{I},1} \in \mathbb{C}^{N(N-1)}$ as the collection of all *off-diagonal* elements in $\mathbf{H}_{\mathsf{srd}}^{(1)}$, i.e., multiuser interference observed at destinations induced by $\mathscr{C}_{\mathsf{local}}$. Combining these, the condition for zero-forcing the total multi-user interference at N destinations, is given by

$$\mathbf{Z}_{\mathcal{I},2} \cdot \mathbf{g}^{(2)} + \mathbf{z}_{\mathcal{I},1} = \mathbf{0}. \tag{5.15}$$

The solvability of (5.15) depends on the rank of $\mathbf{Z}_{\mathcal{I},2}$. If $N_{\mathsf{r},2} \geq N(N-1)$, (5.15) has a solution, whereas multiuser interference can not nulled for $N_{\mathsf{r},2} < N(N-1)$. Here and in the sequel, we call the case with $N_{\mathsf{r},2} = N(N-1)$ relays as the *minimum relay configuration* for $\mathscr{C}_{\mathsf{global}}$ to satisfy the complete interference cancellation at the destinations.

The solution of (5.15) for the case of $N_{\mathsf{r},2} = N(N-1)$ is

$$\mathbf{g}_{\mathsf{zf}}^{(2)} = -\mathbf{Z}_{\mathcal{I},2}^{-1} \cdot \mathbf{z}_{\mathcal{I},1}. \tag{5.16}$$

The computations of $\mathbf{Z}_{\mathcal{I},2}$ and $\mathbf{z}_{\mathcal{I},1}$ require global CSI within $\mathscr{C}_{\mathsf{global}}$ and the knowledge of off-diagonal elements of $\mathbf{H}_{\mathsf{srd}}^{(1)}$, respectively. Note that these channel information have been aforestated for necessary channel information in $\mathscr{C}_{\mathsf{global}}$ for perfect distributed orthogonalization by means of HRP.

Given that P_{r} is fixed, the solution in (5.16) also predefines the individual sum transmission power of each cluster, i.e., P_{global} and P_{local}. The following theorem summarizes this predefined power allocation.

Theorem 5.3.1: Given that $N_{\mathsf{r},2} = N(N-1)$, P_{r} is fixed and the corresponding power constraints are satisfied with equality, there is only one unique cluster transmit power tuple

$$(P_{\mathsf{local}}, P_{\mathsf{global}}) = \left(\frac{P_{\mathsf{r}}}{1+\alpha}, \frac{\alpha P_{\mathsf{r}}}{1+\alpha}\right)$$

where

$$\alpha = \left(\mathbf{Z}_{\mathcal{I},2}^{-1} \cdot \breve{\mathbf{z}}_{\mathcal{I},1}\right)^H \cdot \left(\sigma_{n_r}^2 \mathbf{I}_{N_{r,2}} + P_s\left(\mathbf{H}^{(2)}(\mathbf{H}^{(2)})^H\right) \odot \mathbf{I}_{N_{r,2}}\right) \cdot \left(\mathbf{Z}_{\mathcal{I},2}^{-1} \cdot \breve{\mathbf{z}}_{\mathcal{I},1}\right),$$

and $\breve{\mathbf{z}}_{\mathcal{I},1} = \mathbf{z}_{\mathcal{I},1}/\sqrt{P_{\text{local}}}$ is the normalized interference vector chosen such that $\mathscr{C}_{\text{local}}$ transmits with unit power.

Proof: The proof follows immediately from the definition of the instantaneous transmission power of $\mathscr{C}_{\text{global}}$:

$$\begin{aligned} P_{\text{global}} &= (\mathbf{g}_{\text{zf}}^{(2)})^H \left(\sigma_{n_r}^2 \mathbf{I}_{N_{r,2}} + P_s\left(\mathbf{H}^{(2)}(\mathbf{H}^{(2)})^H\right) \odot \mathbf{I}_{N_{r,2}}\right) \mathbf{g}_{\text{zf}}^{(2)} \\ &= P_{\text{local}} \cdot \left(\mathbf{Z}_{\mathcal{I},2}^{-1} \cdot \breve{\mathbf{z}}_{\mathcal{I},1}\right)^H \cdot \left(\sigma_{n_r}^2 \mathbf{I}_{N_{r,2}} + P_s\left(\mathbf{H}^{(2)}(\mathbf{H}^{(2)})^H\right) \odot \mathbf{I}_{N_{r,2}}\right) \cdot \left(\mathbf{Z}_{\mathcal{I},2}^{-1} \cdot \breve{\mathbf{z}}_{\mathcal{I},1}\right) \\ &= \alpha P_{\text{local}}, \end{aligned} \tag{5.17}$$

where in the second line we inserted the definition of $\mathbf{g}_{\text{zf}}^{(2)}$ from (5.16) and substitute $\breve{\mathbf{z}}_{\mathcal{I},1}$ instead of $\mathbf{z}_{\mathcal{I},1}/\sqrt{P_{\text{local}}}$. Substitution of (5.17) into $P_{\text{local}} + P_{\text{global}} = P_{\text{total}}$ concludes the proof. □

To sum up, with only $N(N-1)$ relays in $\mathscr{C}_{\text{global}}$, which is one relay less than the case in Section 2.4.1, the $\mathscr{C}_{\text{global}}$ can null the interference observed through both clusters. Besides, we benefit from the contributions of local cluster relays without any expense for local CSI dissemination. When compared with a network of all relays stacked in $\mathscr{C}_{\text{local}}$, we drastically improve our beamforming performance by only disseminating $N(N-1)$ relays' local CSI. On the other hand, considering the other extreme case of a network, where all relays are required to exchange local CSI, we intrinsically lose performance due to loss in coherence. However, note our gain in terms of required channel use for CSI dissemination: Now only $N(N-1)$ relays exchanges local CSI instead of N_r, which can be crucial impact in a large AF network of $N_r \gg N(N-1)$. The related performance measures will be later quantified with computer simulations in Section 5.3.5.

Remark 5.3.1: Observe that the structures of the ZF condition (5.15) and the corresponding ZF gain vector (5.16) are identical with that of the out-of-cluster interference mitigation scheme in Section 2.4.3 and the optional second-hop transmission for conventional multiuser relaying in Section 2.4.1. In these sections, we did not further optimize the relay gain vectors for the case of excess number of relays on top of the minimum required for interference cancellation. In the following, we will consider several optimization schemes specific to the problem at hand, and these mathematical formulations can be immediately adapted for the aforementioned two scenarios as well.

Remark 5.3.2: Although we have been assuming that there are no directs link between S-D pairs in the second hop, in case that the effect of direct links would be preferred to be considered in the system, the zero-forcing condition could be modified immediately to $\mathbf{Z}_{\mathcal{I},2} \cdot \mathbf{g}^{(2)} + \mathbf{z}_{\mathcal{I},1} + \mathbf{d}_{\mathcal{I}} = \mathbf{0}$, where $\mathbf{d}_{\mathcal{I}} \in \mathbb{C}^{N(N-1)}$ collects the off-diagonal elements, i.e., interference terms, of the direct source-destination channel matrix $\mathbf{D}$ as defined in Section 2.3.2.2.

5.3.2. Distributed Beamforming Designs for Hierarchical Relaying

In the previous section, we have presented sufficient conditions for ZF interference cancellation and the corresponding relay gain allocation for $\mathcal{C}_{\text{global}}$ with $N_{\text{r},2} = N(N-1)$ relays. However, these do not answer the question of how to choose $\mathbf{G}^{(2)}$ when there are not enough number of relays in $\mathcal{C}_{\text{global}}$ or how to efficiently benefit from excess degrees of freedom when there are more relays in $\mathcal{C}_{\text{global}}$ than necessary. To this end, we shift our focus to further designs of $\mathbf{G}^{(2)}$ in this section. We first focus on ZF beamforming based on the instantaneous SNR and design gain allocation strategies for the sake of maxmin fairness and sum rate maximizations. Afterwards, we extend the optimization framework to the case of general beamforming without the ZF constraint.

5.3.2.1. Zero-forcing Beamforming Designs for $\mathscr{C}_{\text{global}}$

Having more than $N(N-1)$ relays in $\mathscr{C}_{\text{global}}$ results in infinitely many solutions for (5.15). In other words, the nullspace of $\mathbf{Z}_{\mathcal{I},2}$ contains more than the trivial solution, i.e., all zero vector. Defining

$$\mathbf{V}_2 \in \mathbb{C}^{N_{\text{r},2} \times N_{\text{r},2} - N(N-1)} \triangleq \text{null}\{\mathbf{Z}_{\mathcal{I},2}\},$$

any vector $\mathbf{g}_{\text{v}}^{(2)} \in \mathbb{C}^{N_{\text{r},2} - N(N-1)}$ lying in the nullspace $\mathbf{V}_2$ can be added to a particular solution $\mathbf{g}_{\text{o}}^{(2)} \in \mathbb{C}^{N_{\text{r},2}} = -\mathbf{Z}_{\mathcal{I},2}^{\dagger} \cdot \breve{\mathbf{z}}_{\mathcal{I},1}$ of the condition (5.15), such that

$$\mathbf{Z}_{\mathcal{I},2}\mathbf{g}_{\text{zf}}^{(2)} + \mathbf{z}_{\mathcal{I},1} = \mathbf{Z}_{\mathcal{I},2}\Big(\mathbf{g}_{\text{o}}^{(2)} + \mathbf{V}_2\mathbf{g}_{\text{v}}^{(2)}\Big) + \mathbf{z}_{\mathcal{I},1} = \mathbf{0}. \quad (5.18)$$

Hence, in addition to cancelling the multiuser interference, the gain factors can be further optimized for chosen figures of merit such as sum rate, outage rate improvement or fairness.

On the other hand, the vectors $\mathbf{g}_{\mathrm{o}}^{(2)}$, $\mathbf{g}_{\mathrm{v}}^{(2)}$ that are satisfying (5.18), enforce a specific set of feasible cluster transmit power tuples, i.e., $(P_{\mathsf{local}}, P_{\mathsf{global}})$, which is stated by the following theorem.

Theorem 5.3.2: Define $f_{P_{\mathsf{global}}}(\mathbf{g}_{\mathrm{v}}^{(2)}, \mathbf{z}_{\mathcal{I},1})$ as the instantaneous transmit power function of $\mathscr{C}_{\mathsf{global}}$ with $\mathbf{g}_{\mathrm{v}}^{(2)}$ and $\mathbf{z}_{\mathcal{I},1}$ as the arguments. Given that $N_{\mathsf{r},2} > N(N-1)$, and P_{r} is fixed, the feasible cluster transmit powers are bounded as

$$0 \leq P_{\mathsf{local}} \leq \frac{P_{\mathsf{r}}}{1+\eta} \quad \text{and} \quad \frac{\eta P_{\mathsf{r}}}{1+\eta} \leq P_{\mathsf{global}} \leq P_{\mathsf{r}} - P_{\mathsf{local}}, \tag{5.19}$$

where η is the value of the function $f_{P_{\mathsf{global}}}(\mathbf{g}_{\mathrm{v}}^{(2)}, \mathbf{z}_{\mathcal{I},1})$ evaluated at

$$\mathbf{g}_{\mathrm{v}}^{(2)} := \mathbf{g}_{\mathrm{v,min}}^{(2)} = -(\mathbf{V}_2^H \mathbf{M}_2 \mathbf{V}_2)^{-1} \mathbf{V}_2^H \mathbf{M}_2 \mathbf{g}_{\mathrm{o}}^{(2)},$$

$\mathbf{z}_{\mathcal{I},1} := \check{\mathbf{z}}_{\mathcal{I},1}$. The matrix $\mathbf{M}_2 = P_{\mathsf{s}}\Big(\mathbf{H}^{(2)}(\mathbf{H}^{(2)})^H\Big) \odot \mathbf{I}_{N_{\mathsf{r},2}} + \sigma_{n_{\mathsf{r}}}^2 \mathbf{I}_{N_{\mathsf{r},2}}$ is adapted from a related definition in (3.22).

Proof: Since $\mathbf{g}_{\mathrm{o}}^{(2)}$ is an any particular solution to (5.15) for given channel conditions and P_{local}, the instantaneous transmission power of $\mathscr{C}_{\mathsf{global}}$ is a function of both $\mathbf{g}_{\mathrm{v}}^{(2)}$ and $\mathbf{z}_{\mathcal{I},1}$, and is expressed as

$$\begin{aligned} f_{P_{\mathsf{global}}}(\mathbf{g}_{\mathrm{v}}^{(2)}, \mathbf{z}_{\mathcal{I},1}) &= (\mathbf{g}_{\mathrm{o}}^{(2)} + \mathbf{V}_2\mathbf{g}_{\mathrm{v}}^{(2)})^H \mathbf{M}_2 (\mathbf{g}_{\mathrm{o}}^{(2)} + \mathbf{V}_2\mathbf{g}_{\mathrm{v}}^{(2)}) \\ &= (\mathbf{g}_{\mathrm{v}}^{(2)})^H \mathbf{V}_2^H \mathbf{M}_2 \mathbf{V}_2 \mathbf{g}_{\mathrm{v}}^{(2)} + 2\mathrm{Re}\Big\{(\mathbf{g}_{\mathrm{v}}^{(2)})^H \mathbf{V}_2^H \mathbf{M}_2 \mathbf{g}_{\mathrm{o}}^{(2)}\Big\} + (\mathbf{g}_{\mathrm{o}}^{(2)})^H \mathbf{M}_2 \mathbf{g}_{\mathrm{o}}^{(2)}, \end{aligned}$$

where $\mathbf{z}_{\mathcal{I},1}$ is not depicted explicitly, but determines $\mathbf{g}_{\mathrm{o}}^{(2)}$ through (5.16). The matrix $\mathbf{M}_2$ is positive semidefinite by definition; hence for given $\mathbf{z}_{\mathcal{I},1}$, the function $f_{P_{\mathsf{global}}}$ is convex and reaches its minimum value with an argument of

$$\mathbf{g}_{\mathrm{v,min}}^{(2)} = -(\mathbf{V}_2^H \mathbf{M}_2 \mathbf{V}_2)^{-1} \mathbf{V}_2^H \mathbf{M}_2 \mathbf{g}_{\mathrm{o}}^{(2)}.$$

Incorporating this, it can be shown that

$$f_{P_{\mathsf{global}}}(\mathbf{g}_{\mathrm{v,min}}^{(2)}, \mathbf{z}_{\mathcal{I},1}) = P_{\mathsf{local}} f_{P_{\mathsf{global}}}(\mathbf{g}_{\mathrm{v,min}}^{(2)}, \check{\mathbf{z}}_{\mathcal{I},1}) \leq P_{\mathsf{global}},$$

where the latter upper bound comes from the instantaneous sum transmit power constraint imposed on $\mathscr{C}_{\mathsf{global}}$. Next, we define $\eta \triangleq f_{P_{\mathsf{global}}}(\mathbf{g}_{\mathrm{v,min}}^{(2)}, \check{\mathbf{z}}_{\mathcal{I},1})$ for further notational simplicity. We know that if ηP_{local} is smaller than or equal to the design parameter

P_{global}, this leads to the conclusion that there is at least one $\mathbf{g}_{\text{v}}^{(2)}$ that satisfies the instantaneous power constraint. Summing up, we have the following constraints to satisfy

$$\eta P_{\text{local}} \leq P_{\text{global}}, \tag{5.20}$$

$$P_{\text{global}} + P_{\text{local}} = P_{\text{r}}, \tag{5.21}$$

$$P_{\text{global}}, P_{\text{local}} \geq 0. \tag{5.22}$$

Combining and solving the inequalities (5.20-5.22) together, the bounds in (5.19) can be written immediately. □

Signal-to-Noise Ratio Employing the zero-forcing gain vector $\mathbf{g}^{(2)} = \mathbf{g}_{\text{o}}^{(2)} + \mathbf{V}_2\mathbf{g}_{\text{v}}^{(2)}$ for $\mathscr{C}_{\text{global}}$, the instantaneous received (intended) signal power at the ith destination terminal is written through (5.14) as

$$P_{\text{d}_i,\text{signal}}^{\text{zf}} = P_{\text{s}}\Big(\underbrace{(\mathbf{g}_{\text{o}}^{(2)} + \mathbf{V}_2\mathbf{g}_{\text{v}}^{(2)})^H(\mathbf{h}_{ir}^{(2)} \odot \mathbf{f}_{ri}^{(2)})^*(\mathbf{h}_{ir}^{(2)} \odot \mathbf{f}_{ri}^{(2)})^T(\mathbf{g}_{\text{o}}^{(2)} + \mathbf{V}_2\mathbf{g}_{\text{v}}^{(2)})}_{\text{contribution of } \mathscr{C}_{\text{global}}}$$
$$+\underbrace{2\,\text{Re}\Big\{(\mathbf{g}_{\text{o}}^{(2)} + \mathbf{V}_2\mathbf{g}_{\text{v}}^{(2)})^H(\mathbf{h}_{ir}^{(2)} \odot \mathbf{f}_{ri}^{(2)})^*\mathbf{H}_{\text{srd}}^{(1)}[i,i]\Big\}}_{\text{non-coherent joint effect of } \mathscr{C}_{\text{local}} \text{ and } \mathscr{C}_{\text{global}}} + \underbrace{|\mathbf{H}_{\text{srd}}^{(1)}[i,i]|^2}_{\text{contribution of } \mathscr{C}_{\text{local}}}\Big) \tag{5.23}$$

where impact of each cluster on the received signal power is explicitly noted. As the chosen $\mathbf{g}^{(2)} = \mathbf{g}_{\text{o}}^{(2)} + \mathbf{V}_2\mathbf{g}_{\text{v}}^{(2)}$ nulls the interference completely, the corresponding interference terms vanish, i.e.,

$$P_{\text{d}_i,\text{int}}^{\text{zf}} = \sum_{j=1,j\neq i}^{N} P_{\text{s}}\Big((\mathbf{g}_{\text{o}}^{(2)} + \mathbf{V}_2\mathbf{g}_{\text{v}}^{(2)})^H(\mathbf{h}_{jr}^{(2)} \odot \mathbf{f}_{ri}^{(2)})^*(\mathbf{h}_{jr}^{(2)} \odot \mathbf{f}_{ri}^{(2)})^T(\mathbf{g}_{\text{o}}^{(2)} + \mathbf{V}_2\mathbf{g}_{\text{v}}^{(2)})$$
$$+2\,\text{Re}\Big\{(\mathbf{g}_{\text{o}}^{(2)} + \mathbf{V}_2\mathbf{g}_{\text{v}}^{(2)})^H(\mathbf{h}_{jr}^{(2)} \odot \mathbf{f}_{ri}^{(2)})^*\mathbf{H}_{\text{srd}}^{(1)}[i,j]\Big\} + |\mathbf{H}_{\text{srd}}^{(1)}[i,j]|^2\Big) = 0.$$

Note once again that in case of having only $N(N-1)$ relays in $\mathscr{C}_{\text{global}}$, $\mathbf{V}_2$ turns out to be an all-zero matrix, and the term $\mathbf{V}_2\mathbf{g}_{\text{v}}^{(2)}$ drops from the corresponding equations. On other hand, the total noise power at the ith destination is composed of destination local noise plus noise amplified through both $\mathscr{C}_{\text{global}}$ and $\mathscr{C}_{\text{global}}$, i.e.,

$$P_{\text{d}_i,\text{noise}}^{\text{zf}} = \underbrace{\sigma_{n_{\text{d}}}^2}_{\text{destination}} + \underbrace{\tilde{\sigma}_{n_r,i}^2}_{\mathscr{C}_{\text{local}}} + \underbrace{\sigma_{n_r}^2(\mathbf{g}_{\text{o}}^{(2)} + \mathbf{V}_2\mathbf{g}_{\text{v}}^{(2)})^H\tilde{\mathbf{\Gamma}}_i(\mathbf{g}_{\text{o}}^{(2)} + \mathbf{V}_2\mathbf{g}_{\text{v}}^{(2)})}_{\mathscr{C}_{\text{global}}},$$

where $\tilde{\mathbf{\Gamma}}_i = \mathbf{\Gamma}_i^H\mathbf{\Gamma}_i$, $\mathbf{\Gamma}_i = \text{diag}\{\mathbf{f}_{ri}^{(2)}\}$ as defined in Section 3.3.1, and $\tilde{\sigma}_{n_r,i}^2$ is assumed to denote the effective relay-amplified noise impact through $\mathscr{C}_{\text{local}}$ at ith destination.

Finally, we can write the resultant instantaneous SNR at the ith destination as

$$\mathrm{SNR}_i = \frac{P^{\mathrm{zf}}_{\mathsf{d}_i,\mathrm{signal}}}{P^{\mathrm{zf}}_{\mathsf{d}_i,\mathrm{noise}}}. \tag{5.24}$$

For any given feasible power allocation pair $(P_{\mathsf{local}}, P_{\mathsf{global}})$, the additional degrees of freedom in the nullspace $\mathbf{V}_2$, i.e., due to the excess number of relays in $\mathscr{C}_{\mathsf{global}}$, can be used to optimize the relay gain vector $\mathbf{g}_{\mathsf{v}}^{(2)}$ for the sake of SNR-related figures of merit, e.g., fairness between S-D pairs, sum rate or outage rate. However, as the corresponding optimization problems will be dealing with the exact SNR expression, each $\mathscr{C}_{\mathsf{global}}$ relay further needs know full $\mathbf{H}^{(1)}_{\mathsf{srd}}$ instead of only its off-diagonal elements. Furthermore, the effect of amplified noise through $\mathscr{C}_{\mathsf{local}}$, i.e., $\tilde{\sigma}^2_{n_{\mathsf{r}},i}$, should be estimated at each destination, and later the compound vector $\boldsymbol{\sigma}_{n_{\mathsf{r}}} \in \mathbb{R}^N \triangleq [\tilde{\sigma}^2_{n_{\mathsf{r}},1} \cdots \tilde{\sigma}^2_{n_{\mathsf{r}},N}]^T$ should fed back to the $\mathscr{C}_{\mathsf{global}}$. This noise estimation is not considered in the EFP of the protocol in Section 5.3.1.1, and hence, the duration of EFP, L_1, should be extended appropriately.

Max-Min Beamforming Following the corresponding simplification steps for general max-min beamforming problem in (3.10), the max-min beamforming formulation with ZF condition for the current hierarchical relaying scenario can be written as

$$\mathcal{P}^{\mathrm{cluster}}_{\mathrm{mm-zf}}: \quad \max_{\mathbf{g}_{\mathsf{v}}^{(2)},\tau\ge 0} \ \tau \quad \textbf{subject to} \quad \mathrm{SNR}_i \ge \tau \ \forall i$$
$$(\mathbf{g}_{\mathsf{o}}^{(2)} + \mathbf{V}_2\mathbf{g}_{\mathsf{v}}^{(2)})^H\mathbf{M}_2(\mathbf{g}_{\mathsf{o}}^{(2)} + \mathbf{V}_2\mathbf{g}_{\mathsf{v}}^{(2)}) \le P_{\mathsf{global}}.$$

After substituting the SNR definition of (5.24), and making some algebraic manipulations, $\mathcal{P}^{\mathrm{cluster}}_{\mathrm{mm-zf}}$ becomes,

$$\max_{\mathbf{g}_{\mathsf{v}}^{(2)},\tau\ge 0} \ \tau \quad \textbf{subject to} \quad (\mathbf{g}_{\mathsf{v}}^{(2)})^H\breve{\mathbf{A}}_i\mathbf{g}_{\mathsf{v}}^{(2)} + 2\,\mathrm{Re}\{(\mathbf{g}_{\mathsf{v}}^{(2)})^H\breve{\mathbf{a}}_i\} + \breve{a}_i \ge 0 \ \forall i$$
$$(\mathbf{g}_{\mathsf{v}}^{(2)})^H\tilde{\mathbf{M}}_2\mathbf{g}_{\mathsf{v}}^{(2)} + 2\,\mathrm{Re}\{(\mathbf{g}_{\mathsf{v}}^{(2)})^H\tilde{\mathbf{m}}_2\} + \tilde{m}_2 \le P_{\mathsf{global}}, \tag{5.25}$$

where

$$\tilde{\mathbf{M}}_2 = \mathbf{V}_2^H\mathbf{M}_2\mathbf{V}_2,\ \tilde{\mathbf{m}}_2 = \mathbf{V}_2^H\mathbf{M}_2\mathbf{g}_{\mathsf{o}}^{(2)},\ \tilde{m}_2 = (\mathbf{g}_{\mathsf{o}}^{(2)})^H\mathbf{M}_2\mathbf{g}_{\mathsf{o}}^{(2)},$$
$$\breve{\mathbf{A}}_i = \mathbf{V}_2^H\Big((\mathbf{h}_{ir}^{(2)} \odot \mathbf{f}_{ri}^{(2)})^*(\mathbf{h}_{ir}^{(2)} \odot \mathbf{f}_{ri}^{(2)})^T - \tau\sigma^2_{n_{\mathsf{r}}}\tilde{\mathbf{\Gamma}}_i\Big)\mathbf{V}_2,$$
$$\breve{\mathbf{a}}_i = \mathbf{V}_2^H\Big((\mathbf{h}_{ir}^{(2)} \odot \mathbf{f}_{ri}^{(2)})^*(\mathbf{h}_{ir}^{(2)} \odot \mathbf{f}_{ri}^{(2)})^T\mathbf{g}_{\mathsf{o}}^{(2)} + (\mathbf{h}_{ir}^{(2)} \odot \mathbf{f}_{ri}^{(2)})^*\cdot\mathbf{H}^{(1)}_{\mathsf{srd}}[i,i] - \tau\sigma^2_{n_{\mathsf{r}}}\tilde{\mathbf{\Gamma}}_i\mathbf{g}_{\mathsf{o}}^{(2)}\Big),$$
$$\breve{a}_i = (\mathbf{g}_{\mathsf{o}}^{(2)})^H\Big((\mathbf{h}_{ir}^{(2)} \odot \mathbf{f}_{ri}^{(2)})^*(\mathbf{h}_{ir}^{(2)} \odot \mathbf{f}_{ri}^{(2)})^T - \tau\sigma^2_{n_{\mathsf{r}}}\tilde{\mathbf{\Gamma}}_i\Big)\mathbf{g}_{\mathsf{o}}^{(2)} + |\mathbf{H}^{(1)}_{\mathsf{srd}}[i,i]|^2$$
$$-\tau(\tilde{\sigma}^2_{\mathsf{r},i} + \sigma^2_{n_{\mathsf{d}}}) + 2\,\mathrm{Re}\{(\mathbf{g}_{\mathsf{o}}^{(2)})^H(\mathbf{h}_{ir} \odot \mathbf{f}_{ri})^*\cdot\mathbf{H}^{(1)}_{\mathsf{srd}}[i,i]\}.$$

Next, defining $\hat{\mathbf{G}}_{\mathrm{v}}^{(2)} \triangleq \mathbf{g}_{\mathrm{v}}^{(2)}(\mathbf{g}_{\mathrm{v}}^{(2)})^H$ and incorporating the *trace relation* for quadratic functions in (5.25), we obtain

$$\max_{\hat{\mathbf{G}}_{\mathrm{v}}^{(2)}, \mathbf{g}_{\mathrm{v}}^{(2)}, \tau \geq 0} \quad \tau \quad \textbf{subject to} \quad \mathrm{Tr}\left(\breve{\mathbf{A}}_i \hat{\mathbf{G}}_{\mathrm{v}}^{(2)}\right) + 2\,\mathrm{Re}\{(\mathbf{g}_{\mathrm{v}}^{(2)})^H \breve{\mathbf{a}}_i\} + \breve{a}_i \geq 0 \ \forall i$$
$$\mathrm{Tr}\left(\tilde{\mathbf{M}}_2 \hat{\mathbf{G}}_{\mathrm{v}}^{(2)}\right) + 2\,\mathrm{Re}\{(\mathbf{g}_{\mathrm{v}}^{(2)})^H \tilde{\mathbf{m}}_2\} + \tilde{m}_2 \leq P_{\mathrm{global}},$$
$$\mathbf{g}_{\mathrm{v}}^{(2)}(\mathbf{g}_{\mathrm{v}}^{(2)})^H = \hat{\mathbf{G}}_{\mathrm{v}}^{(2)}, \ \hat{\mathbf{G}}_{\mathrm{v}}^{(2)} \succeq 0. \tag{5.26}$$

The equality constraint in (5.26) introduces the requirement that the positive semidefinite matrix $\hat{\mathbf{G}}_{\mathrm{v}}^{(2)}$ is rank-1, and this renders the problem non-convex. Hence, we apply SDR on this constraint, with which $\mathcal{P}_{\mathrm{mm-zf}}^{\mathrm{cluster}}$ is relaxed to the following quasi-convex problem

$$\tilde{\mathcal{P}}_{\mathrm{mm-zf}}^{\mathrm{cluster}}: \max_{\hat{\mathbf{G}}_{\mathrm{v}}^{(2)}, \mathbf{g}_{\mathrm{v}}^{(2)}, \tau \geq 0} \quad \tau \quad \textbf{subject to} \quad \mathrm{Tr}\left(\breve{\mathbf{A}}_i \hat{\mathbf{G}}_{\mathrm{v}}^{(2)}\right) + 2\,\mathrm{Re}\{(\mathbf{g}_{\mathrm{v}}^{(2)})^H \breve{\mathbf{a}}_i\} + \breve{a}_i \geq 0 \ \forall i$$
$$\mathrm{Tr}\left(\tilde{\mathbf{M}}_2 \hat{\mathbf{G}}_{\mathrm{v}}^{(2)}\right) + 2\,\mathrm{Re}\{(\mathbf{g}_{\mathrm{v}}^{(2)})^H \tilde{\mathbf{m}}_2\} + \tilde{m}_2 \leq P_{\mathrm{global}},$$
$$\begin{bmatrix} \hat{\mathbf{G}}_{\mathrm{v}}^{(2)} & \mathbf{g}_{\mathrm{v}}^{(2)} \\ (\mathbf{g}_{\mathrm{v}}^{(2)})^H & 1 \end{bmatrix} \geq 0, \ \hat{\mathbf{G}}_{\mathrm{v}}^{(2)} \succeq 0.$$

The relaxed problem $\tilde{\mathcal{P}}_{\mathrm{zf-mm}}^{\mathrm{cluster}}$ is equivalent to a SDP feasibility check problem when τ is given a priori, and can be efficiently solved by any SDP tool [77, 116]. Thus, we use a bisection algorithm very similar to the one in **Algorithm 2**, where the feasibility check constraints therein are replaced by the constraints of $\tilde{\mathcal{P}}_{\mathrm{zf-mm}}^{\mathrm{cluster}}$. Nevertheless, the optimal result of $\tilde{\mathcal{P}}_{\mathrm{zf-mm}}^{\mathrm{cluster}}$ is a lower bound to the global maxima of $\mathcal{P}_{\mathrm{zf-mm}}^{\mathrm{cluster}}$ due to the SDR. For further details on the tightness of the SDR, interested reader is referred to [7, 41, 58].

Sum Rate Maximization The sum rate maximization problem with the ZF condition is written similar to (3.6) as

$$\mathcal{P}_{\mathrm{sum-zf}}^{\mathrm{cluster}}: \max_{\mathbf{g}_{\mathrm{v}}^{(2)}} \sum_{i=1}^{N} \log_2(1 + \mathrm{SNR}_i)$$
$$\textbf{subject to} \quad (\mathbf{g}_{\mathrm{o}}^{(2)} + \mathbf{V}_2 \mathbf{g}_{\mathrm{v}}^{(2)})^H \mathbf{M}_2 (\mathbf{g}_{\mathrm{o}}^{(2)} + \mathbf{V}_2 \mathbf{g}_{\mathrm{v}}^{(2)}) \leq P_{\mathrm{global}}.$$

Observe that $\mathcal{P}_{\mathrm{sum-zf}}^{\mathrm{cluster}}$ is mathematically in the same structure with that of both problems $\mathcal{P}_{\mathrm{sum}}$ and $\mathcal{P}_{\mathrm{sum-zf}}$ employed for the distributed networks with single-relay-cluster. That is to say,

- objective function of each is a logarithmic sum of several fractions of quadratic functions,
- all problems have quadratic inequality constraints,
- and the optimization variable is a complex vector with a corresponding size, e.g., $\mathbf{g}, \mathbf{g}_{\mathsf{v}}, \mathbf{g}_{\mathsf{v}}^{(2)}$.

In other words, although the input data is different for each, they are numerically solved by following the same steps of logarithmic barrier method introduced in Section 3.2.1.1. Hence, for the sake of brevity, we refer to Section 3.2.1.1 for details, and in the following give only the unconstrained version of the maximization problem $\mathcal{P}_{\mathsf{sum-zf}}^{\mathsf{cluster}}$:

$$\mathbf{g}_{\mathsf{v}}^{(2)\star}(\nu) = \underset{\mathbf{g}_{\mathsf{v}}^{(2)}}{\mathbf{argmin}} \; -\nu \cdot \sum_{i=1}^{N} \log_2(1 + \mathrm{SNR}_i) - \log_e\Big(P_{\mathsf{r}} - (\mathbf{g}_{\mathsf{o}}^{(2)} + \mathbf{V}_2\mathbf{g}_{\mathsf{v}}^{(2)})^H \mathbf{M}_2 (\mathbf{g}_{\mathsf{o}}^{(2)} + \mathbf{V}_2\mathbf{g}_{\mathsf{v}}^{(2)})\Big),$$

where SNR_i is as defined in (5.24). The related necessary derivative expressions can be calculated through the definitions given in Appendix A.2 for generalized quadratic expressions.

Remark 5.3.3: Up until here we have assumed that the gain allocation is optimized for a given fixed cluster power allocation pair $(P_{\mathsf{local}}, P_{\mathsf{global}})$. This fixed power allocation assures that each of C_{local} relays is completely independent from the rest of the network except the first power assignment. However, constricting C_{local} relays' independence, an optimization could also be performed over the feasible set of $(P_{\mathsf{local}}, P_{\mathsf{global}})$ as defined by the Theorem 5.3.2. Although the corresponding formulation is not discussed in this thesis, we report here that such a search for the best cluster power tuple improves the information rate performances only slightly, and hence, dropped here for the sake of brevity.

5.3.2.2. Extension to General Beamforming Designs for $\mathscr{C}_{\mathsf{global}}$

When we drop the ZF constraint and allow some interference at the destinations terminals, the optimization search space for the relay gain vector of $\mathscr{C}_{\mathsf{global}}$ expands. Consequently, the optimal value of any figure of merit trivially improves or at least performs as good as the case with multiuser interference nulling. On the other hand, without the ZF constraint, the beamforming design is free from the minimum number of relays (spatial dimensions) requirement in $\mathscr{C}_{\text{global}}$. Moreover, although marginal (see Section 3.2.3), ZF causes a performance-penalty. Despite the above disadvantages of ZF-beamforming, the problems without the ZF constraint has more optimization (computational) complexity since the size of the complex search vector increased, i.e., instead of $2(N_{\mathsf{r},2} - N(N-1))$ real variables of $\mathbf{g}_{\mathsf{v}}^{(2)}$, we optimize over $2N_{\mathsf{r},2}$ real variables of $\mathbf{g}^{(2)}$. In order to address all of these issues and also to generalize the presented hierarchical framework, we extend the previous findings to the general beamforming designs for $\mathscr{C}_{\text{global}}$ in the following. Note that both beamforming approaches require feedbacks of full $\mathbf{H}_{\mathsf{srd}}^{(1)}$ and variances of the noise amplified through $\mathscr{C}_{\text{local}}$.

The instantenous SINR of the ith S-D link induced by the general beamforming vector $\mathbf{g}^{(2)}$ is written as

$$\text{SINR}_i = \frac{P_{\mathsf{d}_i,\text{signal}}}{P_{\mathsf{d}_i,\text{int}} + P_{\mathsf{d}_i,\text{noise}}}, \tag{5.27}$$

$$\begin{aligned}
P_{\mathsf{d}_i,\text{signal}} &= P_{\mathsf{s}}\Big((\mathbf{g}^{(2)})^H(\mathbf{h}_{ir}^{(2)} \odot \mathbf{f}_{ri}^{(2)})^*(\mathbf{h}_{ir}^{(2)} \odot \mathbf{f}_{ri}^{(2)})^T\mathbf{g}^{(2)} \\
&\qquad +2\text{Re}\Big\{(\mathbf{g}^{(2)})^H(\mathbf{h}_{ir}^{(2)} \odot \mathbf{f}_{ri}^{(2)})^*\mathbf{H}_{\mathsf{srd}}^{(1)}[i,i]\Big\} + |\mathbf{H}_{\mathsf{srd}}^{(1)}[i,i]|^2\Big), \\
P_{\mathsf{d}_i,\text{int}} &= \sum_{j=1,j\neq i}^{N} P_{\mathsf{s}}\Big((\mathbf{g}^{(2)})^H(\mathbf{h}_{jr}^{(2)} \odot \mathbf{f}_{ri}^{(2)})^*(\mathbf{h}_{jr}^{(2)} \odot \mathbf{f}_{ri}^{(2)})^T\mathbf{g}^{(2)} \\
&\qquad +2\text{Re}\Big\{(\mathbf{g}^{(2)})^H(\mathbf{h}_{jr}^{(2)} \odot \mathbf{f}_{ri}^{(2)})^*\mathbf{H}_{\mathsf{srd}}^{(1)}[i,j]\Big\} + |\mathbf{H}_{\mathsf{srd}}^{(1)}[i,j]|^2\Big), \\
P_{\mathsf{d}_i,\text{noise}} &= \sigma_{n_{\mathsf{d}}}^2 + \tilde{\sigma}_{n_{\mathsf{r}},i}^2 + \sigma_{n_{\mathsf{r}}}^2(\mathbf{g}^{(2)})^H\tilde{\mathbf{\Gamma}}_i\mathbf{g}^{(2)}.
\end{aligned}$$

Max-min Beamforming The general max-min beamforming formulation for the current clustered network scenario is given as

$$\mathcal{P}_{\text{mm}}^{\text{cluster}}: \quad \max_{\mathbf{g}^{(2)},\tau\geq 0} \ \tau \quad \textbf{subject to} \quad \text{SINR}_i \geq \tau, \ (\mathbf{g}^{(2)})^H\mathbf{M}_2\mathbf{g}^{(2)} \leq P_{\text{global}}.$$

Skipping the auxiliary derivation lines, we apply the same steps of converting $\mathcal{P}_{\text{mm-zf}}^{\text{cluster}}$ to $\tilde{\mathcal{P}}_{\text{mm-zf}}^{\text{cluster}}$, i.e., substituting the SINR definition of (5.27), defining $\hat{\mathbf{G}}^{(2)} \triangleq \mathbf{g}^{(2)}(\mathbf{g}^{(2)})^H$, introducing the trace relation, and employing SDR, we obtain the relaxed SDP formulation

$$\begin{aligned}
\tilde{\mathcal{P}}_{\text{mm}}^{\text{cluster}}: \quad \max_{\hat{\mathbf{G}}^{(2)},\mathbf{g}^{(2)},\tau\geq 0} \ \tau \quad \textbf{subject to} \quad & \text{Tr}\Big(\breve{\mathbf{A}}_i\hat{\mathbf{G}}^{(2)}\Big) + 2\,\text{Re}\{(\mathbf{g}^{(2)})^H\breve{\mathbf{a}}_i\} + \breve{a}_i \geq 0 \ \forall i \\
& \text{Tr}\Big(\mathbf{M}_2\hat{\mathbf{G}}^{(2)}\Big) \leq P_{\text{global}},
\end{aligned}$$

$$\begin{bmatrix} \hat{\mathbf{G}}^{(2)} & \mathbf{g}^{(2)} \\ (\mathbf{g}^{(2)})^H & 1 \end{bmatrix} \succeq 0, \;\; \hat{\mathbf{G}}^{(2)} \succeq 0,$$

where the definitions of $\breve{\mathbf{A}}_i$, $\breve{\mathbf{a}}_i$, and $\breve{a}_i$ modify to

$$\breve{\mathbf{A}}_i = (\mathbf{h}_{ir}^{(2)} \odot \mathbf{f}_{ri}^{(2)})^*(\mathbf{h}_{ir}^{(2)} \odot \mathbf{f}_{ri}^{(2)})^T - \tau\Bigg(\sum_{j=1,j\neq i}^{N} ((\mathbf{h}_{jr}^{(2)} \odot \mathbf{f}_{ri}^{(2)})^*(\mathbf{h}_{jr}^{(2)} \odot \mathbf{f}_{ri}^{(2)})^T) + \sigma_{n_r}^2 \tilde{\mathbf{\Gamma}}_i\Bigg),$$

$$\breve{\mathbf{a}}_i = (\mathbf{h}_{ir}^{(2)} \odot \mathbf{f}_{ri}^{(2)})^* \cdot \mathbf{H}_{\mathsf{srd}}^{(1)}[i,i] - \tau \sum_{j=1,j\neq i}^{N} (\mathbf{h}_{jr}^{(2)} \odot \mathbf{f}_{ri}^{(2)})^* \cdot \mathbf{H}_{\mathsf{srd}}^{(1)}[i,j],$$

$$\breve{a}_i = |\mathbf{H}_{\mathsf{srd}}^{(1)}[i,i]|^2 - \tau\Bigg(\sum_{j=1,j\neq i}^{N} |\mathbf{H}_{\mathsf{srd}}^{(1)}[i,j]|^2 + \tilde{\sigma}_{\mathsf{r},i}^2 + \sigma_{n_\mathsf{d}}^2\Bigg).$$

Sum Rate Maximization The general sum rate maximization problem for the current clustered network scenario is formulated as

$$\mathcal{P}_{\mathsf{sum}}^{\mathsf{cluster}}: \;\; \max_{\mathbf{g}^{(2)}} \sum_{i=1}^{N} \log_2(1+\mathrm{SINR}_i) \;\; \textbf{subject to} \;\;\;\; (\mathbf{g}^{(2)})^H \mathbf{M}_2 \mathbf{g}^{(2)} \leq P_{\mathrm{global}},$$

which is efficiently solved with the logarithmic barrier method introduced in Section 3.2.1.1 (see the related section of the previous case with ZF constraint).

Recall that the performances of $\mathcal{P}_{\mathrm{mm}}^{\mathsf{cluster}}$ and $\mathcal{P}_{\mathsf{sum}}^{\mathsf{cluster}}$ are upper bounds to that of $\mathcal{P}_{\mathrm{mm-zf}}^{\mathsf{cluster}}$ and $\mathcal{P}_{\mathsf{sum-zf}}^{\mathsf{cluster}}$, respectively. Nevertheless, it can be concluded that there occurs a performance-complexity trade-off between these two beamforming approaches.

5.3.3. Robust Beamforming against Imperfect Feedback

In the previous sections we have considered several gain allocations schemes for $\mathscr{C}_{\mathrm{global}}$, where we either provide sum rate improvement or introduce fairness (outage improvement) to the system while efficiently managing the multiuser interference. Throughout all these designs, we have essentially assumed perfect feedback from the destinations to $\mathscr{C}_{\mathrm{global}}$. Since these feedbacks, i.e., the equivalent channel matrix $\mathbf{H}_{\mathsf{srd}}^{(1)}$ and the vector $\boldsymbol{\sigma}_{n_r}$ of noise variances amplified through $\mathscr{C}_{\mathrm{local}}$, are crucial for interference management, the amount of performance degradation with imperfect feedback emerges as an interesting question. In this section, we aim at designing a robust beamformer that minimizes any possible performance loss that may be caused by the uncertainty in the feedback.

This uncertainty may be the result of an erroneous feedback channel and/or outdated feedback knowledge.

It is assumed that the uncertainty set $\mathcal{U}$ consists all possible realizations of $\mathbf{H}_{\mathsf{srd}}^{(1)}$ and $\boldsymbol{\sigma}_{n_r}$. We adapt a worst-case (max-min) robust optimization approach as proposed in [9, 12, 97], where the optimal decision variable satisfies all possible realizations of the defined uncertainty set. We choose a max-min approach rather than a sum performance, because max-min beamforming's focus of interest, which is the outage regime, is in line with that of worst-case robust approaches. The robust counterpart of the max-min approach is formulated as

$$\mathcal{P}^{\text{fb}}_{\text{robust}}: \quad \max_{\mathbf{g}^{(2)},\tau\geq 0} \ \tau \quad \textbf{subject to} \quad \text{SINR}_i \geq \tau, \ \forall(\mathbf{H}_{\mathsf{srd}}^{(1)}, \boldsymbol{\sigma}_{n_r}) \in \mathcal{U}, \\ (\mathbf{g}^{(2)})^H \mathbf{M}_2 \mathbf{g}^{(2)} \leq P_{\text{global}}, \ i \in \{1, \ldots, N\},$$

where the superscript "fb" denotes "feedback".

In essence, the $\mathscr{C}_{\text{local}}$-amplified relay noise variances are ineffective for interference cancellation, but they rather affect the precision of the optimality of the chosen figure of merit. We can interpret this as that they are effective on the size of the formed beams but not on their direction. Moreover, their impact gradually vanishes out in the mid-to-high average SNR regime. Henceforth, in order to simplify the further analysis, we assume that the $\mathscr{C}_{\text{global}}$ is supplied with perfect knowledge of these noise variances and that the only source of uncertainty is $\mathbf{H}_{\mathsf{srd}}^{(1)}$.

With this simplification, the uncertainty set $\mathcal{U}$ boils down to be $\mathcal{U}_{\mathbf{H}}$, and is defined as the following ellipsoidal uncertainty set

$$\mathcal{U}_{\mathbf{H}} = \left\{ \mathbf{H}_{\mathsf{srd}}^{(1)} = \mathbf{H}_{\mathsf{srd},0}^{(1)} + \sum_{q=1}^{Q_{\mathbf{H}}} \xi_{\mathbf{H},q} \mathbf{H}_{\mathsf{srd},q}^{(1)} \ : \ \|\boldsymbol{\xi}_{\mathbf{H}}\| \leq \rho_{\mathbf{H}} \right\}, \tag{5.28}$$

where $Q_{\mathbf{H}}, \rho_{\mathbf{H}} \in \mathbb{R}_+$, and $\boldsymbol{\xi}_{\mathbf{H}} \triangleq [\xi_{\mathbf{H},1} \cdots \xi_{\mathbf{H},Q_{\mathbf{H}}}]^T$. Consequently, the $\mathcal{P}^{\text{fb}}_{\text{robust}}$ modifies to

$$\tilde{\mathcal{P}}^{\text{fb}}_{\text{robust}}: \quad \max_{\mathbf{g}^{(2)},\tau\geq 0} \ \tau \quad \textbf{subject to} \quad \text{SINR}_i \geq \tau, \ \forall \mathbf{H}_{\mathsf{srd}}^{(1)} \in \mathcal{U}_{\mathbf{H}}, \\ (\mathbf{g}^{(2)})^H \mathbf{M}_2 \mathbf{g}^{(2)} \leq P_{\text{global}}, \ i \in \{1, \ldots, N\}.$$

As the power constraint is independent of the feedbacks, in the following we focus on the SINR constraint and build its computational tractable robust counterpart.

Substituting the definition of uncertain data in (5.28) into the ith SINR constraint as modeled in the first constraint of $\tilde{\mathcal{P}}_{\mathrm{mm}}^{\mathrm{cluster}}$, we obtain

$$\mathrm{Tr}\left(\breve{\mathbf{A}}_i\hat{\mathbf{G}}^{(2)}\right) + 2\,\mathrm{Re}\Bigg\{(\mathbf{g}^{(2)})^H\Bigg((\mathbf{h}_{ir}^{(2)}\odot\mathbf{f}_{ri}^{(2)})^* \cdot \Big(\mathbf{H}_{\mathsf{srd},0}^{(1)}[i,i] + \sum_{q=1}^{Q_{\mathbf{H}}}\xi_{\mathbf{H},q}\mathbf{H}_{\mathsf{srd},q}^{(1)}[i,i]\Big)$$
$$-\tau\sum_{j=1,j\neq i}^{N}(\mathbf{h}_{jr}^{(2)}\odot\mathbf{f}_{ri}^{(2)})^*\Big(\mathbf{H}_{\mathsf{srd},0}^{(1)}[i,j] + \sum_{q=1}^{Q_{\mathbf{H}}}\xi_{\mathbf{H},q}\mathbf{H}_{\mathsf{srd},q}^{(1)}[i,j]\Big)\Bigg)\Bigg\} + \Big|\mathbf{H}_{\mathsf{srd},0}^{(1)}[i,i] + \sum_{q=1}^{Q_{\mathbf{H}}}\xi_{\mathbf{H},q}\mathbf{H}_{\mathsf{srd},q}^{(1)}[i,i]\Big|^2$$
$$-\tau\Bigg(\sum_{j=1,j\neq i}^{N}\Big|\mathbf{H}_{\mathsf{srd},0}^{(1)}[i,j] + \sum_{q=1}^{Q_{\mathbf{H}}}\xi_{\mathbf{H},q}\mathbf{H}_{\mathsf{srd},q}^{(1)}[i,j]\Big|^2 + \tilde{\sigma}_{n_{\mathsf{r}},i}^2 + \sigma_{n_{\mathsf{d}}}^2\Bigg) \geq 0.$$

After some involved algebraic manipulations, the SINR constraint can be written as a quadratic function of $\boldsymbol{\xi}_{\mathbf{H}}$ as

$$\varPhi_i + 2\,\mathrm{Re}\left\{\boldsymbol{\xi}_{\mathbf{H}}^T\breve{\mathbf{h}}_{\mathsf{srd},i}^{(1)}\right\} + \boldsymbol{\xi}_{\mathbf{H}}^T\breve{\mathbf{h}}_{\mathsf{srd},i}^{(1)}(\breve{\mathbf{h}}_{\mathsf{srd},i}^{(1)})^H\boldsymbol{\xi}_{\mathbf{H}}^* \geq 0, \tag{5.29}$$

where

$$\breve{\mathbf{h}}_{\mathsf{srd},i}^{(1)} = \left((\mathbf{g}^{(2)})^H(\mathbf{h}_{ir}^{(2)}\odot\mathbf{f}_{ri}^{(2)})^* + (\mathbf{H}_{\mathsf{srd},0}^{(1)}[i,i])^*\right)\cdot\left[\mathbf{H}_{\mathsf{srd},1}^{(1)}[i,i]\cdots\mathbf{H}_{\mathsf{srd},Q_{\mathbf{H}}}^{(1)}[i,i]\right]^T$$
$$-\tau\sum_{j=1,j\neq i}^{N}\left((\mathbf{g}^{(2)})^H(\mathbf{h}_{jr}^{(2)}\odot\mathbf{f}_{ri}^{(2)})^* + (\mathbf{H}_{\mathsf{srd},0}^{(1)}[i,j])^*\right)\cdot\left[\mathbf{H}_{\mathsf{srd},1}^{(1)}[i,j]\cdots\mathbf{H}_{\mathsf{srd},Q_{\mathbf{H}}}^{(1)}[i,j]\right]^T$$

$$\breve{\mathbf{h}}_{\mathsf{srd},i}^{(1)} = \left[\mathbf{H}_{\mathsf{srd},1}^{(1)}[i,i]\cdots\mathbf{H}_{\mathsf{srd},Q_{\mathbf{H}}}^{(1)}[i,i]\right]^T - \tau\sum_{j=1,j\neq i}^{N}\left[\mathbf{H}_{\mathsf{srd},1}^{(1)}[i,j]\cdots\mathbf{H}_{\mathsf{srd},Q_{\mathbf{H}}}^{(1)}[i,j]\right]^T$$

$$\varPhi_i = \mathrm{Tr}\left(\breve{\mathbf{A}}_i\hat{\mathbf{G}}^{(2)}\right) + 2\mathrm{Re}\Big\{(\mathbf{g}^{(2)})^H(\mathbf{h}_{ir}^{(2)}\odot\mathbf{f}_{ri}^{(2)})^*\cdot\mathbf{H}_{\mathsf{srd},0}^{(1)}[i,i] - \tau\sum_{j=1,j\neq i}^{N}(\mathbf{h}_{jr}^{(2)}\odot\mathbf{f}_{ri}^{(2)})^*\mathbf{H}_{\mathsf{srd},0}^{(1)}[i,j]\Big\}$$
$$+\Big|\mathbf{H}_{\mathsf{srd},0}^{(1)}[i,i]\Big|^2 - \tau\sum_{j=1,j\neq i}^{N}\Big|\mathbf{H}_{\mathsf{srd},0}^{(1)}[i,j]\Big|^2 - \tau\left(\tilde{\sigma}_{n_{\mathsf{r}},i}^2 + \sigma_{n_{\mathsf{d}}}^2\right).$$

As in similar robust problems in Section 4.3, we have two quadratic functions of $\boldsymbol{\xi}_{\mathbf{H}}$, i.e., (5.29) and $\|\boldsymbol{\xi}_{\mathbf{H}}\| \leq \rho_{\mathbf{H}}$, where we want the fulfillment of $\|\boldsymbol{\xi}_{\mathbf{H}}\| \leq \rho_{\mathbf{H}}$ to imply the non-negativity of (5.29). To this end, we apply the *S-lemma* on these two quadratic constraints and express them in terms of the following matrix inequality

$$\begin{bmatrix} \varPhi_i & (\breve{\mathbf{h}}_{\mathsf{srd},i}^{(1)})^H \\ \breve{\mathbf{h}}_{\mathsf{srd},i}^{(1)} & \breve{\mathbf{h}}_{\mathsf{srd},i}^{(1)}(\breve{\mathbf{h}}_{\mathsf{srd},i}^{(1)})^H \end{bmatrix} - \tilde{\zeta}_i\begin{bmatrix} \rho_{\mathbf{H}}^2 & \mathbf{0} \\ \mathbf{0} & -\mathbf{I} \end{bmatrix} \succeq 0, \tag{5.30}$$

for some $\tilde{\zeta}_i \in \mathbb{R}_+$. If the positive semi-definiteness constraint (5.30) is satisfied for a given τ, an uncertainty size $\rho_{\mathbf{H}}$, and a relay gain vector $\mathbf{g}^{(2)}$, this is interpreted as that none of the realizations of the uncertainty can lead an SINR worse than τ. This clause can be simply checked through an SDP feasibility problem through the SDP toolbox YALMIP [26,77,123].

Finally, combining all derivations above with the power constraint as modeled in the second constraint of $\tilde{\mathcal{P}}_{\text{mm}}^{\text{cluster}}$, we write the final SDP for $\tilde{\mathcal{P}}_{\text{robust}}^{\text{fb}}$ as

$$\begin{aligned}\tilde{\mathcal{P}}_{\text{robust}}^{\text{fb}}: \quad & \max_{\hat{\mathbf{G}}^{(2)},\mathbf{g}^{(2)},\tau\geq 0} \quad \tau \\ & \text{subject to} \quad \begin{bmatrix} \Phi_i & (\breve{\mathbf{h}}_{\mathsf{srd},i}^{(1)})^H \\ \breve{\mathbf{h}}_{\mathsf{srd},i}^{(1)} & \breve{\mathbf{h}}_{\mathsf{srd},i}^{(1)}(\breve{\mathbf{h}}_{\mathsf{srd},i}^{(1)})^H \end{bmatrix} - \tilde{\zeta}_i \begin{bmatrix} \rho_{\mathbf{H}}^2 & \mathbf{0} \\ \mathbf{0} & -\mathbf{I} \end{bmatrix} \succeq 0, \\ & \qquad \begin{bmatrix} \hat{\mathbf{G}}^{(2)} & \mathbf{g}^{(2)} \\ (\mathbf{g}^{(2)})^H & 1 \end{bmatrix} \geq 0, \ \operatorname{Tr}\left(\mathbf{M}_2\hat{\mathbf{G}}^{(2)}\right) \leq P_{\text{global}}, \\ & \qquad \hat{\mathbf{G}}^{(2)} \succeq 0, \ \tilde{\zeta}_i \in \mathbb{R}_+, \ i \in \{1,\ldots,N\},\end{aligned}$$

which can be efficiently solved with the bisection method combined with SDP feasibility checks at each iteration (see Section 3.2.2 for further details of the iterative numerical algorithm). The problem $\tilde{\mathcal{P}}_{\text{robust}}^{\text{fb}}$ assures that for any realizations of $\mathbf{H}_{\mathsf{srd}}^{(1)}$ the minimum SINR over all links will be at least equal to the optimized worst-case robust SINR value $\tau^\star$. Nevertheless, note that $\tau^\star$ is a lower bound to the maximal achievable value due to the relaxation applied on $\hat{\mathbf{G}}^{(2)}$(see the second constraint).

Worst-case performance of a given gain vector Another mission of the derived robust counterpart for the SINR in (5.30) is that we can use it to compute the minimum (over all links) worst-case SINR for a given relay gain allocation. In other words, instead of searching for the gain vector leading the best worst-case performance, we can search for the worst-case performance for a given gain vector. We have already noted that if the robust counterpart of SINR is active, i.e., the linear inequality (5.30) is feasible for a given τ, this means that τ is an achievable SINR for all uncertainty realizations. By searching for the maximal τ with which (5.30) is still active, we can find the worst-case SINR performance of the given gain vector for the ith S-D link. Incorporating all SINR constraints, we can also find the minimum worst-case SINR performance for a given relay gain allocation vector $\mathbf{g}^{(2)}$ (assuming that it fulfills given power constraint) through the quasi-convex SDP problem:

$$\tilde{\mathcal{P}}_{\text{robust}}^{\text{fb-worst}}: \quad \max_{\tau\geq 0} \ \tau \ \text{subject to} \ \begin{bmatrix} \Phi_i & (\breve{\mathbf{h}}_{\mathsf{srd},i}^{(1)})^H \\ \breve{\mathbf{h}}_{\mathsf{srd},i}^{(1)} & \breve{\mathbf{h}}_{\mathsf{srd},i}^{(1)}(\breve{\mathbf{h}}_{\mathsf{srd},i}^{(1)})^H \end{bmatrix} - \tilde{\zeta}_i \begin{bmatrix} \rho_{\mathbf{H}}^2 & \mathbf{0} \\ \mathbf{0} & -\mathbf{I} \end{bmatrix} \succeq 0, \quad (5.31)$$

$$i = 1, \dots, N,$$

which can be solved with bisection algorithm combined with SDP feasibility checks.

5.3.4. Relay Selection for Clusters

The previously designed gain allocations assume that the cluster sets $\mathscr{C}_{\mathsf{local}}$ and $\mathscr{C}_{\mathsf{global}}$ are predefined and fixed. The assignment of relays to clusters can be inherently dependent on geographical conditions of the communication area or topological conditions of different relays with respect to each other. Hence, assigning relays which are in the close proximity of each other to the same cluster would be an efficient and realistic approach for minimizing CSI exchange overhead. Such a fixed (and/or random) cluster assignment may not suffice to achieve the sum and outage rate (diversity) performances of conventional coherent multiuser relaying with network-wide global CSI requirement. However, as we assume a slow fading environment, the estimated instantaneous channel conditions can be incorporated to decide on the best cluster structures to increase the performance of chosen beamforming strategy. In the following, we consider two figures of merit for relay selection: sum rate, i.e. $\mathrm{R}_{\mathsf{sum}} = \sum_{i=1}^{N} \mathrm{R}_i$, and minimum S/D link rate, i.e. $\mathrm{R}_{\mathsf{min}} = \min_i \mathrm{R}_i$, aiming at outage rate improvement.

5.3.4.1. Achievable Gains through Relay Selection

The optimal relay selection for clusters requires a master node, which has the global CSI throughout the relay network. In this centralized method, the master node considers all possible cluster formations for $\mathscr{C}_{\mathsf{global}}$, which are in total $\binom{N_{\mathsf{r}}}{N_{\mathsf{r},2}}$. Having the global CSI, the master node can calculate the gain allocation vector, and the corresponding supplied link rates for all possible cluster formations. If relay selection aims at maximizing sum rate, the selection process is written as

$$\mathrm{R}^{\mathrm{rs}}_{\mathsf{sum}}(\mathcal{H}) = \max_{\{\mathcal{C}_{\mathrm{local}},\mathcal{C}_{\mathrm{global}}\}\subset\mathscr{C}_{\mathrm{network}}} \min_{1\leq i\leq N} \mathrm{R}_{\mathsf{sum}}(\mathcal{H}, \mathcal{C}_{\mathrm{local}}, \mathcal{C}_{\mathrm{global}}), \tag{5.32}$$

where $\mathcal{H}$ denotes the whole set of channel realizations and the superscript "rs" denotes "relay selection". If the minimum link rate is desired to be maximized, the selection process becomes

$$\mathrm{R}^{\mathrm{rs}}_{\mathsf{min}}(\mathcal{H}) = \max_{\{\mathcal{C}_{\mathrm{local}},\mathcal{C}_{\mathrm{global}}\}\subset\mathscr{C}_{\mathrm{network}}} \min_{1\leq i\leq N} \mathrm{R}_i(\mathcal{H}, \mathcal{C}_{\mathrm{local}}, \mathcal{C}_{\mathrm{global}}). \tag{5.33}$$

The sum rate $\mathrm{R}_{\mathrm{sum}}(\mathcal{H}, \mathcal{C}_{\mathrm{local}}, \mathcal{C}_{\mathrm{global}})$ or the ith link rate $\mathrm{R}_i(\mathcal{H}, \mathcal{C}_{\mathrm{local}}, \mathcal{C}_{\mathrm{global}})$ is calculated for given $\mathcal{H}$ and cluster sets $\mathcal{C}_{\mathrm{local}}, \mathcal{C}_{\mathrm{global}}$. For relay gain allocation, either directly (5.16) or any of the appropriate beamforming designs proposed in Section 5.3.2 is employed.

Summing up, the master node chooses the best candidates for $\mathscr{C}_{\mathrm{global}}$ using either of (5.32) or (5.33), and the rest of the relays are assigned to $\mathscr{C}_{\mathrm{local}}$. Although optimal, such kind of a global search algorithm is highly complex, and hence, it can only serve as an impractical upper bound for relay selection. Moreover, assuming a master node with global relay network CSI is nothing, but converting the system to conventional coherent multiuser AF relaying with global CSI.

Effective Diversity Gain The max-min relay selection procedure inherently aims at the outage rate regime, and hence, indirectly at the effective distributed diversity gain. In the sequel, we quantify the effective diversity gain as the slope of the outage probability versus average SNR curve (see the related Section 3.2.4). Employing a fixed and random sets $\mathcal{C}_{\mathrm{local}}, \mathcal{C}_{\mathrm{global}}$ with $|\mathcal{C}_{\mathrm{global}}| = N(N-1)$ would result in no diversity, i.e., diversity of 1, independent of the number of relays in $\mathscr{C}_{\mathrm{local}}$. One of the reasons for such a dramatic diversity loss is naturally the assumption of $|\mathcal{C}_{\mathrm{global}}| = N(N-1)$, where all available spatial degrees of freedom is spent to null the multiuser interference. It is trivial to obtain outage improvement (and hence effective diversity) by increasing the number of relays in $\mathscr{C}_{\mathrm{global}}$ and employing a max-min beamformer. Alternatively here, we invoke relay selection for clusters in order to recover the loss in distributed diversity with a practical CSI dissemination overhead.

However, even with $|\mathcal{C}_{\mathrm{global}}| = N(N-1)$, one may expect additional diversity emerging from the local relay clusters as their interference effect is cancelled by $\mathscr{C}_{\mathrm{global}}$. Nevertheless, the limiting fact is that the signal contribution of each relay in $\mathscr{C}_{\mathrm{local}}$ has two parts: coherent and non-coherent. Specifically, let us focus on the intended signal contribution of any kth relay in $\mathscr{C}_{\mathrm{local}}$ at the ith destination, which is written with (5.12) as

$$\xi h_{k,i} \mathbf{h}_{\mathsf{s}k}^{H} \mathbf{f}_{k\mathsf{d}}^{*} f_{i,k} s_i = \underbrace{\xi |h_{k,i}|^2 |f_{i,k}|^2 s_i}_{\text{coherent}} + \underbrace{\xi h_{k,i} f_{i,k} \sum_{j=1, j \neq i}^{N} h_{k,j}^{*} f_{j,k}^{*} s_i}_{\text{non-coherent}}, \tag{5.34}$$

where ξ is a real factor of power scaling. In (5.34) the uncontrolled non-coherent additions can have any direction and hence, ruin the coherent contributions. To sum up, we invoke relay selection for clusters in order to recover the losses in distributed effective diversity with a practical CSI dissemination overhead.

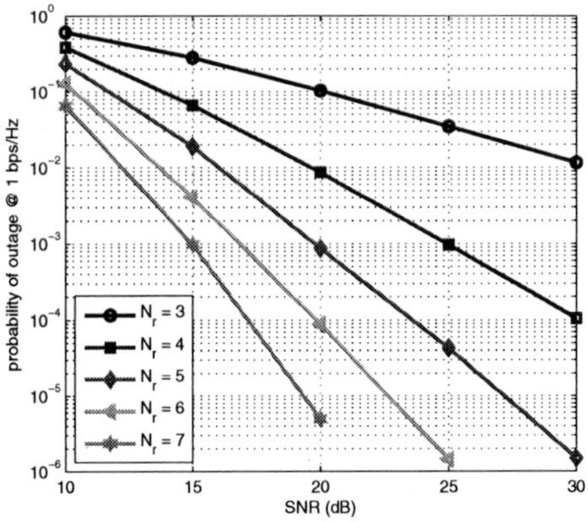

Figure 5.9.: Outage probability at 1 bps/Hz vs. SNR for $N = 2$ S-D links, and $N_{\mathsf{r},2} = 2$ (minimum relay configuration for $\mathscr{C}_{\text{global}}$). The centralized relay selection for clusters is used to maximize the minimum S/D link rate, i.e., the selection process in (5.33).

Introducing the cluster sets $\mathcal{C}_{\text{local}}, \mathcal{C}_{\text{global}}$ in (3.1), the outage probability definition modifies to

$$P_{\text{outage}}(\mathcal{H}, \mathcal{C}_{\text{local}}, \mathcal{C}_{\text{global}}) = \Pr\Big(\min_{1 \le i \le N} \mathrm{R}_i(\mathcal{H}, \mathcal{C}_{\text{local}}, \mathcal{C}_{\text{global}}) \le \mathrm{R}_{\text{outage}}\Big), \tag{5.35}$$

where the gain vector is excluded as it is uniquely determined by given sets. Further, incorporating relay selection for clusters, the outage probability definition modifies to

$$P^{\text{rs}}_{\text{outage}}(\mathcal{H}) = \Pr\Big(\max_{\{\mathcal{C}_{\text{local}}, \mathcal{C}_{\text{global}}\} \subset \mathscr{C}_{\text{network}}} \min_{1 \le i \le N} \mathrm{R}_i(\mathcal{H}, \mathcal{C}_{\text{local}}, \mathcal{C}_{\text{global}}) \le \mathrm{R}_{\text{outage}}\Big). \tag{5.36}$$

With this optimal search over possible relay cluster sets, we maximize the worst-link rate for given $\mathcal{H}$, and hence minimize the "instantaneous" outage probability. We conjecture without any explicit spatial diversity gain proofs defined for infinite average SNR that such a selection diversity can be turned into full effective distributed diversity. This claim translates as searching over $\binom{N_{\mathsf{r}}}{N_{\mathsf{r},2}}$ possible combinations result in an effective diversity of $N_{\mathsf{r}} - N(N-1)$, i.e., we spent $N(N-1)$ degrees of freedom out of available N_{r} for interference mitigation and the rest for outage improvement.

Numerical Simulations In order to verify the aforementioned claim, we present computer simulation results in Fig. 5.9. We plot outage probability for $\mathrm{R}_{\text{outage}} = 1\text{bps/Hz}$

versus average SNR curves for $N = 2$ S-D links and $N_{r,2} = 2$ global cluster relays. We simulate a set of $N_r \in \{3, 4, 5, 6, 7\}$, where we expect to observe an effective diversity gains from 1 to 5, respectively. It is observed in Fig. 5.9 that adding more relays to the cluster C_{local}, which does not induce any overhead to the system, provides additional degrees of freedom to the system, and this is then converted to a distributed diversity gain through relay selection. All outage probability curves confirm our claim of full effective diversity $N_r - N(N-1)$. However, we note that the relay selection does not provide any outage improvement for the special case that there are $N_r = N(N-1)+1$ available relays and the best $N(N-1)$ of these are aimed to be chosen, i.e., $|\mathcal{C}_{local}| = 1$. More specifically, all of the $N(N-1)+1$ possible configurations result exactly the same transmission rates for each S/D link, i.e., the clustered scheme boils down to be a single cluster network in Section 3.2. This fact is revealed by noting that fixing any "one" relay's gain to a constant (as it is in local cluster), the optimal zero-forcing gain vector only differs by a scaling factor.

Remark 5.3.4: For any $N_{r,2} \neq N(N-1)$ and given cluster sets, the $\mathbf{g}^{(2)}$ can be optimized to maximize the minimum link rate as explained in the related sections. This would improve the outage rate as relay selection does. Whereas either through the optimization of relay gains or relay selection, the maximum achievable effective diversity is restricted with maximum available degrees of freedom remaining from interference cancellation.

5.3.4.2. A Simple Relay Selection Scheme

The optimal relay cluster searches fundamentally assume that global network CSI is available at a master node, which contradicts with the essential motivation of the proposed hierarchical relaying protocol. To this end, we propose here a simple and efficient decentralized protocol, which assigns the relays to clusters iteratively for any $N_{r,2}$ in $\mathscr{C}_{global}$. A slightly modified version of the two-phase communication protocol presented in Section 5.3.1 is employed for this selection.

Initially, we assume that all N_r relays in the network are members of $\mathscr{C}_{local}$ cluster and each relay has perfect *local CSI*, i.e. $\mathcal{C}_{local} := \mathscr{C}_{network}$, $\mathcal{C}_{global} := \emptyset$. Following a two-hop training sequence transmission procedure, the destinations estimate the corresponding row of the equivalent channel $\mathbf{H}_{srd}$ of the whole relay network. The perfectly estimated full $\mathbf{H}_{srd}$ is later fed back to all relays in $\mathscr{C}_{local}$.

Each relay $r_k \in \mathcal{C}_{local}$ can determine the *reduced* two-hop equivalent matrix through all the relays except itself, i.e. $\mathcal{C}_{local} \backslash \{r_k\}$, by simply subtracting its own local CSI information from the fed back $\mathbf{H}_{\mathsf{srd}}$, i.e. $\hat{\mathbf{H}}_{\mathsf{srd},k} = \mathbf{H}_{\mathsf{srd}} - g_k \mathbf{h}_{\mathsf{s}k} \mathbf{f}_{k\mathsf{d}}^T$. Incorporating $\hat{\mathbf{H}}_{\mathsf{srd},k}$, each single relay in $\mathscr{C}_{\mathsf{local}}$ computes its gain factor independently from others such that it is the only member of $\mathscr{C}_{\mathsf{global}}$, i.e. $N_{\mathsf{r},2} = 1$, and the rest of the relays are in $\mathscr{C}_{\mathsf{local}}$, i.e. $N_{\mathsf{r},1} = N_{\mathsf{r}} - 1$. While determining this gain factor, the relays employ the general beamforming design of either $\tilde{\mathcal{P}}_{\mathrm{mm}}^{\mathsf{cluster}}$ or $\mathcal{P}_{\mathsf{sum}}^{\mathsf{cluster}}$ depending on the chosen figure of merit, i.e. maximizing the minimum link rate or sum rate, respectively. Moreover, the relays use the same pre-defined cluster power constraints. Consequently, the relays calculate the value of the pre-determined figure of merit, and transmit this value to the master destination, which can be any one destination terminal. In this transmission, the relays benefit from a prioritized carrier sense multiple access (P-CSMA) protocol, so that only one single channel use is required. The master destination informs the relay with the best value for the chosen figure of merit, say r_{k_1}. Henceforth, the corresponding relay broadcasts its local CSI to all other relays in $\mathcal{C}_{\mathsf{local}} := \mathcal{C}_{\mathsf{local}} \backslash \{\mathsf{r}_{k_1}\}$. To sum up, at the end of the first iteration, the first member of $\mathscr{C}_{\mathsf{global}}$ is chosen and its local CSI is shared with the rest of the relays.

In the second iteration, each relay $\mathsf{r}_k \in \mathcal{C}_{\mathsf{local}}$ individually determines the equivalent channel of the relays in $\mathcal{C}_{\mathsf{local}} \backslash \{\mathsf{r}_k\}$ by subtracting the combination of its own and the previously chosen r_{k_1}'s local CSI information from $\mathbf{H}_{\mathsf{srd}}$, i.e.

$$\hat{\mathbf{H}}_{\mathsf{srd},k} = \mathbf{H}_{\mathsf{srd}} - g_k \mathbf{h}_{\mathsf{s}k} \mathbf{f}_{k\mathsf{d}}^T - g_{k_1} \mathbf{h}_{\mathsf{s}k_1} \mathbf{f}_{k_1\mathsf{d}}^T.$$

Next, r_k computes its relay gain factor and the corresponding value for the chosen figure of merit, assuming that $\mathscr{C}_{\mathsf{global}}$ is composed of itself and r_{k_1}, i.e., $N_{\mathsf{r},2} = 2$ and the rest of the relays are in $\mathscr{C}_{\mathsf{local}}$, i.e. $N_{\mathsf{r},1} = N_{\mathsf{r}} - 2$. As soon as the computed values for the desired figure of merit are transmitted to the master destination through a P-CSMA protocol, the best relay in $\mathcal{C}_{\mathsf{local}}$ is chosen, say r_{k_2}. After updating the relay sets $\mathcal{C}_{\mathsf{global}} := \mathcal{C}_{\mathsf{global}} \cup \{\mathsf{r}_{k_2}\}$ and $\mathcal{C}_{\mathsf{local}} := \mathcal{C}_{\mathsf{local}} \backslash \{\mathsf{r}_{k_2}\}$, R_2 disseminates its local CSI to $\mathscr{C}_{\mathsf{local}}$.

Following similar steps, we iterate until $N_{\mathsf{r},2}$ relays are chosen for the $\mathscr{C}_{\mathsf{global}}$, and then we stop the relay selection algorithm. The overall relay selection algorithm is summarized in Algorithm 4. Notice that at the end of the relay selection protocol, the clusters are chosen, the global CSI within $\mathscr{C}_{\mathsf{global}}$ is attained, and the necessary equivalent channel knowledge of the $\mathscr{C}_{\mathsf{local}}$ is readily provided to the relays of $\mathscr{C}_{\mathsf{global}}$. At this point we would like to highlight the followings:

1. Employing simple figures of merit that are related with S(I)NR, the proposed relay selection algorithm is efficient in terms of channel occupancy, i.e., it only needs

Algorithm 4 Iterative Relay Selection Scheme

initiate Set $\mathcal{C}_{\text{local}} = \mathscr{C}_{\text{network}}$, $\mathcal{C}_{\text{global}} = \emptyset$ and $\ell = 1$.
Estimate $\mathbf{H}_{\text{srd}}$ through $\mathscr{C}_{\text{local}}$ as in EFP and feed it back to $\mathscr{C}_{\text{local}}$.

repeat In the ℓth iteration:

Step 1. All $\mathsf{r}_k \in \mathcal{C}_{\text{local}}$ computes the reduced equivalent channel:

$$\hat{\mathbf{H}}_{\text{srd},k} \Leftarrow \mathbf{H}_{\text{srd}} - g_k \mathbf{h}_{\text{s}k} \mathbf{f}_{k\text{d}}^T - \sum_{\rho \in \mathcal{C}_{\text{global}}} g_\rho \mathbf{h}_{\text{s}\rho} \mathbf{f}_{\rho\text{d}}^T.$$

Step 2. Each $\mathsf{r}_k \in \mathcal{C}_{\text{local}}$ individually solves $\tilde{\mathcal{P}}_{\text{mm}}^{\text{cluster}}$ or $\mathcal{P}_{\text{sum}}^{\text{cluster}}$ as if clusters are formed as $\tilde{\mathcal{C}}_{\text{local}} = \mathcal{C}_{\text{local}} \backslash \mathsf{r}_k$ and $\tilde{\mathcal{C}}_{\text{global}} = \mathscr{C}_{\text{network}} \backslash \tilde{\mathcal{C}}_{\text{local}}$.

Step 3. All relays simultaneously transmit the optimal value of $\tilde{\mathcal{P}}_{\text{mm}}^{\text{cluster}}$ or $\mathcal{P}_{\text{sum}}^{\text{cluster}}$ to the master destination through a P-CSMA protocol.

Step 4. The destination chooses the best relay, say $\mathsf{r}_{k_e ll}$, and broadcasts its index number.

Step 5. Update $\mathcal{C}_{\text{local}} \Leftarrow \mathcal{C}_{\text{local}} \backslash \{\mathsf{r}_{k_e ll}\}$ and $\mathcal{C}_{\text{global}} \Leftarrow \mathcal{C}_{\text{global}} \cup \{\mathsf{r}_{k_e ll}\}$.

Step 6. $\mathsf{r}_{k_e ll}$ disseminates its local CSI to $\mathscr{C}_{\text{local}}$.

until the required number of $\mathscr{C}_{\text{global}}$ relays is chosen.

$N_{\text{r},2}$ relay-to-destination, destination-to-relay and relay-to-relay channel uses, respectively.

2. If $N_{\text{r},2} \neq N(N-1)$, the relays need to know the variances of the amplified noise through other relays, in order to compute the beamformers efficiently. However, since each relay only knows its effect on the amplified noise at the destinations, but not the others', such a global noise variance knowledge would require additional channel uses for noise estimation and feedback to the relays. Hence, we assume that S(I)NR expressions can be optimized dropping the unknown noise terms, which would result in very marginal performance losses especially for low relay noise levels.

3. Once the relay selection is performed as explained above, all necessary channel information has already been supplied to the relays. Hence, the actual data transmission can start immediately by skipping the EFP in the original protocol.

5.3.5. Performance Results

5.3.5.1. Simulation Setup

In this section, we discuss the performance of the proposed hierarchical relaying protocol and the effect of relay selection for clusters through Monte-Carlo simulations. As before, we used YALMIP [77] to solve the designed semidefinite problems. The elements of the channel matrices $\mathbf{H}^{(1)}$, $\mathbf{H}^{(2)}$, $\mathbf{F}^{(1)}$ and $\mathbf{F}^{(2)}$ are i.i.d. Rayleigh fading coefficients with zero mean and unit-variance. Furthermore, a local phase reference is assumed throughout the network. We fix $P_\mathsf{s} = 1$ and assume equal relay/destination noise variances $\sigma^2 \triangleq \sigma^2_{n_\mathsf{r}} = \sigma^2_{n_\mathsf{d}}$. Unless otherwise stated, we fix $P_\mathsf{r} = NP_\mathsf{s}$. Further, the power allocation to individual relay clusters are determined according to Theorems 5.3.1 and 5.3.2.

5.3.5.2. Simulation Results

Performance with Relay Selection: We first study the average sum rate performance of our proposal and compare it to that of global CSI requiring ZF solutions. Fig. 5.11 shows the average sum rate performance versus the average SNR for $N_\mathsf{r} = 15$. While simulating the clustered protocol, which we denote in the figures as HRP, we assume minimum relay configuration for $\mathscr{C}_{\text{global}}$. That is, $N_{\mathsf{r},1} = N(N-1)$ and all remaining relays are in $\mathscr{C}_{\text{local}}$, i.e., $N_{\mathsf{r},2} = 15 - N(N-1)$. We consider two cases for HRP in which we either choose the members of $\mathscr{C}_{\text{global}}$ randomly or apply relay selection (RS). In the relay selection, adapting the assignment of relay clusters to the estimated instantaneous channel coefficients, we perform an optimal global search over all relays to find the best configuration that maximizes the instantaneous sum rate. We compare the clustered protocol with the conventional single-relay-cluster protocol, where we consider the relay gain decisions ZF-1 and ZF-2 introduced in Section 2.5. In both reference designs, all 15 relays are employed.

In Fig. 5.10(a) we consider a scenario of $N = 2$ S-D pairs. We immediately realize that HRP without relay selection provides a substantial improvement over ZF-1, whereas it falls behind of ZF-2 in performance. In essence, it may not be fair to compare ZF-2 with HRP, since ZF-2 employs the excess relays efficiently through a smart gain choice of maximum ratio combining. As both HRP and ZF-1 are constrained with spatial dimensions, and directly choose a random (or the only available) relay gain vector out of the nullspace, we believe that such a comparison would make more sense. Never-

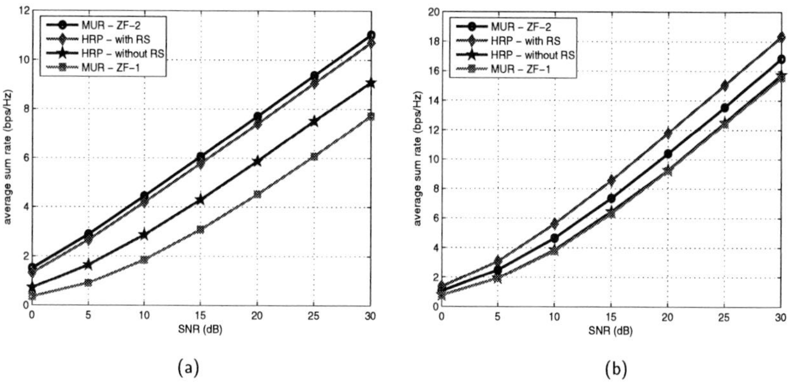

Figure 5.10.: Average sum rate vs. average SNR for $N_r = 15$ relays: (a) $N = 2$ S-D pairs, (b) $N = 4$ S-D pairs.

theless, applying optimal relay selection for clusters in HRP, the sum rate performance approaches that of ZF-2. Note also the considerable impact of relay selection on HRP, e.g., an improvement of almost 2 bps/Hz.

As we increase the number of S-D pairs to $N = 4$, the respective performances of different design choices change behaviour. For example, HRP and ZF-1 performs almost identical for all SNR regime. The basic reason is that now there are only 3 relays in $\mathscr{C}_{\text{local}}$, i.e., minimum relay configuration for $\mathscr{C}_{\text{global}}$ is 12 relays for $N = 4$, whose additional array gain impact is limited at the destinations. Note that a scenario of $(N_{r,1}, N_{r,2}) = (1, 12)$ for HRP would result nothing but exactly the same performance with ZF-1. On the other hand, a more interesting result is that HRP with relay selection considerably outperforms ZF-2, which is a contradiction to the behaviour of the $N = 2$ case.

In summary, we report that HRP act as a powerful alternative for ZF-1. Moreover, when augmented with relay selection, its performance competes with that of ZF-1. Essentially, recall the fundamental advantage of HRP that while single-relay-cluster based designs require all 15 relays to disseminate local CSI, HRP needs only 2 (or 12) relays for $N = 2$ (or 4) pairs to distribute local CSI. Although we admit that relay selection requires additional CSI knowledge, this overhead can be substantially reduced through the proposed simple relay selection method in Section 5.3.4.2. We further study

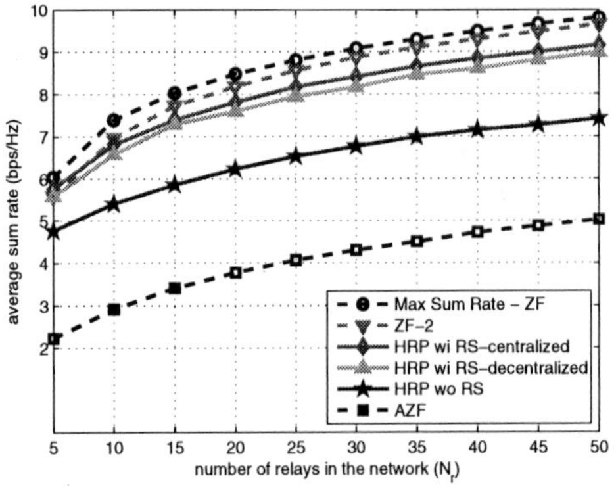

Figure 5.11.: Average sum rates vs. number of relays N_r for $N = 2$ S-D links, $N_{r,2} = 2$, SNR = 20 dB.

the performance of relay selection in the next figure.

Fig. 5.11 provides an overview of the comparative sum rate performances for single- and multi-relay clusters networks. In addition to the previous reference schemes, the asymptotic zero-forcing relaying with only local CSI requirement per relay (no CSI dissemination overhead), serves as a lower-bound for the sum rate. As an upper bound, we consider the conventional single-relay-cluster multiuser relaying with network-wide global CSI, where the ZF relay gains are chosen such that sum rate is maximized, i.e., through $\mathcal{P}_{\text{sum-zf}}$ in Section 3.2.1.2. Moreover, three variants of HRP are simulated: without relay selection, with centralized relay selection (global brute-force search) and with decentralized relay selection (proposed iterative relay selection method). We consider a scenario of $N = 2$ S-D pairs and hence, the proposed heterogeneous relaying protocol uses $N_{r,2} = 2$ relays in $\mathscr{C}_{\text{global}}$, and the remanining $N_r - 2$ relays in $\mathscr{C}_{\text{local}}$.

The figure depicts that using only the minimum number of relays necessary to orthogonalize S/D links, HRP without RS, i.e., random and fixed cluster assignment, halves the gap between the two bounding reference systems with almost no CSI exchange with respect to coherent multiuser relaying with network-wide global CSI. Applying an exhaustive search for relay selection, the sum rate performance of HRP approaches close to that of single-relay-cluster proposals. Nevertheless, as pointed out before, such an exhaustive search suffers from local CSI exchange overhead. Fortunately, the proposed decentralized selection scheme (green curve) shows a prosperous performance to achieve the sum rate measures of the centralized search. Besides that "almost" optimal relay

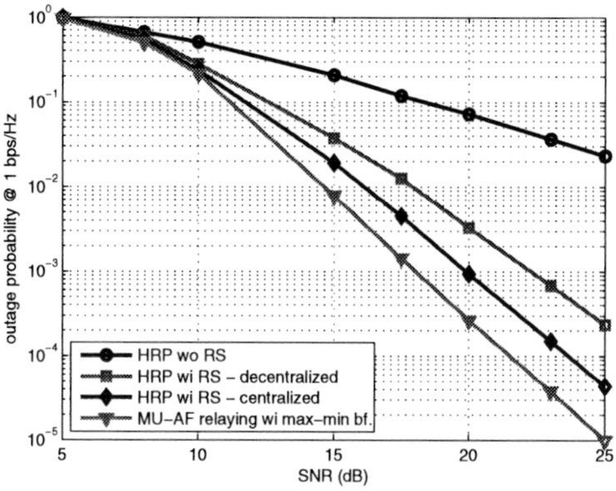

Figure 5.12.: Outage probability at 1 bps/Hz vs. average SNR for $N = 2$ S-D links, and $N_{r,2} = 2$, $N_r = 5$ relays.

clusters are found in a distributed manner, only two iterations ($N(N-1)$ in general) are needed for relay selection. In contrary to the centralized relay selection approach, only one relay ($N(N-1)-1$ in general) out of N_r disseminates its local CSI to the others. Comparing specifically the sum rate performances of conventional multiuser relaying and the proposed HRP with decentralized relay selection (green-curve), we notice that the latter performs almost 1 bps/Hz worse than the former for all N_r values. However, the CSI dissemination overhead for the former scales with N_r^2, whereas independent of N_r, HRP requires only a single relay to broadcasts its local CSI to the others. We shortly conclude that HRP releases the multiuser network from paying spatial deficiency penalty for each additional relay due to the overhead of CSI dissemination.

In the previous figure, we have shown that the proposed suboptimal relay selection scheme performs quite satisfactory to improve sum rate with a very limited overhead. On the other hand, we have also shown in Fig. 5.9 that relay selection with centralized search provides full diversity. Hence, the next question to answer is how the suboptimal relay selection scheme performs in terms of outage. In Fig. 5.12, we study the outage probability at 1 bps/Hz for a scenario of $N = 2$ S-D pairs and $N_r = 5$ relaying in the network. Consequently, we expect a maximum effective diversity gain of three. As references, we plot the outage performances of the optimal centralized relay selection scheme and a random relay cluster choice. We observe that our decentralized relay selection method achieves an effective diversity (see the slope of the outage curve) that only slightly differs from that of the centralized approach. However, note the difference at the outage probability for a given target average SNR, where we are faced with an array gain loss. On the other hand, we also present the outage performance of the conventional single-relay-cluster scheme employing *max-min* ZF beamforming with all $N_r = 5$ relays (Section 3.2.1.2). It is noticed that although both relaying architectures achieve the same (maximum achievable) effective diversity, there is a parallel shift for the curve of HRP, i.e. higher outage probabilities, due to the loss in full coherence in between all the relays in the network.

Distributed Beamforming: Considering our HRP proposal, we have derived sum rate and max-min rate oriented beamforming solutions with and without ZF constraint for gain allocation of $\mathscr{C}_{\text{global}}$ throughout Section 5.3.2. We already know from the discussions and similar simulations in the previous chapters that max-min type gain allocations provides fairness in link rates over sum-rate maximization and that removing the ZF constraint provides additional improvement on the rate performances. Instead, our aim here is rather to study the trade-off between employing the clusters $\mathscr{C}_{\text{local}}$ and $\mathscr{C}_{\text{global}}$.

In Fig. 5.13, we try to answer the following question: How many relays in $\mathscr{C}_{\text{local}}$ can compensate the removal of relays from $\mathscr{C}_{\text{global}}$? To this end, we consider a scenario of $N = 2$ S-D pairs, $N_{r,1} = 8$ relays in $\mathscr{C}_{\text{local}}$ and $N_{r,2} = 2$ relays in $\mathscr{C}_{\text{global}}$, i.e. minimum relay configuration for $\mathscr{C}_{\text{global}}$. Further, we assume that the total relay network power P_r is shared equally between clusters. We consider 8 additional relays to join the network. We compare the sum rate performance of two different cases: all 8 relays joins either to $\mathscr{C}_{\text{local}}$ or to $\mathscr{C}_{\text{global}}$ one by one. The sum rate for excess relays in $\mathscr{C}_{\text{global}}$ is computed through the general sum-rate beamforming $\mathcal{P}_{\text{sum}}^{\text{cluster}}$. As depicted in Fig. 5.13, 8 additional relays in $\mathscr{C}_{\text{local}}$ improves the sum rate for about 0.8 bps/Hz (from about 5.4 to 6.2bps/Hz), whereas collecting these relays within $\mathscr{C}_{\text{global}}$ provides a sum rate gain of 2.4 bps/Hz. This result can be interpreted from two different aspects. First, adding all relays to $\mathscr{C}_{\text{local}}$ provides sum rate improvement without any need for CSI dissemination or for informing the other relays. However, the impact on sum rate is rather limited, since the full coherence within relays can not be achieved. Secondly, although further CSI dissemination is required, adding relays to $\mathscr{C}_{\text{global}}$ provides substantial gains. Another interesting question to answer is that how many relays are needed in $\mathscr{C}_{\text{local}}$ to compensate the sum rate performance coming with each additional relay in $\mathscr{C}_{\text{global}}$. The answer to this question is given in the right sub-figure of Fig. 5.13. Let us focus on a specific example: In order to achieve the performance of 4 additional relays in $\mathscr{C}_{\text{global}}$, i.e.,

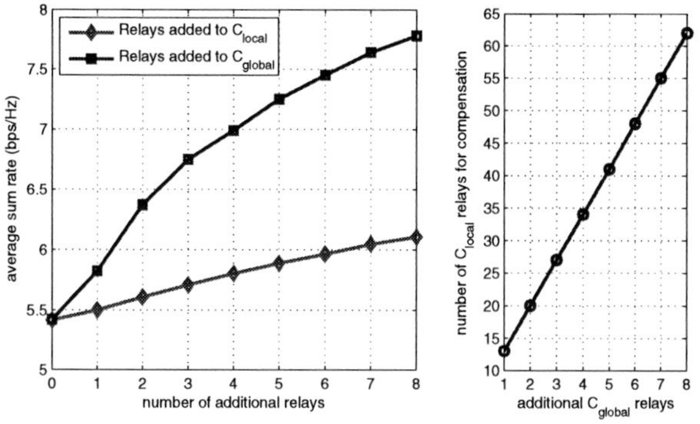

Figure 5.13.: Average sum rate trade-off between adding relays to $\mathscr{C}_{\text{local}}$ or to $\mathscr{C}_{\text{global}}$.

$N_{\mathsf{r},1} = 8$ and $N_{\mathsf{r},2} = 6$, we need to have 34 relays in $\mathscr{C}_{\text{local}}$, i.e. $N_{\mathsf{r},1} = 34$ and $N_{\mathsf{r},2} = 2$, meaning that we need to add 26 additional relays to $\mathscr{C}_{\text{local}}$. Having a closer look at the respective curve, we observe that for each additional relay in $\mathscr{C}_{\text{global}}$ we need to add 7 relays to $\mathscr{C}_{\text{local}}$.

In Fig. 5.14 we modify our scenario such that we fix the total number of relays within the network to $N_{\mathsf{r}} = 15$ and compare the sum rate performance of different $(N_{\mathsf{r},1}, N_{\mathsf{r},2})$ relay configurations. Initially we assume that there are $N_{\mathsf{r},2} = 2$ relays in $\mathscr{C}_{\text{global}}$, and all the remaining of relays are in $\mathscr{C}_{\text{local}}$. We sequentially take one relay out of $\mathscr{C}_{\text{local}}$ and add it to $\mathscr{C}_{\text{global}}$. Hence, at the end we reach an empty $\mathscr{C}_{\text{local}}$ and a full $\mathscr{C}_{\text{global}}$, which boils down to the conventional single-cluster multiuser relaying proposal with network-wide global CSI. As upper and lower bounds we consider the cases of Sum Rate Max with global CSI and AZF with local CSI through all $N_{\mathsf{r}} = 15$ relays, respectively. Focusing first on the minimum relay configuration case $(N_{\mathsf{r},2} = 2)$, we once again observe that with only two relays exchanging CSI, the sum rate is substantially improved with respect to the lower bound. Recall that the feasible set of cluster power allocations is described with Theorem 5.3.2. We simulate two feasible sets for cluster power allocations $(P_{\text{local}}, P_{\text{global}})$, where the clusters either share the total power $P_{\mathsf{r}} = 2$ equally or the global cluster is favored through $(P_{\text{local}}, P_{\text{global}}) = (0.5, 1.5)$. For both cases, there is a rate jump from the relay configuration $(N_{\mathsf{r},1} = 1, N_{\mathsf{r},1} = 14)$ to $(N_{\mathsf{r},1} = 0, N_{\mathsf{r},1} = 15)$. This due to the aforementioned fact that the pre-set cluster powers are not optimal to maximize the sum rate. In other words, there is still a room to optimize the cluster power allocations per instantaneous channel realization, which we omit not to introduce a power dependence

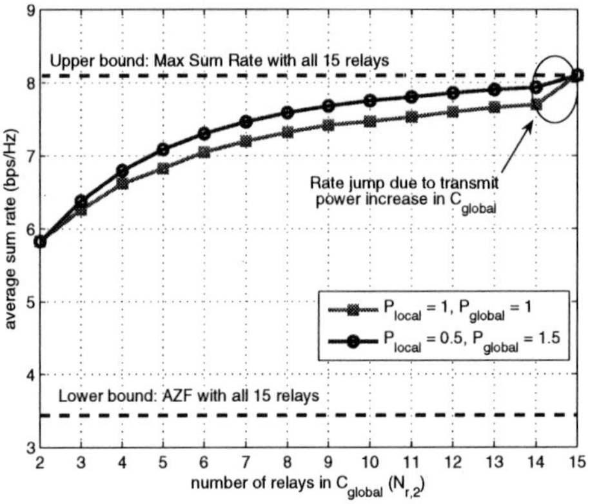

Figure 5.14.: Average sum rate comparison between different relay cluster ($\mathscr{C}_{\text{local}}$,$\mathscr{C}_{\text{global}}$) configurations.

in between clusters at each time that propagation channel changes. In summary, this figure provides a trade-off between the two clusters, through which we can decide the relay configurations for each cluster to achieve a desired network sum rate.

Robust Optimization: In all of the above discussions, the feedback from destinations plays a crucial real in terms of efficient interference mitigation. Specifically, the off-diagonal elements of the fedback two-hop equivalent channel of $\mathscr{C}_{\text{local}}$, $\mathbf{H}_{\mathsf{srd}}^{(1)}$, determine the accuracy of multiuser interference cancellation, and the equivalent $\mathscr{C}_{\text{local}}$ amplified relay noise vector $\boldsymbol{\sigma}_{n_{\mathsf{r}}}$ affects the optimality of the beamforming solutions. In order to account any possible imperfection with these feedbacks, in Section 5.3.3, we have designed a worst-case robust max-min beamformer $\tilde{\mathcal{P}}_{\text{robust}}^{\text{fb}}$, where we assumed $\mathbf{H}_{\mathsf{srd}}^{(1)}$ as the only source of imperfection. In Fig. 5.15, we investigate the impact of this robust design on the minimum worst-case SINRs (the minimum is taken over S-D links).

As a non-robust benchmark, we use the max-min beamforming problem without ZF given by $\tilde{\mathcal{P}}_{\text{mm}}^{\text{cluster}}$, where we ignore the presence of uncertainty. The minimum worst-case SINR through this non-robust gain design can be calculated through $\tilde{\mathcal{P}}_{\text{robust}}^{\text{fb-worst}}$ in (5.31). We consider a network composed of $N = 2$ S-D pairs and $N_{\mathsf{r}} = 10$ relays. We set $Q_{\mathbf{H}} = 1$ and fix the perturbation to be equal to the nominal data. Further, we scale each realization as explained in Section 4.3.4 so that $\rho_{\mathbf{H}} = 0$ and $\rho_{\mathbf{H}} = 1$ correspond to an equal size of perturbation as the nominal data and no uncertainty, respectively.

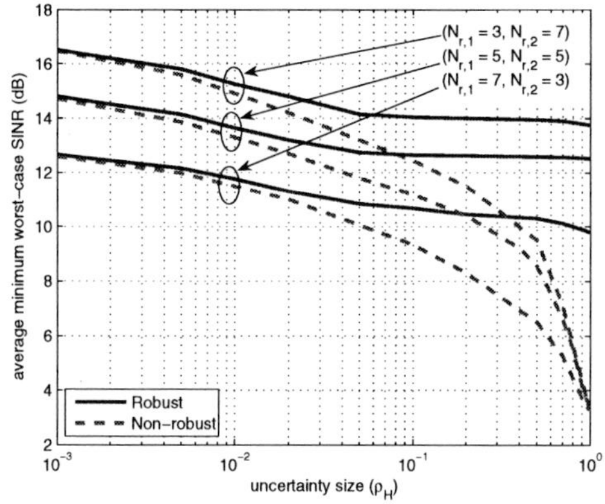

Figure 5.15.: Average minimum worst-case SINR vs. the size of uncertainty $\rho_{\mathbf{H}}$ for three sets of relay configurations: $(N_{r,1}, N_{r,2}) = (7,3), (5,5)$ or $(3,7)$.

Note that the impact capability of the uncertainty here is different than that of considered in Section 4.3.4 for single-cluster coherent AF relaying. In order to get more insight on the elaborations for Fig. 5.15, let us briefly focus on the noiseless intended received signal expressions at the ith destination for both cases:

$$(\mathbf{h}_{ir}^{(2)} \odot \mathbf{f}_{ri}^{(2)})^T \mathbf{g}^{(2)} + \underbrace{\mathbf{H}_{\mathsf{srd}}^{(1)}}_{\text{source of uncertainty}} \qquad \text{from (5.14)}, \tag{5.37}$$

$$\underbrace{(\mathbf{h}_{ir} \odot \mathbf{f}_{ri})}_{\text{source of uncertainty}}^T \mathbf{g} \qquad \text{from (3.20)}. \tag{5.38}$$

We see that the uncertainty is additive to the received signal in (5.37), whereas it is multiplicative to the gain vector decision in (5.38). In other words, while the uncertainty can null, in the extreme case, the nominal data in (5.38), it is not that probable for the uncertainty in (5.37) to do so, through it is dependent on the direction of the equivalent channels and the power constraints. Nevertheless, it is natural to expect that the impact of uncertainty is rather limited here.

We simulate three sets of curves, where each corresponds to one specific relay cluster configuration, i.e., $(N_{r,1}, N_{r,2}) = (3,7), (5,5), (7,3)$. We observe that at the extreme limit of $\rho_{\mathbf{H}} \rightarrow 1$, the non-robust performance of all configurations goes to a sink value at around 3 dB, which correspond to 10-15 dB worst-case performance loss with respect to the no uncertainty case. However, the worst-case results of the robust proposal are noticed to be quite resistant against the uncertainty for almost all $\rho_{\mathbf{H}}$ choices. In essence, beyond $\rho_{\mathbf{H}} = 0.1$ the average minimum worst-case SINR for all cluster configurations tend to saturate. This can be interpreted as follows. As the impact of the uncertainty is limited, the robust design can choose the relay gains such that they cancel all contribution from the local cluster (even the positive signal power contribution). Since there is still the *per se* contribution of the global cluster, the corresponding saturation SINR values are attained in the worst-case. This claim is also supported with the saturation levels of different cluster configurations. The more relay there are in $\mathscr{C}_{\mathsf{global}}$, the higher is the worst-case SINR, e.g., 14, 12.5, 10 dB for $(N_{\mathsf{r},1}, N_{\mathsf{r},2}) = (3,7), (5,5), (7,3)$, respectively. In summary, the robust proposal provides the network to benefit from the local cluster whenever the feedback is reliable, and further protects it from the destructive impacts of uncertain feedback.

5.4. Concluding Remarks

In this chapter, our perspective for system design has shifted from sufficiency to cost-efficiency. To this end, we exploited relay clustering through two different relaying protocols with a common target of providing "affordable" distributed spatial multiplexing in large relay networks.

We have been promoting coherent multiuser relaying since it provides several MIMO gains in a distributed manner through simple and low-cost AF relays. The primary assumption behind these gains is that CSI dissemination throughout the network is managed effectively without harming the spectral efficiency. Even assuming that spectral efficiency is not a problem any more, there is the impact of topological conditions, which are commonly neglected in theoretical considerations. That is, some relays may be geographically separated from each other such that they can not exchange local CSI.

Our first proposal addressed such a scenario where the relays are grouped into smaller clusters. The clusters are formed by the relays which are in the close proximity of each other. Hence, the topological conditions can easily form the cluster structures. Each of these clusters is unaware of the rest of the relay network and only employs *cluster-wide* global CSI. They are homogeneous in the sense that each treats the multiple S-D pairs independently and identically, i.e., CSI knowledge per cluster, employed relay gain allocations and transmit power constraints are similar. As the information flow in between clusters is cut off, the clustered network inherently sacrifices from array and diversity gains. However, we have shown that the network can recover its outage performance simply through cluster and time specific phase offsets applied at the relays, which does not require any additional cost or CSI knowledge. Finally, we have analyzed the impact of cluster formations on the CSI dissemination overhead, the transmit energy consumption and the average link-rate performance. We have proved that the most uniform distribution of the relays to clusters maximizes the average link-rate performance and minimizes the induced overhead.

Nevertheless, we enforced each of these homogeneous clusters to *per se* cancel all multiuser interference by using the available spatial degrees of freedom within the cluster. In other words, we require each cluster to have at least minimum relay configuration, which still stands as a threat against spectral efficiency for rather large N.

In a second part, we changed our view of clustering with an ultimate aim of complete independence for as much as relay in the network. That is, starting from the very basic setup of AZF relaying, we compose a very large cluster $\mathscr{C}_{\mathsf{local}}$, in which each relay is an independent sub-cluster of its own and requires only local CSI per relay. We have already seen that AZF gain allocation design performs quite efficient (in terms of array gain), if we can erase the interference that they cause at the non-intended destinations (see the performance of ZF-2 in general). To this end, we introduced an eraser cluster $\mathscr{C}_{\mathsf{global}}$ to the network with a primary mission of cancelling multiuser interference. The sufficient CSI information for this cluster to null multiuser interference has been found to be global CSI within this single cluster plus two-hop equivalent channel coefficients of all non-intended signals at the destinations established through $\mathscr{C}_{\mathsf{local}}$ relays, i.e., $\mathbf{H}^{(1)}_{\mathsf{srd}}$. After determining the structure of the whole network, we have focused on distributed beamforming designs for $\mathscr{C}_{\mathsf{global}}$, where we have also addressed the imperfectness of the critical feedback channel. Moreover, we have identified relay selection for clusters as an efficient means to achieve full effective diversity, which is quantified by the number of excess spatial dimensions left after interference cancellation. Relay selection has been further exploited to approach very close to the sum rate performances of similar multiuser relaying schemes requiring *network-wide* global CSI.

Such a perfect independence within $\mathscr{C}_{\mathsf{local}}$ provides an efficient adaptability to the network variations, such that any relay can join or leave the network without informing any other or a master node. Since $\mathbf{H}^{(1)}_{\mathsf{srd}}$ is estimated at the destinations from time to time, any change with it due to the variation of number of relays, will be tracked within the next estimation cycle.

To sum up, both clustering protocols provide an efficient trade-off between information transfer performance and CSI dissemination overhead in large AF networks. Henceforth, we can enjoy interference-free multiuser communication, which is now enabled to be affordable, high-rate and reliable.

Part II.

Multiuser MIMO Decode-and-Forward Relaying

Chapter 6

Decode-and-Forward Relaying: From One- to Two-Way Protocol

This chapter shifts our focus to decode-and-forward based relaying protocols. Our perspective follows the practical nature of wireless multiuser networks, where all terminals within the network transmit and receive information packets. That is, there occurs a bidirectional traffic pattern. Two-way relaying has been recently proposed to exploit spectral efficiency in such bidirectional data transfer links. Hence, for the sake of an inclusive and comparative study, we consider both one- and two-way DF relaying protocols in this chapter, and extend the respective current theory to the case of multiple antenna equipped nodes.

6.1. Introduction

In the first part of this thesis we have focused on the amplify-and-forward relay channel, where the relay amplifies its received signal according to an average or peak power constraint, and the decoding takes place only at the intended final destination. Although AF relaying has been commonly promoted for its transparency to coding/modulation and low-complexity due to the absence of decoding/re-encoding processes, it has a main drawback of forwarding not only desired signal but also *amplified* noise. Here and in the following chapters of the thesis, we shift our focus to decode-and-forward relaying, in which the received signal is decoded and re-encoded before it is forwarded to the destination. The DF relaying avoids relay noise amplification, but causes error propagation because the transmit signal of the source is subject to successive encoding/decoding processes until it reaches the destination.

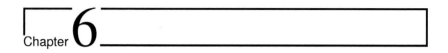

Independent of the applied relaying scheme, the half-duplex two-hop relay channel suffers from reduced spectral efficiency since the relay can not transmit and receive concurrently through the same physical channel. Specifically in a two-hop system, the half-duplex loss appears as a pre-log factor $\frac{1}{2}$ in the end-to-end achievable rate expression. This has been also reported with the analysis of diversity-multiplexing trade-off (DMT), which characterize the slow fading performance through outage formulations. In [2, 4], it has been proved that the optimal DMT of half-duplex two-hop one-way relay channel is limited by a maximal multiplexing gain of $\frac{1}{2}$ in the absence of source-to-destination direct link. In order to achieve this trade-off, respective AF and DF protocols have been further investigated.

Consider a wireless communication network of two terminals and a single relay, where the terminals want to exchange information via the relay with a purpose of range-extension, i.e., there is no source-to-destination direct link. If the conventional half-duplex relaying schemes are used, the relay successively establishes two unidirectional links for each information flow direction and requires four channel-uses in total. We refer this relaying protocol as *one-way relaying* in the sequel. However, it has been shown in [102] that the spectral efficiency of such a bidirectional link can be immediately improved with a protocol called *two-way relaying*, which only requires two-channel uses. Two-way relaying composed of two phases assuming that each phase occupies a single channel-use. In the first phase, both terminals transmit simultaneously via a multiple access scenario to the relay, whereas in the second phase, the relay broadcasts a combination of the previously received signals back to the terminals. The key idea of two-way relaying is the *self-interference cancellation*. Since the terminals know their own transmitted signal in the first phase, they can subtract the back-propagated self-interference prior to decoding in the second phase. Note that the aforementioned pre-log factor $\frac{1}{2}$ is still effective on each information flow direction, but since the same physical channel is used to establish the two connections, the sum rate is substantially higher when compared with its one-way counterpart. Fig. 6.1 shows the traffic patterns for both relaying protocols.

Here and in the sequel of this thesis, we focus on how to extend the aforementioned relaying protocols so, that the terminals and the relay can benefit from multiple antennas through transmission and reception. Further, we aim at designing efficient MIMO one- and two-way relaying protocols, with which we can exploit transmit CSI at the nodes.

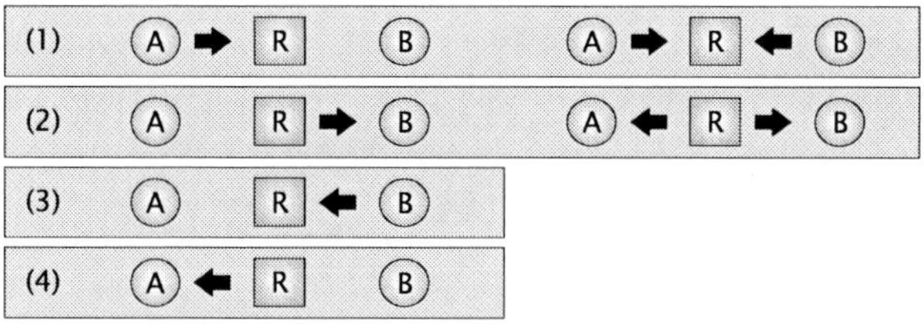

Figure 6.1.: Traffic patterns for one- and two-way relaying protocols.

State-of-the-Art - *Two-Way Relaying:* One of the first works that aimed for recovering the spectral efficiency loss in one-way relay channel, was for a cellular relaying setup, where each of the M base-to-mobile station link is aided by a dedicated relay [83]. The base station occupies M orthogonal channels to transmit to M mobiles with the corresponding relays. In the $(M+1)$th channel, all relays forward their received signal causing interference at the mobiles. It has been shown that the pre-log factor of the capacity of a single base-to-mobile link becomes $\frac{M}{M+1}$ instead of $\frac{1}{2}$.

The two-way communication channel was first studied by Shannon [111], where two terminals were assumed to transmit concurrently to each other *without* a relay in between. In [74, 136], a three phase protocol in combination with network coding approach [48, 143], has been considered for establishing a bidirectional relay link in between two terminals. In the consecutive first two channel uses (phases), the terminals transmit to the relay via orthogonal channels. After decoding each received signal perfectly, the relay combines both information sequence on bit-level through XOR additions and broadcasts it back to the terminals in the third phase. The terminals apply self-interference cancellation through XORing the decoded information with the previously transmitted bit sequence to the relay. With such a three-phase protocol, the pre-log factor with respect to the sum-rate is increased to $\frac{2}{3}$ from $\frac{1}{2}$.

Eventually, Rankov *et. all.* proposed the *two-way relaying protocol* with AF relay processing in [99], and showed that the number of required channel uses can be further reduced to two. The proposal has been extended to DF relaying in [100], where superposition coding (SPC) is used at the relay to combine the two received signal from the terminals. On the other hand, assuming SPC based DF relaying, optimal time sharing between the two signaling phases has been found in [89], and relay selection was addressed for two-way relaying in [88]. Exploring more on the achievable limits of two-way relaying, the achievable rate region for the full-duplex relay case is investigated in [101]. There, a DF relaying based on block Markov superposition coding and CF relaying based on Wyner-Ziv source coding were proposed. Reference [90] identified the SPC not to be the optimal combining scheme for the half-duplex channel, where the capacity region of the broadcast phase in terms of the minimal probability of error has been determined. Broadcasting a common message obtained through network coding (XORing) has been claimed to be inferior in general, which was further shown to achieve the capacity for certain distributions of the common message. Finally in [66], the constraint of two-phase communication was removed and three different DF relaying protocols have been considered with various numbers of phases (two to four). The corresponding performance bounds for each has been derived in terms of achievable sum rate.

Recently, a novel half-duplex relaying scheme called denoise-and-forward (DNF) has been proposed in [68, 96]. The DNF scheme consists of two phases, where the terminals simultaneously transmit to the relay in the first phase. The relay does not decode the signals from the terminals, but it maps the received signals into symbols from a discrete constellation and broadcasts back in the second phase. That is to say, the DNF considers the two-way problem on the practical implementation level and focuses on finite alphabet symbol mapping and constellation designs. It has been shown in [68] that DNF outperforms its practical opponents like three-phase network coding and four-phase one-way relaying.

State-of-the-Art - *MIMO DF Relaying:* The capacity scaling laws of MIMO one-way two-hop relay networks have been derived in [23], and it was extended to two-way relaying in [124]. In [75, 114, 131], a MIMO network composed of one S-D pair and a single relay has been considered. Such a scenario is commonly referred as the *MIMO relay channel* in the literature, where the S-D direct link is assumed to be reasonably strong in comparison with the link through the relay. The capacity of this conventional two-hop MIMO relay channel has been investigated in [131] and the corresponding upper and lower bounds have been provided. In [75], the authors specifically focused on DF relaying, and improved the lower bound for the capacity provided by [131], through precoding methods such as superposition coding and dirty-paper coding. Similar precoding strategies with partial or full decoding at the relay has been also considered in [114].

The above mentioned scenario has been adapted in [52] with differences that the number of assisting relay nodes was increased to two, and the S-D direct link was neglected. Two different MIMO precoding techniques have been investigated. Either a common message is broadcasted to the relays, which is later forwarded to the destination employing distributed space time coding, or the common message is divided into two parts, each of which is transmitted through one specific relay. On the other hand, in [30], MIMO relay channel is modified to have multiple single-antenna destinations. Assuming that the source uses DPC and the single MIMO relay linearly amplifies and forwards the received signal, a joint source precoder and relay gain matrix design was performed (refer to [30, 122] for further references on MIMO AF relaying).

The two-way broadcast capacity has been extended to the multiple antennas case in [137], where the characterization of the capacity region follows a random coding approach. In [138], channel correlation in two-way broadcast channel has been addressed for the special case of single-antenna terminals. Further, space time coding has been combined with network coding in [92] for a potential application in multi-hop MIMO networks.

Contribution of this Chapter We consider two MIMO terminals exchanging information via a single MIMO relay node, either using four-phase one-way relaying or two-phase two-way relaying. In the context of one-way relaying, we present an overview of MIMO signaling through the four phases, and provide expressions for end-to-end information transfer rates for given transmit signal covariances. Further, we characterize the capacity region of the bidirectional link established with one-way relaying, in terms of time sharing parameters of each phase.

Our main contribution takes place in the second part, where we extend the two-way relaying to multiple antenna equipped nodes. We primarily focus on network coding based signal combination at the relay, whereas we briefly present superposition encoding for the sake of completeness. After providing the principles of MIMO two-way relaying, we characterize capacity regions for each signaling phase and the overall two-phase region for given transmit signal covariances. It has been conjectured in [137] that when bit-level XOR precoding is used to combine the two information sequence from the terminals, the downlink rates from the relay to the terminals are restricted with the worst relay-to-terminal channel, i.e., it is nothing but multicasting a common message. However, we propose a variant of the XOR precoding approach, with which unbalanced downlink rates can be supported, and hence, the full MIMO two-way broadcast capacity region derived in [137], is achieved. The key idea to this is to apply *zero-padding per symbol* to the information sequence to be sent to the weak terminal, prior to XOR addition. Moreover, we provide *a priori* information about this process to the corresponding terminal for efficient decoding. Finally, we conclude this chapter by presenting a comparative study between different types of signal combinations at the relay for two-way relaying.

Outline of this Chapter The rest of this chapter is organized as follows. After presenting the system model in the beginning of Section 6.2, we present the four-phase one-way MIMO relaying protocol in Section 6.2.1, where we give achievable rate expressions for each phase in details. In Section 6.2.2, we focus on MIMO two-way relaying and present the signal models for both of the two phases. Next, the proposed signal combination schemes at the relay are respectively explained in Section 6.2.2.1 and 6.2.2.2.

6.2. Principles of MIMO DF Relaying

We consider a wireless relay network where two terminals A and B are communicating with each other via one DF relay R. Motivating the employment of the relay with range extension, we assume that there is no direct link between A and B. Each node in the network is equipped respectively with N_{A}, N_{B}, N_{R} antennas in both transmit and receive modes. It is assumed that the relay is half-duplex, i.e., it can not transmit and receive simultaneously.

The matrices describing the frequency flat fading channels from A to R and from B to R are denoted by $\mathbf{H}_{\mathsf{A}} \in \mathbb{C}^{N_{\mathsf{R}} \times N_{\mathsf{A}}}$ and $\mathbf{H}_{\mathsf{B}} \in \mathbb{C}^{N_{\mathsf{R}} \times N_{\mathsf{B}}}$, respectively. We assume channel reciprocity such that the channel matrices from R to A and B are represented by $\mathbf{H}_{\mathsf{A}}^T$ and $\mathbf{H}_{\mathsf{B}}^T$, respectively. Furthermore, considering a slow-fading environment, the channel matrices are assumed to stay constant within a transmission cycle duration, and succeeding realizations of the propagation channels are statistically independent (block fading).

Both terminals have data to transmit to the other. The terminal A wants to transmit a bit sequence $\mathbf{x}_{\mathsf{A}}$ to the terminal B, where the elements of the bit sequence vector are either 1 or 0. The bit sequence $\mathbf{x}_{\mathsf{A}}$ is later coded to a complex transmit signal vector of $\mathbf{x}_{\mathsf{A}} \rightarrow \mathbf{s}_{\mathsf{A}} \in \mathbb{C}^{N_{\mathsf{A}}}$. Likewise, the bit sequence that the terminal B wants to transmit to the terminal A and the corresponding coded signal vector are denoted by $\mathbf{x}_{\mathsf{B}}$ and $\mathbf{s}_{\mathsf{B}} \in \mathbb{C}^{N_{\mathsf{B}}}$, respectively. Hence, the signal covariance matrices of terminals A and B are then defined as $\mathbf{\Lambda}_{\mathsf{A}} \triangleq \mathbb{E}\{\mathbf{s}_{\mathsf{A}}\mathbf{s}_{\mathsf{A}}^H\}$ and $\mathbf{\Lambda}_{\mathsf{B}} \triangleq \mathbb{E}\{\mathbf{s}_{\mathsf{B}}\mathbf{s}_{\mathsf{B}}^H\}$, respectively. We impose an average transmit signal power constraint on each data-exchanging terminal in the network such that

$$\mathrm{Tr}(\mathbf{\Lambda}_{\mathsf{n}}) \leq P_{\mathsf{n}}, \quad \mathsf{n} \in \{\mathsf{A},\mathsf{B}\}.$$

The terminals employ one of the two possible relaying protocol options: one- or two-way relaying. In the following we present the transmit and receive signal models and the achievable rate expressions for both protocols.

6.2.1. One-Way Relaying Protocol

As both terminals have data to share with the other, the total transmission cycle will last for four time-slots (or equivalently channel-uses), where we devote the first two time-slots for the transmission from A to B, and the last two for the transmission from B to A. In the first time-slot, A transmits $\mathbf{s}_{\mathsf{A}}$ to R, and the $N_{\mathsf{R}} \times 1$ received signal at R is

$$\mathbf{r}_{\mathsf{R}} = \mathbf{H}_{\mathsf{A}}\mathbf{s}_{\mathsf{A}} + \mathbf{n}_{\mathsf{R}}, \tag{6.1}$$

where the relay noise vector $\mathbf{n}_{\mathsf{R}}$ is spatio-temporally white circularly symmetric complex Gaussian noise sequence with zero-mean and covariance matrix $\mathbb{E}\{\mathbf{n}_{\mathsf{R}}\mathbf{n}_{\mathsf{R}}^H\} = \sigma_{n_{\mathsf{R}}}^2\mathbf{I}_{N_{\mathsf{R}}}$.

Assuming Gaussian codebooks are used for coding at the transmitter, the maximal information transfer rate from A to R is theoretically given by [117]

$$\mathrm{R}_{\mathsf{AR}}^{\rightarrow} \leq \mathrm{I}_{\mathsf{AR}}^{\rightarrow} = \log_2\left|\mathbf{I}_{N_{\mathsf{R}}} + \frac{1}{\sigma_{n_{\mathsf{R}}}^2}\mathbf{H}_{\mathsf{A}}\mathbf{\Lambda}_{\mathsf{A}}\mathbf{H}_{\mathsf{A}}^H\right|, \tag{6.2}$$

where the one-way directing arrow "$\rightarrow$" at the superscript of rate notation indicates that it is valid for one-way relaying protocol. Here and in the sequel, I is used for denoting the mutual information, whereas R denotes the allocated rate which can be maximally I. After perfect decoding of the received signal, the relay re-encodes the received bit sequence to the relay transmit signal vector $\mathbf{x}_{\mathsf{A}} \rightarrow \mathbf{s}_{\mathsf{R}} \in \mathbb{C}^{N_{\mathsf{R}}}$ and sends it to the terminal B in the second time-slot. Correspondingly, $\mathbf{\Lambda}_{\mathsf{R}} \triangleq \mathbb{E}\{\mathbf{s}_{\mathsf{R}}\mathbf{s}_{\mathsf{R}}^H\}$ denotes the covariance of the relay transmit signal, and we impose an average transmit power constraint on $\mathbf{s}_{\mathsf{R}}$, i.e., $\mathrm{Tr}(\mathbf{\Lambda}_{\mathsf{R}}) \leq P_{\mathsf{R}}$. Next, the $N_{\mathsf{B}} \times 1$ received signal vector at B is written as

$$\mathbf{r}_{\mathsf{B}} = \mathbf{H}_{\mathsf{B}}^T\mathbf{s}_{\mathsf{R}} + \mathbf{n}_{\mathsf{B}}, \tag{6.3}$$

where $\mathbf{n}_\mathsf{B} \sim \mathcal{CN}(\mathbf{0}, \sigma_{n_\mathsf{B}}^2 \mathbf{I}_{N_\mathsf{B}})$ similar to the relay noise in (6.1). Hence, the maximal achievable transfer rate from R to B is

$$\vec{R}_{\mathsf{RB}} \leq \vec{I}_{\mathsf{RB}} = \log_2 \left| \mathbf{I}_{N_\mathsf{B}} + \frac{1}{\sigma_{n_\mathsf{B}}^2} \mathbf{H}_\mathsf{B}^T \mathbf{\Lambda}_\mathsf{R} \mathbf{H}_\mathsf{B}^* \right|. \tag{6.4}$$

As we do not allow for data buffering, the maximal information transfer rate between terminals A and B is limited by the weakest link, i.e.,

$$\vec{R}_{\mathsf{AB}} \leq \mathbf{min}\{\alpha_1 \vec{R}_{\mathsf{AR}}, \alpha_2 \vec{R}_{\mathsf{RB}}\}, \tag{6.5}$$

where the terms α_1, α_2 are the normalized time share allocated to the first and the second time-slots respectively, where $0 \leq \alpha_1, \alpha_2 \leq 1$.

After the data transmission of the terminal A is accomplished, the next two time-slots are occupied for the data transmission from B to A. Since the transmission structure is symmetric and the signal expressions for B's transmission are very similar to that of A, we drop further details and directly present the maximal information transfer rate from B and A:

$$\vec{R}_{\mathsf{BA}} \leq \mathbf{min}\{\alpha_3 \vec{R}_{\mathsf{BR}}, \alpha_4 \vec{R}_{\mathsf{RA}}\}, \tag{6.6}$$

where

$$\vec{R}_{\mathsf{BR}} \leq \vec{I}_{\mathsf{BR}} = \log_2 \left| \mathbf{I}_{N_\mathsf{R}} + \frac{1}{\sigma_{n_\mathsf{R}}^2} \mathbf{H}_\mathsf{B} \mathbf{\Lambda}_\mathsf{B} \mathbf{H}_\mathsf{B}^H \right| \tag{6.7}$$

$$\vec{R}_{\mathsf{RA}} \leq \vec{I}_{\mathsf{RA}} = \log_2 \left| \mathbf{I}_{N_\mathsf{A}} + \frac{1}{\sigma_{n_\mathsf{A}}^2} \mathbf{H}_\mathsf{A}^T \mathbf{\Lambda}_\mathsf{R} \mathbf{H}_\mathsf{A}^* \right|, \tag{6.8}$$

and α_3, α_4 are the normalized time share allocated to the third and the fourth time-slots, respectively. Note the time-slot allocation constraint that $\sum_{i=1}^4 \alpha_i = 1$. Hence, the overall capacity region for exchanging information through one-way relaying is the convex hull of all rate pairs $(\vec{R}_{\mathsf{AB}}, \vec{R}_{\mathsf{BA}})$ satisfying (6.5) and (6.6), i.e.,

$$\mathcal{C}_{\text{one-way}} = \left\{ (\vec{R}_{\mathsf{AB}}, \vec{R}_{\mathsf{BA}}) \in \mathbb{R}_+ \; : \; \begin{array}{c} \vec{R}_{\mathsf{AB}} \leq \mathbf{min}\{\alpha_1 \vec{R}_{\mathsf{AR}}, \alpha_2 \vec{R}_{\mathsf{RB}}\} \\ \vec{R}_{\mathsf{BA}} \leq \mathbf{min}\{\alpha_3 \vec{R}_{\mathsf{BR}}, \alpha_4 \vec{R}_{\mathsf{RA}}\} \\ \sum_{i=1}^4 \alpha_i = 1, \quad \alpha_i \in \mathbb{R}_+ \end{array} \right\}. \tag{6.9}$$

Finally, the sum rate of the exchanged data between the terminals A and B by using one-way relaying protocol, is computed as $\vec{R}_{\text{sum}} = \vec{R}_{\mathsf{AB}} + \vec{R}_{\mathsf{BA}}$.

6.2.2. Two-Way Relaying Protocol

In contrary with the previous one-way protocol, data exchange of the two terminals is accomplished only in two time-slots with two-way relaying protocol. In the first time-slot, both terminals simultaneously transmit to the relay R using the same physical channel. Hence, the received signal at the relay is written as

$$\mathbf{r}_{\mathsf{R}} = \mathbf{H}_{\mathsf{A}}\mathbf{s}_{\mathsf{A}} + \mathbf{H}_{\mathsf{B}}\mathbf{s}_{\mathsf{B}} + \mathbf{n}_{\mathsf{R}}. \tag{6.10}$$

As there occurs a multiple access scenario at the relay, the communication in this first time-slot will be called as MAC phase in the sequel. We require the relay to perfectly decode the received signal and extract both intended bit sequences $\mathbf{x}_{\mathsf{A}}$ and $\mathbf{x}_{\mathsf{B}}$. Note that for perfect decoding, the relay needs to have sufficient number of antennas [119]. Assuming that the transmit signals are drawn from independent Gaussian codebooks, the achievable rates $\mathrm{R}_{\mathsf{AR}}^{\leftrightarrow}$ (from A to R) and $\mathrm{R}_{\mathsf{BR}}^{\leftrightarrow}$ (from B to R) are defined by the following three expressions [49]

$$\mathrm{R}_{\mathsf{AR}}^{\leftrightarrow} \leq \mathrm{I}_{\mathsf{AR}}^{\leftrightarrow} = \log_2 \left| \mathbf{I}_{N_{\mathsf{R}}} + \frac{1}{\sigma_{n_{\mathsf{R}}}^2} \mathbf{H}_{\mathsf{A}} \mathbf{\Lambda}_{\mathsf{A}} \mathbf{H}_{\mathsf{A}}^H \right| \tag{6.11}$$

$$\mathrm{R}_{\mathsf{BR}}^{\leftrightarrow} \leq \mathrm{I}_{\mathsf{BR}}^{\leftrightarrow} = \log_2 \left| \mathbf{I}_{N_{\mathsf{R}}} + \frac{1}{\sigma_{n_{\mathsf{R}}}^2} \mathbf{H}_{\mathsf{B}} \mathbf{\Lambda}_{\mathsf{B}} \mathbf{H}_{\mathsf{B}}^H \right| \tag{6.12}$$

$$\mathrm{R}_{\mathsf{AR}}^{\leftrightarrow} + \mathrm{R}_{\mathsf{BR}}^{\leftrightarrow} \leq \mathrm{I}_{\mathsf{AB,MAC}}^{\leftrightarrow} = \log_2 \left| \mathbf{I}_{N_{\mathsf{R}}} + \frac{1}{\sigma_{n_{\mathsf{R}}}^2} \left(\mathbf{H}_{\mathsf{A}} \mathbf{\Lambda}_{\mathsf{A}} \mathbf{H}_{\mathsf{A}}^H + \mathbf{H}_{\mathsf{B}} \mathbf{\Lambda}_{\mathsf{B}} \mathbf{H}_{\mathsf{B}}^H \right) \right|. \tag{6.13}$$

For given covariance matrices $\mathbf{\Lambda}_{\mathsf{A}}$ and $\mathbf{\Lambda}_{\mathsf{B}}$ fulfilling the transmit power constraints, the achievable rate region for the MAC phase is defined as the convex hull of the set of rate pairs satisfying (6.11), (6.12), (6.13), and given as the following polyhedron (see Fig. 6.2(a))

$$\mathcal{C}_{\mathrm{MAC}}(\mathbf{\Lambda}_{\mathsf{A}}, \mathbf{\Lambda}_{\mathsf{B}}) = \left\{ (\mathrm{R}_{\mathsf{AR}}^{\leftrightarrow}, \mathrm{R}_{\mathsf{BR}}^{\leftrightarrow}) \in \mathbb{R}_+ \;:\; \mathrm{R}_{\mathsf{AR}}^{\leftrightarrow} \leq \mathrm{I}_{\mathsf{AR}}^{\leftrightarrow}, \mathrm{R}_{\mathsf{BR}}^{\leftrightarrow} \leq \mathrm{I}_{\mathsf{BR}}^{\leftrightarrow}, \mathrm{R}_{\mathsf{AR}}^{\leftrightarrow} + \mathrm{R}_{\mathsf{BR}}^{\leftrightarrow} \leq \mathrm{I}_{\mathsf{AB,MAC}}^{\leftrightarrow} \right\}. \tag{6.14}$$

Note that the two-way directing arrow "$\leftrightarrow$" at the superscript of rate notations indicates the two-way relaying protocol.

The relay now needs to combine the two independent information sequence coming from the terminals in such a way that upon receiving the resultant signal, each terminal can subtract the back-propagated self-interference. For this BC phase of the two-way protocol, we consider two coding schemes: superposition coding and bit-wise XOR pre-coding. In the sequel of this thesis, we will focus on the XOR pre-coding in details. However, for the sake of completeness, we also present the basics of the SPC scheme.

6.2.2.1. Superposition Coding

The relay codes both received bit sequences to individual transmit signals $\mathbf{s}_{\mathsf{RA}}$ and $\mathbf{s}_{\mathsf{RB}}$, i.e., $\mathbf{x}_{\mathsf{A}} \rightarrow \mathbf{s}_{\mathsf{RB}} \in \mathbb{C}^{N_{\mathsf{R}}}$ and $\mathbf{x}_{\mathsf{B}} \rightarrow \mathbf{s}_{\mathsf{RA}} \in \mathbb{C}^{N_{\mathsf{R}}}$, with independent Gaussian codebooks. The final relay transmit signal is obtained by superimposing these two independent signals after an appropriate pre-coding. That is to say, the relay broadcast

$$\mathbf{s}_{\mathsf{R}} = \mathbf{W}_{\mathsf{A}}\mathbf{s}_{\mathsf{RA}} + \mathbf{W}_{\mathsf{B}}\mathbf{s}_{\mathsf{RB}}, \tag{6.15}$$

where $\mathbf{W}_{\mathsf{A}} \in \mathbb{C}^{N_{\mathsf{A}} \times N_{\mathsf{R}}}$ and $\mathbf{W}_{\mathsf{B}} \in \mathbb{C}^{N_{\mathsf{B}} \times N_{\mathsf{R}}}$ are the corresponding pre-coding matrices. With (6.15), we establish a *modified* broadcast channel. It is different than the conventional broadcast channel, because upon receiving $\mathbf{s}_{\mathsf{R}}$ each terminal can subtract the unintended part of the signal as it was the corresponding terminal's own transmit signal in the MAC phase, i.e., no interference management necessary. Moreover, it is also different than the multicast channel as there are private messages to individual terminals.

The received signals at the corresponding terminals are

$$\begin{aligned} \mathbf{r}_{\mathsf{A}} &= \underbrace{\mathbf{H}_{\mathsf{A}}^{T}\mathbf{W}_{\mathsf{A}}\mathbf{s}_{\mathsf{RA}}}_{\text{intended signal}} + \underbrace{\mathbf{H}_{\mathsf{A}}^{T}\mathbf{W}_{\mathsf{B}}\mathbf{s}_{\mathsf{RB}}}_{\text{self-interference}} + \mathbf{n}_{\mathsf{A}}, \\ \mathbf{r}_{\mathsf{B}} &= \underbrace{\mathbf{H}_{\mathsf{B}}^{T}\mathbf{W}_{\mathsf{A}}\mathbf{s}_{\mathsf{RA}}}_{\text{self-interference}} + \underbrace{\mathbf{H}_{\mathsf{B}}^{T}\mathbf{W}_{\mathsf{B}}\mathbf{s}_{\mathsf{RB}}}_{\text{intended signal}} + \mathbf{n}_{\mathsf{B}}, \end{aligned} \tag{6.16}$$

where *self-interference* terms can be cancelled completely. For instance, let us focus on terminal A: As it knows $\mathbf{x}_{\mathsf{A}}$, it can obtain $\mathbf{s}_{\mathsf{RB}}$ with the assumption that it has the knowledge of the Gaussian codebook used at the relay. Next, assuming that it further has receive CSI $\mathbf{H}_{\mathsf{A}}^{T}$, and pre-coding matrix $\mathbf{W}_{\mathsf{B}}$ knowledge, it cancels the self-interference part, and decodes the rest of the received signal. A similar operation is performed at the terminal B. Hence, the maximal information rate from R to A and B in the broadcast phase can be respectively written as

$$\overset{\leftrightarrow}{\mathrm{R}}_{\mathsf{Rn}} \leq \overset{\leftrightarrow}{\mathrm{I}}_{\mathsf{Rn}} = \log_2 \left| \mathbf{I}_{N_{\mathsf{R}}} + \frac{1}{\sigma_{n_{\mathsf{n}}}^2} \mathbf{H}_{\mathsf{n}}^{T}\mathbf{W}_{\mathsf{n}}\mathbf{\Lambda}_{\mathsf{Rn}}\mathbf{W}_{\mathsf{n}}^{H}\mathbf{H}_{\mathsf{n}}^{*} \right|, \quad \mathsf{n} \in \{\mathsf{A},\mathsf{B}\}, \tag{6.17}$$

where $\mathbf{\Lambda}_{\mathsf{Rn}} \triangleq \mathbb{E}\{\mathbf{s}_{\mathsf{Rn}}\mathbf{s}_{\mathsf{Rn}}^{H}\}$ denotes the covariance matrix of the corresponding transmit signal. Note that both rate expressions are only coupled via the transmit power constraint at the relay, which is given as

$$\sum_{\mathsf{n}\in\{\mathsf{A},\mathsf{B}\}} \mathrm{Tr}(\mathbf{W}_{\mathsf{n}}\mathbf{\Lambda}_{\mathsf{Rn}}\mathbf{W}_{\mathsf{n}}^{H}) \leq P_{\mathsf{R}}. \tag{6.18}$$

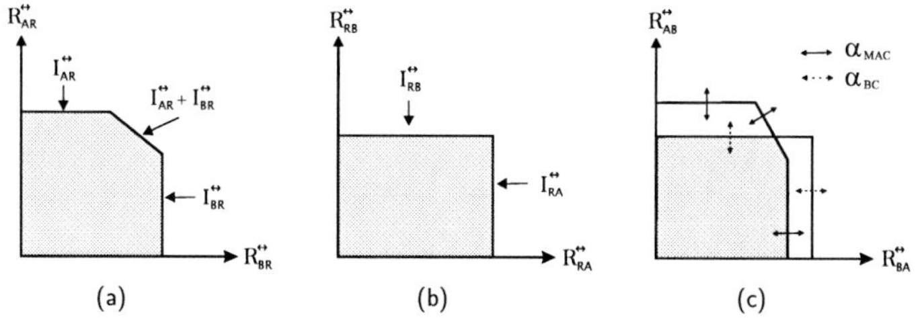

Figure 6.2.: Illustration of achievable rate regions for given covariance matrices: (a) MAC, (b) BC, (c) Overall two-way relaying.

Hence, for the given covariance matrices $\mathbf{\Lambda}_{\mathsf{RA}}$ and $\mathbf{\Lambda}_{\mathsf{RB}}$ fulfilling (6.18), the achievable rate region for the MAC phase is defined as the convex hull of the set of rate pairs satisfying (6.17), and given as the following rectangle (see Fig. 6.2(b))

$$\mathcal{C}_{\mathrm{BC}}^{\mathrm{SPC}}(\mathbf{\Lambda}_{\mathsf{RA}}, \mathbf{\Lambda}_{\mathsf{RB}}) = \left\{(\overleftrightarrow{\mathrm{R}}_{\mathsf{RA}}, \overleftrightarrow{\mathrm{R}}_{\mathsf{RB}}) \;:\; \overleftrightarrow{\mathrm{R}}_{\mathsf{Rn}} \leq \overleftrightarrow{\mathrm{I}}_{\mathsf{Rn}},\; \mathsf{n} \in \{\mathsf{A}, \mathsf{B}\}\right\}. \tag{6.19}$$

Taking the convex hull of all achievable rate pairs $(\overleftrightarrow{\mathrm{R}}_{\mathsf{AB}}, \overleftrightarrow{\mathrm{R}}_{\mathsf{BA}})$ satisfying (6.11), (6.12), (6.13), (6.17), we can write the overall rate region for the two-way relaying with SPC as

$$\mathcal{C}_{\mathrm{two-way}}^{\mathrm{SPC}}(\mathbf{\Lambda}_{\mathsf{A}}, \mathbf{\Lambda}_{\mathsf{B}}, \mathbf{\Lambda}_{\mathsf{RA}}, \mathbf{\Lambda}_{\mathsf{RB}}) = \left\{ (\overleftrightarrow{\mathrm{R}}_{\mathsf{AB}}, \overleftrightarrow{\mathrm{R}}_{\mathsf{BA}}) \in \mathbb{R}_{+} \;:\; \begin{array}{rl} \overleftrightarrow{\mathrm{R}}_{\mathsf{AB}} & \leq \mathbf{min}\{\alpha_{\mathrm{MAC}}\overleftrightarrow{\mathrm{R}}_{\mathsf{AR}}, \alpha_{\mathrm{BC}}\overleftrightarrow{\mathrm{R}}_{\mathsf{RB}}\}, \\ \overleftrightarrow{\mathrm{R}}_{\mathsf{BA}} & \leq \mathbf{min}\{\alpha_{\mathrm{MAC}}\overleftrightarrow{\mathrm{R}}_{\mathsf{BR}}, \alpha_{\mathrm{BC}}\overleftrightarrow{\mathrm{R}}_{\mathsf{RA}}\}, \\ \overleftrightarrow{\mathrm{R}}_{\mathsf{AB}} + \overleftrightarrow{\mathrm{R}}_{\mathsf{BA}} & \leq \alpha_{\mathrm{MAC}}\overleftrightarrow{\mathrm{I}}_{\mathsf{AB,MAC}}, \\ \alpha_{\mathrm{MAC}} + \alpha_{\mathrm{BC}} & = 1, \quad \alpha_{\mathrm{MAC}}, \alpha_{\mathrm{BC}} \in \mathbb{R}_{+} \end{array} \right\} \tag{6.20}$$

where α_{MAC} and α_{BC} denote the fraction of time allocated to the MAC and BC phases respectively, and $\alpha_{\mathrm{MAC}} + \alpha_{\mathrm{BC}} = 1$ (see Fig. 6.2(c)). Note that the achievable two-way rate region is nothing but the intersection of rate regions of MAC and BC phases.

The design and optimization of the pre-coding and covariance matrices are not in the scope of thesis. For further details, we refer the interested reader to [55].

6.2.2.2. XOR Precoding

The XOR precoding scheme combines the two information bit sequences on bit-level prior to encoding. Specifically, the relay applies bitwise XOR operation on both decoded bit sequences $\mathbf{x}_{\mathsf{A}}$ and $\mathbf{x}_{\mathsf{B}}$, i.e.,

$$\mathbf{x}_{\mathsf{A}} \oplus \mathbf{x}_{\mathsf{B}} = \mathbf{x}_{\mathsf{R}}.$$

Next, $\mathbf{x}_\mathsf{R}$ is coded to the relay transmit signal vector $\mathbf{s}_\mathsf{R}$ with a covariance matrix of $\mathbf{\Lambda}_\mathsf{R} \triangleq \mathbb{E}\{\mathbf{s}_\mathsf{R}\mathbf{s}_\mathsf{R}^H\}$. In case that the lengths of $\mathbf{x}_\mathsf{A}$ and $\mathbf{x}_\mathsf{B}$ are not equal, i.e., $\overleftrightarrow{\mathrm{R}}_\mathsf{AR} \neq \overleftrightarrow{\mathrm{R}}_\mathsf{BR}$, then we apply zero-padding to the shorter sequence. For instance, say $\overleftrightarrow{\mathrm{R}}_\mathsf{AR} \geq \overleftrightarrow{\mathrm{R}}_\mathsf{BR}$, which means that the length of the bit sequence $\mathbf{x}_\mathsf{A}$ is longer than $\mathbf{x}_\mathsf{B}$. Hence, we pad zeros to $\mathbf{x}_\mathsf{B}$ prior to XOR addition.

As $\mathbf{s}_\mathsf{R}$ is broadcasted to the terminals, the received signals are given as

$$\mathbf{r}_\mathsf{n} = \mathbf{H}_\mathsf{n}^T\mathbf{s}_\mathsf{R} + \mathbf{n}_\mathsf{n}, \quad \mathsf{n} \in \{\mathsf{A}, \mathsf{B}\}.$$

Upon receiving $\mathbf{r}_\mathsf{n} \in \mathbb{C}^{N_\mathsf{n}}$, each terminal decodes $\mathbf{s}_\mathsf{R}$ to obtain $\mathbf{x}_\mathsf{R}$. The self-interference cancellation is done after the decoding by applying a simple XOR operation, i.e.,

$$\mathbf{x}_\mathsf{R} \oplus \mathbf{x}_\mathsf{A} = \mathbf{x}_\mathsf{B} \quad \text{and} \quad \mathbf{x}_\mathsf{R} \oplus \mathbf{x}_\mathsf{B} = \mathbf{x}_\mathsf{A}.$$

With this interference cancellation, the maximal information rate from R to the terminals in the broadcast phase is written through conventional interference-free MIMO decoding rates as

$$\overleftrightarrow{\mathrm{R}}_\mathsf{Rn} \leq \overleftrightarrow{\mathrm{I}}_\mathsf{Rn} = \log_2\left|\mathbf{I}_{N_\mathsf{n}} + \frac{1}{\sigma_{n_\mathsf{n}}^2}\mathbf{H}_\mathsf{n}^T\mathbf{\Lambda}_\mathsf{R}\mathbf{H}_\mathsf{n}^*\right|, \quad \mathsf{n} \in \{\mathsf{A}, \mathsf{B}\}. \tag{6.21}$$

In the broadcast phase of two-way relaying with XOR precoding, we are broadcasting common information (multicast) $\mathbf{s}_\mathsf{R}$ to the terminals. If we follow the fundamental theory of multicast channel, we need to adjust the rates with respect to the weakest terminal. In other words, we should set $\overleftrightarrow{\mathrm{R}}_\mathsf{RA} = \overleftrightarrow{\mathrm{R}}_\mathsf{RB}$ (see [55] for the case when this equal rate constraint is enforced). However, employing XOR precoding, we differ from the conventional multicast problem such that we can support unbalanced rates for $\mathbf{s}_\mathsf{R}$, and hence, improve the downlink sum rate.

The capacity region $\mathcal{C}_{\mathrm{BC}}^{\mathrm{XOR}}$ for the broadcast phase of two-way relaying with XOR precoding has been derived in [137]. For given covariance matrix $\mathbf{\Lambda}_\mathsf{R}$, it is basically defined similar to $\mathcal{C}_{\mathrm{BC}}^{\mathrm{SPC}}$ in (6.19) by just replacing (6.17) with (6.21). However, a random coding approach was taken there, and no practical implementation has been proposed to achieve the derived capacity. In the following, we propose a coding scheme which achieve the capacity region, and can be easily implemented (see [147]). Basically, the scheme assumes that one of the relay-to-terminal links can support higher rate than the other, and proposes to pad zeros per symbol to the lower-rate bit sequence before XOR operation. Correspondingly, through the decoding at the lower-rate terminal, there needs an a priori decoding knowledge for padded zeros for perfect decoding. Note

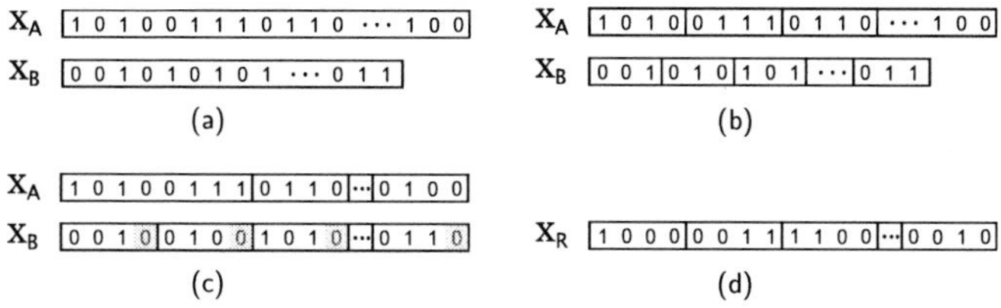

Figure 6.3.: Illustrative example for zero-padding based encoding with detailed implementations steps: (a) Unbalanced information bit sequences, (b) Symbol segmentation, (c) Zero-padding per symbol, (d) Bit-wise XOR operation.

that the zero-padding (per symbol) concept here is different than the zero-padding per sequence applied in the case of unequal bit-sequence through the uplink to the relay. However, both can be incorporated appropriately.

Let us explain this scheme with a conceptual example: In the first phase the relay perfectly decodes the bit sequences $\mathbf{x}_{\mathsf{A}}$ and $\mathbf{x}_{\mathsf{B}}$ from A and B, respectively. It wants to broadcast these to the corresponding terminals in the second phase, i.e., $\mathbf{x}_{\mathsf{A}}$ to B and $\mathbf{x}_{\mathsf{B}}$ to A. Assume that for a given covariance matrix, the link to the terminal A is strictly weaker than that of B, i.e., $\overleftrightarrow{\mathrm{R}}_{\mathsf{RA}} < \overleftrightarrow{\mathrm{R}}_{\mathsf{RB}}$. As the relay can transmit more bits to B than to A, the length of $\mathbf{x}_{\mathsf{A}}$ (L_{A}) is longer than $\mathbf{x}_{\mathsf{B}}$ (L_{B}) for a given block of bits in a certain amount of channel use (see Fig. 6.3(a)). Further assume that these downlink transmission rates are also supported with the achievable rate region of the MAC phase, i.e., the bottleneck is the BC phase.

- **Encoding:** We interprete $\ell_{\mathsf{A}}(\ell_{\mathsf{B}})$ consecutive bits of $\mathbf{x}_{\mathsf{A}}$ ($\mathbf{x}_{\mathsf{B}}$) as a symbol to be encoded by a Gaussian codebook, where $\ell_{\mathsf{A}} = 4$ and $\ell_{\mathsf{B}} = 3$ for our illustrative example in Fig. 6.3(b). The relay pads $\ell_{\mathsf{A}} - \ell_{\mathsf{B}}$ zeros after each ℓ_{B} consecutive bits in $\mathbf{x}_{\mathsf{B}}$, such that the length of $\mathbf{x}_{\mathsf{B}}$ is extended to be L_{A} instead of L_{B} (see Fig. 6.3(c)). After XORing the $\mathbf{x}_{\mathsf{A}}$ and the *extended* $\mathbf{x}_{\mathsf{B}}$ (see Fig. 6.3(d)), we encode the resultant $\mathbf{x}_{\mathsf{R}}$ with a codebook of rate $\mathbf{max}\{\overleftrightarrow{\mathrm{R}}_{\mathsf{RA}}, \overleftrightarrow{\mathrm{R}}_{\mathsf{RB}}\} = \overleftrightarrow{\mathrm{R}}_{\mathsf{RB}}$.

- **Decoding:** At the receive sides, the terminal B employs a regular MIMO decoding with a codebook of size $\overleftrightarrow{\mathrm{R}}_{\mathsf{RB}}$ to obtain $\mathbf{x}_{\mathsf{R}}$ perfectly. On the other hand, the terminal A receives information with an higher rate than its own link can support, i.e., $\overleftrightarrow{\mathrm{R}}_{\mathsf{RA}} < \overleftrightarrow{\mathrm{R}}_{\mathsf{RB}}$ for this example. At this point, we provide an *a priori knowledge* to A for decoding, and inform it with the knowledge that there are some intentional and redundant zeros padded per symbol to its intended bit sequence $\mathbf{x}_\mathsf{B}$. Consequently, for this specific simple example, instead of searching through 2^4 possible codewords it can omit $2^4 - 2^3$ of these, which stand for the additional $\ell_\mathsf{A} - \ell_\mathsf{B}$ bits per symbol and do not bear information for the terminal A. Although A can not fully decode $\mathbf{x}_\mathsf{R}$, it has perfect access to $\mathbf{x}_\mathsf{B}$-related part of it. With this shrinking of the codebook size, the number of symbols that can be transmitted (correspondingly the transmission rate) is reduced.

To sum up, by using appropriate zero padding per symbol and *a priori* knowledge at the corresponding terminal, two-way relaying with XOR precoding can support unbalanced downlink rates. In [55], equal downlink rates were assumed, which achieves the two-way broadcast capacity region at only one point [137]. However, with this scheme, all points of the broadcast capacity region are achieved. A current work [147] has taken the presented approach as a basis, and proposed a finite alphabet encoding/decoding scheme based on quadrature amplitude modulation.

Finally, we can write the overall rate region $\mathcal{C}_{\mathsf{two-way}}^{\mathrm{XOR}}$ for the two-way relaying with XOR precoding by replacing (6.17) with (6.21) in (6.20). In the following section, we study the optimization of the covariance matrix $\mathbf{\Lambda}_\mathsf{R}$ in details for different objective functions.

6.2.2.3. Why prefer XOR precoding?

In the sequel of this thesis, we specifically focus on network coding based signal combination at the relay. Our motivation to prefer XOR-precoding arises from its low-complexity structure and high-rate performance.

First, as it simply implements an XOR addition at bit-level, which can be trivially implemented in hardware circuit, the signal combination at the relay requires almost no further signal-processing. Secondly, we need only a single encoding/decoding chain to encode the XORed bit sequence, whereas SPC based schemes require two independent encoders to code the corresponding information sequence of each terminal.

Considering the sufficiency of self-interference cancellation, we notice a fundamental difference between SPC and XOR-precoding based scheme. In the former scheme, decoding is performed after self-interference cancellation, whereas in the latter, self-interference cancellation is applied on the decoded information. Correspondingly, while the terminals only require *receive CSI* for decoding the XOR precoded signal, they additionally need the precoding matrix of the other terminal (hence, CSI of other terminal) to cancel the corresponding interference in the superposition coded signal (see (6.16)).

In summary, XOR-precoding provides a low-complexity solution in three different aspects:

- implementation,
- encoding/decoding,
- and required information for self-interference cancellation.

We next shift our view from complexity to performance. It has been recently shown in [55] that network coding based schemes outperforms SPC based ones in terms of achievable sum rate. Further, it has been proven in [91] that SPC based schemes are in fact suboptimal, and that an optimal approach follows a joint-coding principle. The intuition behind the performance loss of SPC is that there occurs a transmit power coupling between two diagonalized MIMO channels. That is, two different beam sets (two independent covariances) share a common resource of transmit power, which limits the resultant performance. In contrary, the single covariance of the network coding applied information sequence enjoys the full power utilization by incorporating any correlation between the two MIMO channels.

In addition to the above listed advantages of network coding based signal combination, the recent advances in the design of network codes has raised the popularity of XOR-precoding in two-way applications (see [48,68,96,143] and references therein).

Nevertheless, in spite of all above mentioned relative disadvantages of SPC, it has been recently proposed that the back-propagated self-interference can be better utilized in SPC by employing it for channel estimation in the broadcast phase [146], which can not be extended to XOR-precoding based two-way systems.

Chapter 7

Transmit Covariance Optimization for MIMO DF Relay Networks

In this chapter, we study and compare the impact of transmit CSI on both one- and two-way MIMO relaying protocols. Covariance matrices of the transmit signals are designed such that either the sum or the minimum of terminal-to-terminal rates is maximized. Imperfect CSIT case is also addressed within context of this chapter. In overall, it is shown that two-way relaying achieves a quite substantial improvement in spectral efficiency compared to conventional one-way relaying independent of the availability of CSIT.

7.1. Introduction

In the previous chapter, we have derived the rate expressions and capacity regions for both one- and two-way relaying. However, we have not specified the design of covariance matrices, which strictly depends on the availability of transmit CSI at the terminals/relay. In case that there is no CSIT at the nodes, we set the corresponding covariance to be an identity matrix, i.e., feeding spatially white inputs to the transmit antennas.

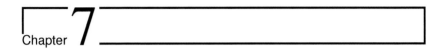

In this chapter, we aim at answering the question of how these end-to-end rate expressions can be optimized if the nodes exploit the knowledge of transmit CSI. Especially, the assumption of CSIT at the relay is a reasonable in practice, since the relay has to estimate the MIMO channels for decoding in the receive mode anyway. So, through the corresponding transmission mode in the reverse direction, this knowledge can be used for precoding by further assuming channel reciprocity and that the bursts are short enough compared to the coherence time of the involved MIMO channels. The importance of CSIT for both single- and multi-user MIMO communications has been reported in various references such as [36,49,60,115,117,125,137].

Contribution of this Chapter The impact of transmit CSI on the capacity of both one- and two-way relaying schemes is studied in this chapter. In the first part, we focus on the capacity region characterization. Later, we consider two commonly employed objective functions in order to optimize the transmit covariance matrices: maximizing the sum rate and maximizing the minimum link rate. As the covariance optimization for one-way relaying is trivially solved through SVD based beamforming combined with water-filling, we rather focus on precoding at the relay for the two-way relaying case. In the literature for MIMO two-way relaying, a suboptimal but trivial approach of precoder optimization is usually taken, where the multiple access and broadcast phases are decoupled. That is, the achievable rates for different phases are independently computed using the well-known schemes proposed in the literature for the general MIMO broadcast and multiple access channels [36,49,115,125]. Later, in order to find the overall two-phase rates, the resultant rates from different phases are cross-checked, i.e., minimum supported rate is chosen. Nevertheless, this decoupling approach is suboptimal since resources can not be fully utilized, i.e., the bottleneck phase can be different for each node/link. Our main contribution with respect to the works such as [137] is that our optimization techniques are based on overall terminal-to-terminal rates, and optimally allocate rate over two phases avoiding decoupling.

In the second part of the chapter, we study the case of imperfect CSIT at the relay for two-way relaying. The mismatch at the channel estimates may arise either from an estimation error or change of channels (outdateness) due to mobility of the nodes. We propose a worst-case robust covariance matrix design, such that both links are served with reliable information rates even in the case of channel uncertainty.

Outline of this Chapter The impact of CSIT on one-way relaying is studied in Section 7.2.1. In Section 7.2.2, we consider the two-way relaying protocol with a close look on efficient precoding schemes at the relay. There, we assume either CSIT at all nodes in the network or only at the relay. Assuming fixed average transmit power constraints, we maximize the sum rate of the data exchanging terminals in Section 7.2.2.1, and later in 7.2.2.2, we introduce max-min fairness to two-way relaying by maximizing the minimum unidirectional (from A to B or vice versa) link rate. We study the performance improvement of employing CSIT at the nodes in Section 7.2.3 through Monte-Carlo simulations, where we also depict the negative impact of decoupling two phases during optimal rate allocation. In Section 7.3, we consider that the CSIT at the relay is uncertain, and we optimize the covariance matrix within defined spherical uncertainty sets for downlink channels. Finally, conclusive remarks are drawn in Section 7.4.

7.2. MIMO DF Relaying with CSIT

7.2.1. One-Way Relaying

When the terminals A and B prefer to apply one-way relaying and have CSIT, the optimization of the covariance matrices $\mathbf{\Lambda}_{\mathsf{A}}$ in (6.2), $\mathbf{\Lambda}_{\mathsf{B}}$ in (6.7), and $\mathbf{\Lambda}_{\mathsf{R}}$ for both directions in (6.4) and (6.8), boil down to be the same with that of point-to-point MIMO capacity maximization optimization. That is, we diagonalize the channel by applying SVD to the corresponding MIMO channel, precoding the transmit signal with the right unitary matrix, and multiplying the received signal at the corresponding node by the hermitian of the left unitary matrix. Assuming fixed transmit power constraint, the maximal point-to-point rate resulting power allocation is obtained by applying water-filling in between the non-zero singular values of the channel [117]. Note that it is reasonable to assume that each node (both terminals and relay) have CSIT knowledge, which is anyway attained during the receive mode of the corresponding node as we readily assume channel reciprocity and slow-fading environment. For instance, during the transmission from A to B, the relay estimates the channel $\mathbf{H}_{\mathsf{A}}$ in the first time slot, and hence, can use its transpose as transmit CSI in the fourth time-slot through the transmission from B to A. Similarly, as the terminal B estimates the channel $\mathbf{H}_{\mathsf{A}}^T$ while receiving information from the relay in the second time-slot, it can use $\mathbf{H}_{\mathsf{A}}$ as transmit CSI in the third time-slot while transmitting to R. However, in case that there is no CSIT at a given node, the corresponding covariance is set to be a diagonal matrix with equal weights for the diagonal elements. To sum up, it can be concluded that it is rather trivial to optimize the covariances for one-way relaying.

Having obtained the maximal transfer rate for each link, we can compute the maximal achievable rate pair $(\mathrm{R}_{\vec{\mathsf{AB}}}, \mathrm{R}_{\vec{\mathsf{BA}}})$ from (6.5) and (6.6). Assuming fixed time-sharing parameters, the capacity region for the rate pair $(\mathrm{R}_{\vec{\mathsf{AB}}}, \mathrm{R}_{\vec{\mathsf{BA}}})$ is nothing but a rectangle. However, we can further optimize the time-allocation parameters α_i such that the sum rate or the minimum link rate is maximized. The sum rate maximization problem is defined as

$$\begin{aligned}\mathcal{P}^{\text{sum}}_{\text{one-way}} \quad : \quad & \max_{\alpha_1,\ldots,\alpha_4,\vec{\mathrm{R}}_{\mathsf{AB}},\vec{\mathrm{R}}_{\mathsf{BA}}} && \vec{\mathrm{R}}_{\mathsf{AB}} + \vec{\mathrm{R}}_{\mathsf{BA}} \\ & \textbf{subject to} && \alpha_1 \vec{\mathrm{R}}^{\star}_{\mathsf{AR}} - \vec{\mathrm{R}}_{\mathsf{AB}} \geq 0, \quad \alpha_2 \vec{\mathrm{R}}^{\star}_{\mathsf{RB}} - \vec{\mathrm{R}}_{\mathsf{AB}} \geq 0 \\ & && \alpha_3 \vec{\mathrm{R}}^{\star}_{\mathsf{BR}} - \vec{\mathrm{R}}_{\mathsf{BA}} \geq 0, \quad \alpha_4 \vec{\mathrm{R}}^{\star}_{\mathsf{RA}} - \vec{\mathrm{R}}_{\mathsf{BA}} \geq 0 \\ & && \sum_{i=1}^{4} \alpha_i = 1, \;\; \vec{\mathrm{R}}_{\mathsf{AB}}, \vec{\mathrm{R}}_{\mathsf{BA}} > 0, \alpha_i > 0, i = 1, \ldots, 4, \end{aligned} \tag{7.1}$$

where the superscript * denotes that the corresponding rate is obtained by optimizing the covariance matrices as explained previously through SVD. The problem $\mathcal{P}^{\text{sum}}_{\text{one-way}}$ is readily modeled as a linear program (LP), and can be solved efficiently with any standard numerical LP solver [77]. On the other hand, replacing the objective function in $\mathcal{P}^{\text{sum}}_{\text{one-way}}$ simply with

$$\max_{\alpha_1,\ldots,\alpha_4,\vec{\mathrm{R}}_{\mathsf{AB}},\vec{\mathrm{R}}_{\mathsf{BA}}} \quad \textbf{min}\,\{\vec{\mathrm{R}}_{\mathsf{AB}}, \vec{\mathrm{R}}_{\mathsf{BA}}\},$$

we can as well optimize the time-sharing parameters for max-min fairness (again through an efficient LP formulation). Consider a special case that there is CSIT available for each of the four transmissions during data exchanging. Such an assumption results in that $\vec{\mathrm{R}}^{\star}_{\mathsf{AR}} = \vec{\mathrm{R}}^{\star}_{\mathsf{RA}}$, $\vec{\mathrm{R}}^{\star}_{\mathsf{BR}} = \vec{\mathrm{R}}^{\star}_{\mathsf{RB}}$. When we aim at max-min fairness, then we trivially obtain $\alpha_1 = \alpha_4$ and $\alpha_2 = \alpha_3$. Consequently, time-share parameters are readily computed as

$$\alpha_1 = \alpha_4 = \frac{1}{2} \cdot \frac{\vec{\mathrm{R}}^{\star}_{\mathsf{RB}}}{\vec{\mathrm{R}}^{\star}_{\mathsf{AR}} + \vec{\mathrm{R}}^{\star}_{\mathsf{RB}}} \quad \text{and} \quad \alpha_2 = \alpha_3 = \frac{1}{2} \cdot \frac{\vec{\mathrm{R}}^{\star}_{\mathsf{AR}}}{\vec{\mathrm{R}}^{\star}_{\mathsf{AR}} + \vec{\mathrm{R}}^{\star}_{\mathsf{RB}}}.$$

7.2.2. Two-Way Relaying

In order to get more insight, let us first identify the impact of CSIT on the achievable two-way rate region which is the intersection of the capacity regions of MAC and BC phases. We consider a simplified case where the time-sharing parameters α_{MAC} and α_{BC} are fixed to be equal to each other. Refer to [89, 145] for considerations on optimal time-sharing between the phases.

MAC Capacity Region As derived for the general MIMO MAC in [49], the capacity region for the MAC phase is

$$\mathcal{C}_{\text{MAC}} = \bigcup_{\substack{\mathbf{\Lambda}_{\mathsf{A}} \succeq 0, \mathbf{\Lambda}_{\mathsf{B}} \succeq 0, \\ \text{Tr}(\mathbf{\Lambda}_{\mathsf{A}}) \leq \text{P}_{\mathsf{A}}, \text{Tr}(\mathbf{\Lambda}_{\mathsf{B}}) \leq \text{P}_{\mathsf{B}}}} \left\{ (\overset{\leftrightarrow}{\text{R}}_{\mathsf{AR}}, \overset{\leftrightarrow}{\text{R}}_{\mathsf{BR}}) \in \mathbb{R}_+ \; : \; \begin{array}{c} \overset{\leftrightarrow}{\text{R}}_{\mathsf{AR}} \leq \overset{\leftrightarrow}{\text{I}}_{\mathsf{AR}}, \; \overset{\leftrightarrow}{\text{R}}_{\mathsf{BR}} \leq \overset{\leftrightarrow}{\text{I}}_{\mathsf{BR}}, \\ \overset{\leftrightarrow}{\text{R}}_{\mathsf{AR}} + \overset{\leftrightarrow}{\text{R}}_{\mathsf{BR}} \leq \overset{\leftrightarrow}{\text{I}}_{\mathsf{AB,MAC}} \end{array} \right\} \quad (7.2)$$

where $\overset{\leftrightarrow}{\text{I}}_{\mathsf{AR}}$, $\overset{\leftrightarrow}{\text{I}}_{\mathsf{BR}}$, and $\overset{\leftrightarrow}{\text{I}}_{\mathsf{AB,MAC}}$ are defined in (6.14). While determining the boundary points of the capacity region, we should pay attention to decoding orders of the terminals. The boundary of the capacity region is in general curved, except at the sum rate point, which is a straight line [49] (Note the difference with respect to no CSIT or given covariances cases). Different set of covariance matrices achieves the points on these curved portions, where each curved portion is the result of a given decoding order. Referring to [49] for further details on information theoretical issues, we propose the following iterative optimization procedure. It is a two stage procedure, where at each stage we assume one of the two possible decoding orders, e.g., either decode B first or decode A first, and correspondingly compute the boundary points of the resultant region. Later, we take a union of these two regions to find the overall MAC capacity region. To summarize:

- **Initialize:** Say we choose to decode B first and A last. Compute the maximal achievable rate for information transfer from A to R independent from B (no interference):

$$\mathcal{P}_{\text{MAC},\mathsf{A}} \; : \; \overset{\leftrightarrow}{\text{R}}{}^{\star}_{\mathsf{AR}} = \max_{\mathbf{\Lambda}_{\mathsf{A}} \succeq 0} \; \log_2 \left| \mathbf{I}_{N_{\mathsf{R}}} + \frac{1}{\sigma^2_{n_{\mathsf{R}}}} \mathbf{H}_{\mathsf{A}} \mathbf{\Lambda}_{\mathsf{A}} \mathbf{H}_{\mathsf{A}}^H \right| \; \textbf{subject to} \; \text{Tr}(\mathbf{\Lambda}_{\mathsf{A}}) \leq P_{\mathsf{A}}.$$

- **Iterate:** Set $\boldsymbol{\tau} := [0, \; \Delta, \; 2\Delta, \; \cdots, \; \overset{\leftrightarrow}{\text{R}}{}^{\star}_{\mathsf{AR}} - \Delta, \; \overset{\leftrightarrow}{\text{R}}{}^{\star}_{\mathsf{AR}}]$, where $\Delta \geq 0$. For each $\tau \in \boldsymbol{\tau}$, solve

$$\begin{aligned} \mathcal{P}_{\text{MAC},\mathsf{AB}} \; : \; \overset{\leftrightarrow}{\text{R}}{}^{\star}_{\mathsf{AB,MAC}}(\tau) = & \max_{\mathbf{\Lambda}_{\mathsf{A}} \succeq 0, \mathbf{\Lambda}_{\mathsf{B}} \succeq 0} \; \log_2 \left| \mathbf{I}_{N_{\mathsf{R}}} + \frac{1}{\sigma^2_{n_{\mathsf{R}}}} (\mathbf{H}_{\mathsf{A}} \mathbf{\Lambda}_{\mathsf{A}} \mathbf{H}_{\mathsf{A}}^H + \mathbf{H}_{\mathsf{B}} \mathbf{\Lambda}_{\mathsf{B}} \mathbf{H}_{\mathsf{B}}^H) \right| \\ & \textbf{subject to} \; \log_2 \left| \mathbf{I}_{N_{\mathsf{R}}} + \frac{1}{\sigma^2_{n_{\mathsf{R}}}} \mathbf{H}_{\mathsf{A}} \mathbf{\Lambda}_{\mathsf{A}} \mathbf{H}_{\mathsf{A}}^H \right| \geq \tau \\ & \qquad \text{Tr}(\mathbf{\Lambda}_{\mathsf{A}}) \leq P_{\mathsf{A}}, \text{Tr}(\mathbf{\Lambda}_{\mathsf{B}}) \leq P_{\mathsf{B}}, \end{aligned}$$

  which maximizes the sum rate assuming that A is decoded last with a minimal rate of τ. Then, the boundary point for given τ is

$$(\overset{\leftrightarrow}{\text{R}}_{\mathsf{AR}}, \overset{\leftrightarrow}{\text{R}}_{\mathsf{BR}}) := (\tau, \mathbf{min}\{\overset{\leftrightarrow}{\text{R}}{}^{\star}_{\mathsf{AB,MAC}}(\tau) - \tau, \overset{\leftrightarrow}{\text{R}}{}^{\star}_{\mathsf{BR}}(\tau)\}), \quad (7.3)$$

where

$$\overset{\leftrightarrow}{\mathrm{R}}^{\star}_{\mathsf{BR}}(\tau) = \log_2 \left| \mathbf{I}_{N_\mathsf{R}} + \frac{1}{\sigma^2_{n\mathsf{R}}} \mathbf{H}_\mathsf{B} \mathbf{\Lambda}^\star_\mathsf{B} \mathbf{H}^H_\mathsf{B} \right|,$$

and $\mathbf{\Lambda}^\star_\mathsf{B}$ is the optimized covariance from $\mathcal{P}_{\mathrm{MAC},\mathsf{AB}}$ for given τ.

Hence, for each τ until $\overset{\leftrightarrow}{\mathrm{R}}^{\star}_{\mathsf{AR}}$, we find another boundary point of the MAC phase capacity for the decoding order where A is decoded last. Note that the smaller Δ is, the better resolution we obtain for the boundary. Both $\mathcal{P}_{\mathrm{MAC},\mathsf{A}}$ and $\mathcal{P}_{\mathrm{MAC},\mathsf{AB}}$ can be efficiently solved through semidefinite programming [26, 77]. In order to gain insight about the corresponding solutions, let us now provide more details on the SDP models.

Implementation of $\mathcal{P}_{\mathrm{MAC},\mathsf{A}}$*:* We first focus on $\mathcal{P}_{\mathrm{MAC},\mathsf{A}}$ and re-formulate it by dropping the logarithm operation as it has no impact on the optimization due to monotonicity:

$$\max_{\mathbf{\Lambda}_\mathsf{A} \succeq 0} \left| \mathbf{I}_{N_\mathsf{R}} + \frac{1}{\sigma^2_{n\mathsf{R}}} \mathbf{H}_\mathsf{A} \mathbf{\Lambda}_\mathsf{A} \mathbf{H}^H_\mathsf{A} \right| \ \textbf{subject to} \ \mathrm{Tr}(\mathbf{\Lambda}_\mathsf{A}) \leq P_\mathsf{A}.$$

Since $\mathbf{\Lambda}_\mathsf{A}$ is constrained to be a positive semidefinite matrix, the arguments of the determinants are also positive semidefinite. However, the determinant constraints are not convex, but can be rendered to be convex through monotonic transformations. The geometric mean of eigenvalues, i.e., $\left(|\mathbf{I}_{N_\mathsf{R}} + \frac{1}{\sigma^2_{n\mathsf{R}}} \mathbf{H}_\mathsf{A} \mathbf{\Lambda}_\mathsf{A} \mathbf{H}^H_\mathsf{A}| \right)^{\frac{1}{N_\mathsf{R}}}$ is a concave function [77, 87]. Moreover, it has been readily implemented by the geomean function of the semidefinite toolbox YALMIP [77], and hence, can be employed in the semidefinite program modeling. Finally, as the power constraint is already represented as an LMI of itself, the problem $\mathcal{P}_{\mathrm{MAC},\mathsf{A}}$ is written in the following final SDP formulation

$$\max_{\mathbf{\Lambda}_\mathsf{A} \succeq 0} \ \mathsf{geomean}\left(\mathbf{I}_{N_\mathsf{R}} + \frac{1}{\sigma^2_{n\mathsf{R}}} \mathbf{H}_\mathsf{A} \mathbf{\Lambda}_\mathsf{A} \mathbf{H}^H_\mathsf{A} \right) \ \textbf{subject to} \ \mathrm{Tr}(\mathbf{\Lambda}_\mathsf{A}) \leq P_\mathsf{A}, \qquad (7.4)$$

and can be efficiently solved with any SDP solver such as YALMIP. Note that as we had dropped the logarithm function and introduced the geomean, the optimal value of (7.4) does not immediately result in $\overset{\leftrightarrow}{\mathrm{R}}^{\star}_{\mathsf{AR}}$. To this end, we first take N_Rth power of the optimal argument of the problem (7.4) and then, compute $\log_2$ of the resultant value. Alternatively, we can directly use $\mathbf{\Lambda}^\star_\mathsf{A}$ so that $\overset{\leftrightarrow}{\mathrm{R}}^{\star}_{\mathsf{AR}} = \log_2 \left| \mathbf{I}_{N_\mathsf{R}} + \frac{1}{\sigma^2_{n\mathsf{R}}} \mathbf{H}_\mathsf{A} \mathbf{\Lambda}^\star_\mathsf{A} \mathbf{H}^H_\mathsf{A} \right|$.

Implementation of $\mathcal{P}_{\mathrm{MAC},\mathsf{AB}}$*:* Next, we examine the solution for the problem $\mathcal{P}_{\mathrm{MAC},\mathsf{AB}}$. Similar to the previous case, we first drop the logarithm operation from the objective function. Note that we can not drop the corresponding logarithm operation from the constraint, since the constraint $\left|\mathbf{I}_{N_\mathsf{R}}+\frac{1}{\sigma^2_{n_\mathsf{R}}}\mathbf{H}_\mathsf{A}\boldsymbol{\Lambda}_\mathsf{A}\mathbf{H}_\mathsf{A}^H\right| \geq \tau$ does not necessarily mean that $\log_2\left|\mathbf{I}_{N_\mathsf{R}}+\frac{1}{\sigma^2_{n_\mathsf{R}}}\mathbf{H}_\mathsf{A}\boldsymbol{\Lambda}_\mathsf{A}\mathbf{H}_\mathsf{A}^H\right| \geq \tau$ holds. Hence, we need to incorporate it with the lower bound τ. By doing so, we obtain the equivalent problem

$$\begin{aligned}
&\max_{\boldsymbol{\Lambda}_\mathsf{A}\succeq 0,\boldsymbol{\Lambda}_\mathsf{B}\succeq 0} \quad \left|\mathbf{I}_{N_\mathsf{R}}+\frac{1}{\sigma^2_{n_\mathsf{R}}}(\mathbf{H}_\mathsf{A}\boldsymbol{\Lambda}_\mathsf{A}\mathbf{H}_\mathsf{A}^H+\mathbf{H}_\mathsf{B}\boldsymbol{\Lambda}_\mathsf{B}\mathbf{H}_\mathsf{B}^H)\right| \\
&\textbf{subject to } \left|\mathbf{I}_{N_\mathsf{R}}+\frac{1}{\sigma^2_{n_\mathsf{R}}}\mathbf{H}_\mathsf{A}\boldsymbol{\Lambda}_\mathsf{A}\mathbf{H}_\mathsf{A}^H\right| \geq 2^\tau,\ \mathrm{Tr}(\boldsymbol{\Lambda}_\mathsf{A})\leq P_\mathsf{A}, \mathrm{Tr}(\boldsymbol{\Lambda}_\mathsf{B})\leq P_\mathsf{B},
\end{aligned}$$

After introducing the geomean functions instead of the determinant operations, we write the final SDP formulation as

$$\begin{aligned}
&\max_{\boldsymbol{\Lambda}_\mathsf{A}\succeq 0,\boldsymbol{\Lambda}_\mathsf{B}\succeq 0} \quad \mathsf{geomean}\Big(\mathbf{I}_{N_\mathsf{R}}+\frac{1}{\sigma^2_{n_\mathsf{R}}}(\mathbf{H}_\mathsf{A}\boldsymbol{\Lambda}_\mathsf{A}\mathbf{H}_\mathsf{A}^H+\mathbf{H}_\mathsf{B}\boldsymbol{\Lambda}_\mathsf{B}\mathbf{H}_\mathsf{B}^H)\Big) \\
&\textbf{subject to } \mathsf{geomean}\Big(\mathbf{I}_{N_\mathsf{R}}+\frac{1}{\sigma^2_{n_\mathsf{R}}}\mathbf{H}_\mathsf{A}\boldsymbol{\Lambda}_\mathsf{A}\mathbf{H}_\mathsf{A}^H\Big) \geq 2^{\frac{\tau}{N_\mathsf{R}}},\ \mathrm{Tr}(\boldsymbol{\Lambda}_\mathsf{A})\leq P_\mathsf{A}, \mathrm{Tr}(\boldsymbol{\Lambda}_\mathsf{B})\leq P_\mathsf{B}, \quad (7.5)
\end{aligned}$$

where we introduced the effect of using geomean to the corresponding lower bound, i.e.,

$$\left|\mathbf{I}_{N_\mathsf{R}}+\frac{1}{\sigma^2_{n_\mathsf{R}}}\mathbf{H}_\mathsf{A}\boldsymbol{\Lambda}_\mathsf{A}\mathbf{H}_\mathsf{A}^H\right| \geq 2^\tau \equiv \mathsf{geomean}\Big(\mathbf{I}_{N_\mathsf{R}}+\frac{1}{\sigma^2_{n_\mathsf{R}}}\mathbf{H}_\mathsf{A}\boldsymbol{\Lambda}_\mathsf{A}\mathbf{H}_\mathsf{A}^H\Big) \geq 2^{\frac{\tau}{N_\mathsf{R}}}.$$

The resultant problem (7.5) is solved with any SDP solver. Note that in order to find the resultant $\overset{\leftrightarrow}{\mathrm{R}}{}^\star_{\mathsf{AB,MAC}}(\tau)$, we should again take care of the logarithm and power effects.

In the second stage of boundary determination, we change the decoding order to decode B last, and perform similar optimization steps as before. Specifically, in the initialization step we compute the maximal achievable rate for information transfer from B to R independent from A; in the iteration steps we maximize the sum rate assuming that B is decoded last with a minimal rate of given τ. Finally, we determine the overall MAC capacity region boundary by taking the union of the two computed sub-boundaries.

BC Capacity Region The broadcast capacity region for two-way relaying is derived in [137], and given as

$$\mathcal{C}_{\mathrm{BC}} = \bigcup_{\boldsymbol{\Lambda}_\mathsf{R}\succeq 0,\mathrm{Tr}(\boldsymbol{\Lambda}_\mathsf{R})\leq P_\mathsf{R}} \left\{(\overset{\leftrightarrow}{\mathrm{R}}_{\mathsf{RA}},\overset{\leftrightarrow}{\mathrm{R}}_{\mathsf{RB}})\in\mathbb{R}_+ \ :\ \overset{\leftrightarrow}{\mathrm{R}}_{\mathsf{RA}}\leq\overset{\leftrightarrow}{\mathrm{I}}_{\mathsf{RA}},\ \ \overset{\leftrightarrow}{\mathrm{R}}_{\mathsf{RB}}\leq\overset{\leftrightarrow}{\mathrm{I}}_{\mathsf{RB}},\right\} \quad (7.6)$$

where $\overleftrightarrow{\mathrm{I}}_{\mathsf{RA}}$ and $\overleftrightarrow{\mathrm{I}}_{\mathsf{RB}}$ are defined in (6.21). The boundary points of this two-way broadcast region can be found by following a very similar iterative optimization procedure to that for MAC phase:

- **Initialize:** Choose any one of the terminals, say A, and compute the maximal achievable rate for information transfer from R to A independent from B:

$$\mathcal{P}_{\mathsf{BC,A}} \;:\; \overleftrightarrow{\mathrm{R}}_{\mathsf{RA}}^{\star} = \max_{\mathbf{\Lambda}_{\mathsf{R}} \succeq 0} \; \log_2 \left| \mathbf{I}_{N_{\mathsf{A}}} + \frac{1}{\sigma_{n_{\mathsf{A}}}^2} \mathbf{H}_{\mathsf{A}}^T \mathbf{\Lambda}_{\mathsf{R}} \mathbf{H}_{\mathsf{A}}^* \right| \;\; \textbf{subject to} \;\; \mathrm{Tr}(\mathbf{\Lambda}_{\mathsf{R}}) \le P_{\mathsf{R}}.$$

- **Iterate:** Set $\boldsymbol{\tau} := [0,\ \Delta,\ 2\Delta,\ \cdots,\ \overleftrightarrow{\mathrm{R}}_{\mathsf{AR}}^{\star} - \Delta,\ \overleftrightarrow{\mathrm{R}}_{\mathsf{RA}}^{\star}]$, where $\Delta \ge 0$. For each $\tau \in \boldsymbol{\tau}$, solve

$$\mathcal{P}_{\mathsf{BC,B}} \;:\; \overleftrightarrow{\mathrm{R}}_{\mathsf{RB}}^{\star}(\tau) = \max_{\mathbf{\Lambda}_{\mathsf{R}} \succeq 0} \; \log_2 \left| \mathbf{I}_{N_{\mathsf{B}}} + \frac{1}{\sigma_{n_{\mathsf{B}}}^2} \mathbf{H}_{\mathsf{B}}^T \mathbf{\Lambda}_{\mathsf{R}} \mathbf{H}_{\mathsf{B}}^* \right|$$

$$\textbf{subject to} \;\; \log_2 \left| \mathbf{I}_{N_{\mathsf{A}}} + \frac{1}{\sigma_{n_{\mathsf{A}}}^2} \mathbf{H}_{\mathsf{A}}^T \mathbf{\Lambda}_{\mathsf{R}} \mathbf{H}_{\mathsf{A}}^* \right| \ge \tau, \;\; \mathrm{Tr}(\mathbf{\Lambda}_{\mathsf{R}}) \le P_{\mathsf{R}}.$$

Then, the boundary point for given τ is

$$(\overleftrightarrow{\mathrm{R}}_{\mathsf{RA}}, \overleftrightarrow{\mathrm{R}}_{\mathsf{RB}}) := (\tau, \overleftrightarrow{\mathrm{R}}_{\mathsf{RB}}^{\star}(\tau)). \tag{7.7}$$

Both $\mathcal{P}_{\mathsf{BC,A}}$ and $\mathcal{P}_{\mathsf{BC,B}}$ can be solved efficiently through semidefinite programming [26, 77] by following similar modeling steps explained previously for MAC boundary.

Two-Way Capacity Region For given time-sharing parameters α_{MAC} and α_{BC}, the achievable rate region for the end-to-end information transfer rate pair $(\overleftrightarrow{\mathrm{R}}_{\mathsf{AB}}, \overleftrightarrow{\mathrm{R}}_{\mathsf{BA}})$ is given by the intersection of MAC and BC phase capacity regions, i.e.,

$$\mathcal{C}_{\mathsf{two-way}}(\alpha_{\mathrm{MAC}}, \alpha_{\mathrm{BC}}) = \mathcal{C}_{\mathrm{MAC}}(\alpha_{\mathrm{MAC}}) \cap \mathcal{C}_{\mathrm{BC}}(\alpha_{\mathrm{BC}}).$$

In Fig. 7.1, we show an example for capacity region boundary of the two-way relay protocol for one specific channel realization. We fix $N_{\mathsf{A}} = N_{\mathsf{B}} = N_{\mathsf{R}} = 2$, $P_{\mathsf{A}} = P_{\mathsf{B}} = P_{\mathsf{R}} = 1$, $\sigma_{n_{\mathsf{A}}}^2 = \sigma_{n_{\mathsf{B}}}^2 = \sigma_{n_{\mathsf{R}}}^2 = 10^{-2}$ (corresponds to an average SNR of 20 dB). Moreover, we set $\alpha_{\mathrm{MAC}} = 0.47$, $\alpha_{\mathrm{BC}} = 0.53$ in order to emphasize the intersection of two capacity regions. The MIMO channels employed in the example are

$$\mathbf{H}_{\mathsf{A}} = \begin{bmatrix} 1.3756 + 1.5238i & -0.1785 - 0.6004i \\ 0.2826 - 0.1480i & 0.3433 + 0.2429i \\ -1.0853 + 0.0659i & 0.1042 + 0.6472i \end{bmatrix},$$

$$\mathbf{H}_{\mathsf{B}} = \begin{bmatrix} 1.3756 + 1.5238i & -0.1785 - 0.6004i \\ 0.2826 - 0.1480i & 0.3433 + 0.2429i \\ -1.0853 + 0.0659i & 0.1042 + 0.6472i \end{bmatrix}.$$

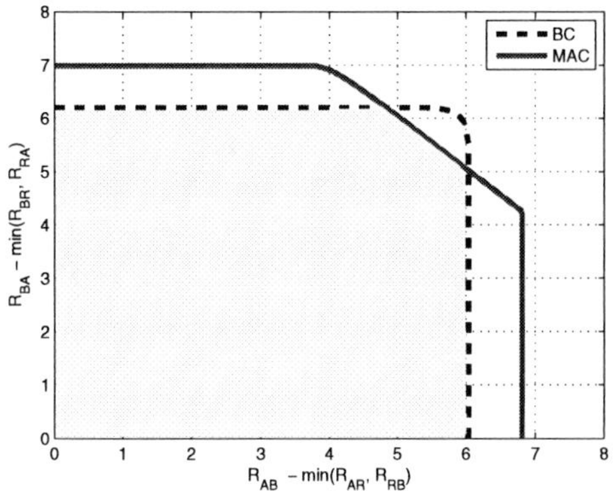

Figure 7.1.: Two-way capacity region is the intersection of MAC and BC phases' capacity regions.

The shaded area in the figure depicts the two-way capacity region for the given channel realizations.

After characterizing the boundary points of the two-way capacity region, we are next interested in finding the optimal boundary points that either maximizes the sum rate or the minimum link rate.

7.2.2.1. Maximizing Sum Rate

Our goal is to find the end-to-end unidirectional rate pair $(\overset{\leftrightarrow}{\mathrm{R}}_{\mathsf{AB}}, \overset{\leftrightarrow}{\mathrm{R}}_{\mathsf{BA}})$ within the capacity region $\mathcal{C}_{\text{two-way}}$ that maximizes the sum rate $\mathrm{R}^{\text{sum}}_{\text{two-way}} = \overset{\leftrightarrow}{\mathrm{R}}_{\mathsf{AB}} + \overset{\leftrightarrow}{\mathrm{R}}_{\mathsf{BA}}$. The general sum rate maximization problem is formulated as

$$\mathcal{P}^{\text{sum}}_{\text{two-way}}: \begin{cases} \max\limits_{\mathbf{\Lambda}_{\mathsf{A}}\succeq 0, \mathbf{\Lambda}_{\mathsf{B}}\succeq 0, \mathbf{\Lambda}_{\mathsf{R}}\succeq 0, \overset{\leftrightarrow}{\mathrm{R}}_{\mathsf{AB}}, \overset{\leftrightarrow}{\mathrm{R}}_{\mathsf{BA}}\geq 0} \quad \overset{\leftrightarrow}{\mathrm{R}}_{\mathsf{AB}} + \overset{\leftrightarrow}{\mathrm{R}}_{\mathsf{BA}} \\ \textbf{subject to} \quad \mathrm{Tr}(\mathbf{\Lambda}_{\mathsf{A}}) \leq P_{\mathsf{A}},\ \mathrm{Tr}(\mathbf{\Lambda}_{\mathsf{B}}) \leq P_{\mathsf{B}},\ \mathrm{Tr}(\mathbf{\Lambda}_{\mathsf{R}}) \leq P_{\mathsf{R}} \\ \overset{\leftrightarrow}{\mathrm{R}}_{\mathsf{AB}} \leq \frac{1}{2}\mathbf{min}\Big\{ \log_2\left|\mathbf{I}_{N_{\mathsf{R}}} + \frac{1}{\sigma^2_{n_{\mathsf{R}}}}\mathbf{H}_{\mathsf{A}}\mathbf{\Lambda}_{\mathsf{A}}\mathbf{H}^H_{\mathsf{A}}\right|, \log_2\left|\mathbf{I}_{N_{\mathsf{B}}} + \frac{1}{\sigma^2_{n_{\mathsf{B}}}}\mathbf{H}^T_{\mathsf{B}}\mathbf{\Lambda}_{\mathsf{R}}\mathbf{H}^*_{\mathsf{B}}\right|\Big\} \\ \overset{\leftrightarrow}{\mathrm{R}}_{\mathsf{BA}} \leq \frac{1}{2}\mathbf{min}\Big\{ \log_2\left|\mathbf{I}_{N_{\mathsf{R}}} + \frac{1}{\sigma^2_{n_{\mathsf{R}}}}\mathbf{H}_{\mathsf{B}}\mathbf{\Lambda}_{\mathsf{B}}\mathbf{H}^H_{\mathsf{B}}\right|, \log_2\left|\mathbf{I}_{N_{\mathsf{A}}} + \frac{1}{\sigma^2_{n_{\mathsf{A}}}}\mathbf{H}^T_{\mathsf{A}}\mathbf{\Lambda}_{\mathsf{R}}\mathbf{H}^*_{\mathsf{A}}\right|\Big\} \\ \overset{\leftrightarrow}{\mathrm{R}}_{\mathsf{AB}} + \overset{\leftrightarrow}{\mathrm{R}}_{\mathsf{BA}} \leq \frac{1}{2}\log_2\left|\mathbf{I}_{N_{\mathsf{R}}} + \frac{1}{\sigma^2_{n_{\mathsf{R}}}}(\mathbf{H}_{\mathsf{A}}\mathbf{\Lambda}_{\mathsf{A}}\mathbf{H}^H_{\mathsf{A}} + \mathbf{H}_{\mathsf{B}}\mathbf{\Lambda}_{\mathsf{B}}\mathbf{H}^H_{\mathsf{B}})\right|, \end{cases} \tag{7.8}$$

where the factor $1/2$ stands for the equal time-sharing between the two phases. To the best of our knowledge, $\mathcal{P}^{\text{sum}}_{\text{two-way}}$, as formulated in (7.8), does not have an immediate trivial solution due to the involved logdet functions which jointly upper and lower bound each other. However, note that the maximal sum rate can always be found through an exhaustive search of the two-way capacity region derived in the previous section.

Here and in the sequel, in order to simplify the analysis, we assume that the terminals A and B do not have transmit CSI. In general, we have assumed that channel matrices stay constant only for a single transmission cycle duration (one MAC plus one BC phase), and that there is no feedback channel from R to A and B. In other words, the transmit CSI can be attained only in receive mode by assuming channel reciprocity. Hence, the assumption of CSIT only at the relay is reasonable in practice, whereas CSIT knowledge at the terminals require either a secure feedback channel or very-slow (immobile) channel environment.

Dropping the CSIT assumption at the terminals, the corresponding transmit covariances become $\mathbf{\Lambda}_{\mathsf{n}} = (P_{\mathsf{n}}/N_{\mathsf{n}}) \cdot \mathbf{I}_{N_{\mathsf{n}}}$ for $\mathsf{n} \in \{\mathsf{A}, \mathsf{B}\}$. Similarly, the MAC phase rate regions reduced to be a simple polyhedron and the sum rate maximization problem in (7.8) modifies to

$$\mathcal{P}^{\text{sum}}_{\text{two-way}} : \begin{cases} \max\limits_{\mathbf{\Lambda}_{\mathsf{R}} \succeq 0, \overleftrightarrow{\mathrm{R}}_{\mathsf{AB}}, \overleftrightarrow{\mathrm{R}}_{\mathsf{BA}} \geq 0} & \overleftrightarrow{\mathrm{R}}_{\mathsf{AB}} + \overleftrightarrow{\mathrm{R}}_{\mathsf{BA}} \\ \textbf{subject to} & \overleftrightarrow{\mathrm{R}}_{\mathsf{AB}} \leq \frac{1}{2} \mathbf{min} \Big\{ \overleftrightarrow{\mathrm{I}}_{\mathsf{AR}}, \log_2 \Big| \mathbf{I}_{N_{\mathsf{B}}} + \frac{1}{\sigma^2_{n_{\mathsf{B}}}} \mathbf{H}^T_{\mathsf{B}} \mathbf{\Lambda}_{\mathsf{R}} \mathbf{H}^*_{\mathsf{B}} \Big| \Big\}, \\ & \overleftrightarrow{\mathrm{R}}_{\mathsf{BA}} \leq \frac{1}{2} \mathbf{min} \Big\{ \overleftrightarrow{\mathrm{I}}_{\mathsf{BR}}, \log_2 \Big| \mathbf{I}_{N_{\mathsf{A}}} + \frac{1}{\sigma^2_{n_{\mathsf{A}}}} \mathbf{H}^T_{\mathsf{A}} \mathbf{\Lambda}_{\mathsf{R}} \mathbf{H}^*_{\mathsf{A}} \Big| \Big\}, \\ & \overleftrightarrow{\mathrm{R}}_{\mathsf{AB}} + \overleftrightarrow{\mathrm{R}}_{\mathsf{BA}} \leq \frac{1}{2} \overleftrightarrow{\mathrm{I}}_{\mathsf{AB,MAC}}, \;\; \mathrm{Tr}(\mathbf{\Lambda}_{\mathsf{R}}) \leq P_{\mathsf{R}}. \end{cases} \tag{7.9}$$

Taking a closer look on (7.9), we observe that we are, in fact, maximizing the sum of the two logdet expressions, which are upper bounded by scalars due to the minimum operations. As we attempt to upper bound a concave function, we face with a non-convex problem formulation. Besides, the current structure of the problem does not let itself for trivial numerical implementation with the available semidefinite tool boxes such as YALMIP [77], SeDuMi [116] or SDPT3 [121]. Determinant incorporating problems can be solved in several different ways with the aforementioned toolboxes, however with some restrictions. If you have a determinant (or sum of determinants) in your objective function, the built-in function logdet of SDPT3 can be used to model the objective. However, this function can be solely used in the objective but not in constraints consisting logdet expressions. On the other hand, as addressed before, determinant expressions can be also modeled via the geometric mean of eigenvalues of the corresponding semidefinite matrix (there is a builtin function geomean in YALMIP). This function has the advantage of being effectively used in both objective and constraints. However, we note that since the modeling via geomean relies on monotonic equivalence, it should not be used in mixed expression in which determinant is combined with other functions. For instance, if sum of two logdet functions is desired to be maximized, each of these can, of course, be modeled through geomean; but the sum of two geomean functions would not give the optimal result due to conversion.

In order to further simplify the problem definition equivalently, we drop the MAC sum constraint $\overleftrightarrow{\mathrm{I}}_{\mathsf{AB},\mathrm{MAC}}$ which does not have impact on the optimal argument of $\mathcal{P}^{\mathrm{sum}}_{\mathrm{two-way}}$. However, after solving the simplified problem, we have to check if the sum of the optimized rate pair fulfills $\overleftrightarrow{\mathrm{I}}_{\mathsf{AB},\mathrm{MAC}}$ or not. Further to facilitate the solution of the simplified $\mathcal{P}^{\mathrm{sum}}_{\mathrm{two-way}}$, we decompose it into four sub-problems

$$\mathcal{P}^{\mathrm{sum}}_{\mathrm{two-way}} = \mathcal{P}^{\mathrm{sum},1}_{\mathrm{two-way}} \cup \mathcal{P}^{\mathrm{sum},2}_{\mathrm{two-way}} \cup \mathcal{P}^{\mathrm{sum},3}_{\mathrm{two-way}} \cup \mathcal{P}^{\mathrm{sum},4}_{\mathrm{two-way}},$$

where

$$\mathcal{P}^{\mathrm{sum},1}_{\mathrm{two-way}} : \begin{cases} \max\limits_{\mathbf{\Lambda}_{\mathsf{R}}\succeq 0} \ \min\left\{\overleftrightarrow{\mathrm{I}}_{\mathsf{AR}}, \log_2\left|\mathbf{I}_{N_{\mathsf{B}}} + \frac{1}{\sigma^2_{n_{\mathsf{B}}}}\mathbf{H}^T_{\mathsf{B}}\mathbf{\Lambda}_{\mathsf{R}}\mathbf{H}^*_{\mathsf{B}}\right|\right\}, \\ \textbf{subject to} \quad \overleftrightarrow{\mathrm{I}}_{\mathsf{BR}} \le \log_2\left|\mathbf{I}_{N_{\mathsf{A}}} + \frac{1}{\sigma^2_{n_{\mathsf{A}}}}\mathbf{H}^T_{\mathsf{A}}\mathbf{\Lambda}_{\mathsf{R}}\mathbf{H}^*_{\mathsf{A}}\right|, \mathrm{Tr}(\mathbf{\Lambda}_{\mathsf{R}}) \le P_{\mathsf{R}}, \end{cases} \tag{7.10}$$

$$\mathcal{P}^{\mathrm{sum},2}_{\mathrm{two-way}} : \begin{cases} \max\limits_{\mathbf{\Lambda}_{\mathsf{R}}\succeq 0} \ \min\left\{\overleftrightarrow{\mathrm{I}}_{\mathsf{BR}}, \log_2\left|\mathbf{I}_{N_{\mathsf{A}}} + \frac{1}{\sigma^2_{n_{\mathsf{A}}}}\mathbf{H}^T_{\mathsf{A}}\mathbf{\Lambda}_{\mathsf{R}}\mathbf{H}^*_{\mathsf{A}}\right|\right\}, \\ \textbf{subject to} \quad \overleftrightarrow{\mathrm{I}}_{\mathsf{AR}} \le \log_2\left|\mathbf{I}_{N_{\mathsf{B}}} + \frac{1}{\sigma^2_{n_{\mathsf{B}}}}\mathbf{H}^T_{\mathsf{B}}\mathbf{\Lambda}_{\mathsf{R}}\mathbf{H}^*_{\mathsf{B}}\right|, \mathrm{Tr}(\mathbf{\Lambda}_{\mathsf{R}}) \le P_{\mathsf{R}}, \end{cases} \tag{7.11}$$

$$\mathcal{P}^{\mathrm{sum},3}_{\mathrm{two-way}} : \begin{cases} \textbf{check feasibility} \ \overleftrightarrow{\mathrm{I}}_{\mathsf{AR}} \le \log_2\left|\mathbf{I}_{N_{\mathsf{B}}} + \frac{1}{\sigma^2_{n_{\mathsf{B}}}}\mathbf{H}^T_{\mathsf{B}}\mathbf{\Lambda}_{\mathsf{R}}\mathbf{H}^*_{\mathsf{B}}\right|, \\ \qquad \overleftrightarrow{\mathrm{I}}_{\mathsf{BR}} \le \log_2\left|\mathbf{I}_{N_{\mathsf{A}}} + \frac{1}{\sigma^2_{n_{\mathsf{A}}}}\mathbf{H}^T_{\mathsf{A}}\mathbf{\Lambda}_{\mathsf{R}}\mathbf{H}^*_{\mathsf{A}}\right|, \\ \qquad \mathrm{Tr}(\mathbf{\Lambda}_{\mathsf{R}}) \le P_{\mathsf{R}}, \mathbf{\Lambda}_{\mathsf{R}} \succeq 0, \end{cases} \tag{7.12}$$

$$\mathcal{P}^{\mathrm{sum},4}_{\mathrm{two-way}} : \begin{cases} \max\limits_{\mathbf{\Lambda}_{\mathsf{R}}\succeq 0} \ \log_2\left|\mathbf{I}_{N_{\mathsf{A}}} + \frac{1}{\sigma^2_{n_{\mathsf{A}}}}\mathbf{H}^T_{\mathsf{A}}\mathbf{\Lambda}_{\mathsf{R}}\mathbf{H}^*_{\mathsf{A}}\right| + \log_2\left|\mathbf{I}_{N_{\mathsf{B}}} + \frac{1}{\sigma^2_{n_{\mathsf{B}}}}\mathbf{H}^T_{\mathsf{B}}\mathbf{\Lambda}_{\mathsf{R}}\mathbf{H}^*_{\mathsf{B}}\right|, \\ \textbf{subject to} \quad \overleftrightarrow{\mathrm{I}}_{\mathsf{BR}} > \log_2\left|\mathbf{I}_{N_{\mathsf{A}}} + \frac{1}{\sigma^2_{n_{\mathsf{A}}}}\mathbf{H}^T_{\mathsf{A}}\mathbf{\Lambda}_{\mathsf{R}}\mathbf{H}^*_{\mathsf{A}}\right|, \\ \qquad \overleftrightarrow{\mathrm{I}}_{\mathsf{BR}} > \log_2\left|\mathbf{I}_{N_{\mathsf{A}}} + \frac{1}{\sigma^2_{n_{\mathsf{A}}}}\mathbf{H}^T_{\mathsf{A}}\mathbf{\Lambda}_{\mathsf{R}}\mathbf{H}^*_{\mathsf{A}}\right|, \ \mathrm{Tr}(\mathbf{\Lambda}_{\mathsf{R}}) \le P_{\mathsf{R}}. \end{cases} \tag{7.13}$$

Note that the time-sharing factor 1/2 has been dropped from all sub-problems since it is ineffective for the optimization. We focus on each sub-problem individually (see the sub-figures 1 to 4 in Fig. 7.2 for the corresponding illustration of each optimization region):

e

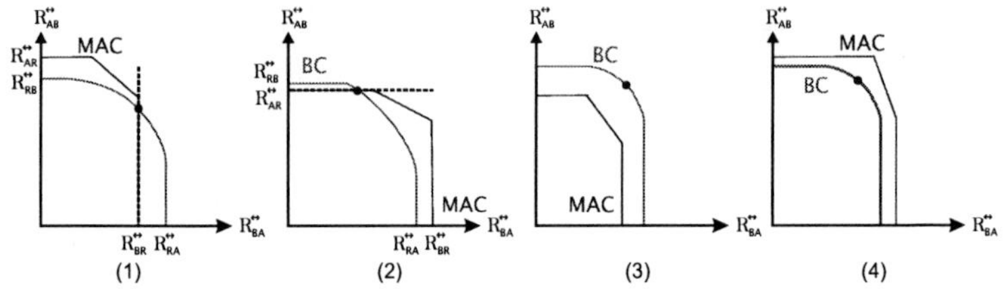

Figure 7.2.: Illustration of optimization region for each decomposed sub-problem for two-way sum rate maximization.

1. The sub-problem $\mathcal{P}^{\text{sum},1}_{\text{two-way}}$ assumes that B-to-R link (MAC) is bottleneck for the transmission from B to A, and hence, aims at utilizing the resources to maximize the transfer rate from A to B while assigning as much as MAC supported rate to the reverse-direction. In the first sub-figure in Fig. 7.2, this corresponds to search on the dashed vertical line. For implementation, the minimum operation in the objective can be dropped and the rate expression representing the transfer from R to B can be maximized directly (see modeling of $\mathcal{P}_{\text{MAC,AB}}$ for further details). However, afterwards we have to check if the optimal value $\overleftrightarrow{R}_{\text{RB}}$ fulfills $\overleftrightarrow{I}_{\text{AR}}$ or not (take the minimum).

2. Similar to $\mathcal{P}^{\text{sum},1}_{\text{two-way}}$, the sub-problem $\mathcal{P}^{\text{sum},2}_{\text{two-way}}$ assumes that A-to-R link (MAC) is bottleneck for the transmission from A to B, and hence, maximizes the transfer rate from B to A. As illustrated in the second sub-figure in Fig. 7.2, we search on the dashed horizontal line. Likewise for implementation, the minimum operation in the objective can be dropped and the rate expression representing the transfer from R to A can be maximized directly (see modeling of $\mathcal{P}_{\text{MACAB}}$ for further details). Note again that the resultant rates should be checked to fulfill $\overleftrightarrow{I}_{\text{BR}}$.

3. The sub-problem $\mathcal{P}^{\text{sum},4}_{\text{two-way}}$ assumes that MAC phase links (A to R and B to R) are bottleneck for both data transfer directions, and verifies this assumption through a standard semidefinite feasibility check.

4. Lastly, the sub-problem $\mathcal{P}^{\text{sum},4}_{\text{two-way}}$ assumes that none of the MAC phase links limits the two-way traffic, but the BC phase links are causing the bottleneck. Hence, it tries to maximize the sum rate within the two-way broadcast capacity region. Unfortunately, in contrary with the previous three, the problem formulation of this sub-problem is non-convex due to the upper bound constraints on rate expressions.

To summarize, an involved way of solving the sum rate maximization problem is to solve the first three sub-problems through SDP, and then to apply an iterative line search approach for the fourth sub-problem (we lower bound one of the link rates with a threshold, maximizes the other, and increment this threshold iteratively). At last, we take the union of the results, i.e., choose the maximum sum rate obtained out of the feasible sub-problems.

Instead, in the following we propose a much simpler optimization procedure, which benefits from a special order of applying the aforementioned sub-problems and incorporates the crossing points of the MAC and BC phase capacity regions. We summarize the steps of the proposed optimization procedure as follows:

- Step 1: We drop the rate upper bound constraints from $\mathcal{P}^{\text{sum},4}_{\text{two-way}}$, and optimize the sum only with power and semi-definiteness constraints, i.e.,

$$\begin{aligned}&\max_{\mathbf{\Lambda}_{\mathsf{R}}\succeq 0} \;\; \log_2\left|\mathbf{I}_{N_{\mathsf{A}}}+\frac{1}{\sigma^2_{n_{\mathsf{A}}}}\mathbf{H}_{\mathsf{A}}^T\mathbf{\Lambda}_{\mathsf{R}}\mathbf{H}_{\mathsf{A}}^*\right|+\log_2\left|\mathbf{I}_{N_{\mathsf{B}}}+\frac{1}{\sigma^2_{n_{\mathsf{B}}}}\mathbf{H}_{\mathsf{B}}^T\mathbf{\Lambda}_{\mathsf{R}}\mathbf{H}_{\mathsf{B}}^*\right|\\&\textbf{subject to} \quad \operatorname{Tr}(\mathbf{\Lambda}_{\mathsf{R}})\le P_{\mathsf{R}},\end{aligned} \tag{7.14}$$

  with which we are finding the maximal sum rate for the BC phase ignoring the MAC bounds on the corresponding rates. This maximization is a concave problem with a unique global maxima, since each rate expression is concave and sum of concave functions are again concave [26]. Each single rate expression in the objective, which has positive semidefinite argument by definition, can be modeled with the built-in function logdet of SDPT3 tool, and hence, can be solved efficiently through SDP. Recall that as we mentioned previously, we can not use geomean function here instead of logdet. Because the optimal covariance maximizing the sum of two geomean does not necessarily maximizes the sum of the logarithm functions, i.e.,

$$\sum_{\mathsf{n}\in\{\mathsf{A},\mathsf{B}\}} \log_2\left|\mathbf{I}_{N_{\mathsf{n}}}+\frac{1}{\sigma^2_{n_{\mathsf{n}}}}\mathbf{H}_{\mathsf{n}}^T\mathbf{\Lambda}_{\mathsf{R}}\mathbf{H}_{\mathsf{n}}^*\right|$$

  is not monotonically equivalent to

$$\sum_{\mathsf{n}\in\{\mathsf{A},\mathsf{B}\}} \mathsf{geomean}\left(\mathbf{I}_{N_{\mathsf{n}}}+\frac{1}{\sigma^2_{n_{\mathsf{n}}}}\mathbf{H}_{\mathsf{n}}^T\mathbf{\Lambda}_{\mathsf{R}}\mathbf{H}_{\mathsf{n}}^*\right)\equiv\sum_{\mathsf{n}\in\{\mathsf{A},\mathsf{B}\}}\left(\left|\mathbf{I}_{N_{\mathsf{n}}}+\frac{1}{\sigma^2_{n_{\mathsf{n}}}}\mathbf{H}_{\mathsf{n}}^T\mathbf{\Lambda}_{\mathsf{R}}\mathbf{H}_{\mathsf{n}}^*\right|\right)^{\frac{1}{N_{\mathsf{n}}}}$$

  in terms of $\mathbf{\Lambda}_{\mathsf{R}}$.

- Step 2: Having obtained the optimal rate pair $(\overleftrightarrow{R}_{RA}^{\star}, \overleftrightarrow{R}_{RB}^{\star})$, check the following clauses:
  - If $\overleftrightarrow{R}_{RA}^{\star} > \overleftrightarrow{I}_{BR}$ and $\overleftrightarrow{R}_{RB}^{\star} > \overleftrightarrow{I}_{AR}$,

    Set $R_{two-way}^{sum} := \overleftrightarrow{I}_{AB,MAC}$ (To determine $\overleftrightarrow{R}_{AB}$ and $\overleftrightarrow{R}_{BA}$ specifically, choose any pair on the MAC region boundary satisfying the sum rate.) Done ✓
  - else if $\overleftrightarrow{R}_{RA}^{\star} > \overleftrightarrow{I}_{BR}$ and $\overleftrightarrow{R}_{RB}^{\star} \leq \overleftrightarrow{I}_{AR}$,

    Set $\overleftrightarrow{R}_{BA} := \overleftrightarrow{I}_{BR}$ and go to Step 3.
  - else if $\overleftrightarrow{R}_{RA}^{\star} \leq \overleftrightarrow{I}_{BR}$ and $\overleftrightarrow{R}_{RB}^{\star} > \overleftrightarrow{I}_{AR}$,

    Set $\overleftrightarrow{R}_{AB} := \overleftrightarrow{I}_{AR}$ and go to Step 4.
  - else if $\overleftrightarrow{R}_{RA}^{\star} + \overleftrightarrow{R}_{RB}^{\star} <= \overleftrightarrow{I}_{AB,MAC}$,

    Set $R_{two-way}^{sum} = \overleftrightarrow{R}_{RA}^{\star} + \overleftrightarrow{R}_{RB}^{\star}$ $(\overleftrightarrow{R}_{AB} := \overleftrightarrow{R}_{RB}^{\star}, \overleftrightarrow{R}_{BA} := \overleftrightarrow{R}_{RA}^{\star})$. Done ✓
  - else

    Set $R_{two-way}^{sum} := \overleftrightarrow{I}_{AB,MAC}$ (To determine $\overleftrightarrow{R}_{AB}$ and $\overleftrightarrow{R}_{BA}$ specifically, choose any pair on the MAC region boundary satisfying the sum rate.) Done ✓
- Step 3: Solve $\mathcal{P}_{two-way}^{sum,1}$ and set $\overleftrightarrow{R}_{AB} = \mathbf{min}\{\overleftrightarrow{I}_{AR}, \overleftrightarrow{R}_{RB}^{\star}\}$. Check if the determined $\overleftrightarrow{R}_{AB}$ and $\overleftrightarrow{R}_{BA}$ fulfills $\overleftrightarrow{I}_{AB,MAC}$. Done ✓
- Step 4: Solve $\mathcal{P}_{two-way}^{sum,2}$ and set $\overleftrightarrow{R}_{BA} = \mathbf{min}\{\overleftrightarrow{I}_{BR}, \overleftrightarrow{R}_{RA}^{\star}\}$. Check if the determined $\overleftrightarrow{R}_{AB}$ and $\overleftrightarrow{R}_{BA}$ fulfills $\overleftrightarrow{I}_{AB,MAC}$. Done ✓

Summing up, we find the maximal two-way sum rate only by solving at most two SDPs with this four-step optimization procedure. Let us now elaborate on the individual steps so that we gain more insight on the optimality of the optimization. After finding the rate pair that maximizes sum rate of BC phase in Step 1, in Step 2 we check if this point represented by the chosen pair is within the MAC phase capacity boundary or not. If yes, which is described by the fourth clause in Step 2, we are done and find the maximal sum rate (see the first sub-figure in Fig. 7.3 for an illustration). If the point is outside the MAC boundary and none of its projections on the axes are satisfied by the individual MAC link constraints (the first clause), MAC sum rate is defined to be the maximal two-way sum rate (see the second sub-figure in Fig. 7.3). If any one of the projections satisfies the corresponding individual MAC link constraint, i.e., one-sided BC rate is supported by one-sided MAC rate, then the point falls in either the

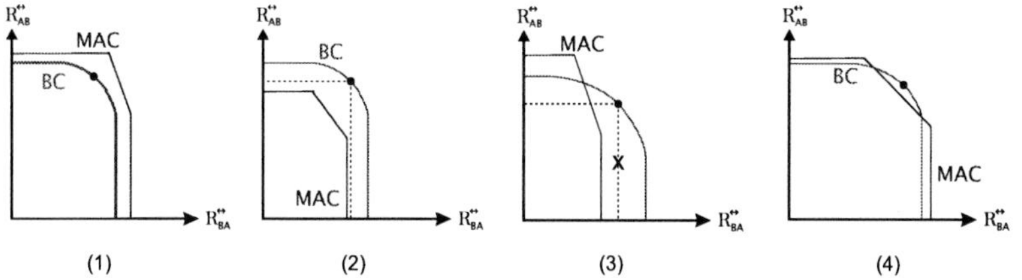

Figure 7.3.: Illustrative elaborations on the proposed sum rate optimization procedure.

second or the third clause (see the third sub-figure in Fig. 7.3), and we need to solve the corresponding sub-problem. Finally, if both of the projections satisfy the corresponding individual MAC link constraint, but the point is still outside the MAC boundary, we are again limited by the MAC sum rate (see the last sub-figure in Fig. 7.3).

7.2.2.2. Maximizing Minimum Link Rate

Two-way rate allocation aiming at maximum sum rate may result in an unfair resource management especially in asymmetric channel conditions. In order to support reliable and sufficient transfer rates for both directions, fairness can be introduced into the rate allocation. We embrace a max-min type fairness, which maximizes the minimum end-to-end unidirectional rate and tries to allocate as much as possible equal rates to each direction. Assuming that CSIT is only available at the relay, the corresponding max-min fairness aiming covariance optimization is formulated as

$$\mathcal{P}^{\text{mm}}_{\text{two-way}} : \left\{ \begin{array}{ll} \max\limits_{\boldsymbol{\Lambda}_{\mathsf{R}}\succeq 0, \overleftrightarrow{\mathrm{R}}_{\mathsf{AB}}, \overleftrightarrow{\mathrm{R}}_{\mathsf{BA}}\geq 0} & \min\{\overleftrightarrow{\mathrm{R}}_{\mathsf{AB}}, \overleftrightarrow{\mathrm{R}}_{\mathsf{BA}}\} \\ \textbf{subject to} & \overleftrightarrow{\mathrm{R}}_{\mathsf{AB}} \leq \frac{1}{2}\min\left\{\overleftrightarrow{\mathrm{I}}_{\mathsf{AR}}, \log_2\left|\mathbf{I}_{N_{\mathsf{B}}} + \frac{1}{\sigma^2_{n_{\mathsf{B}}}}\mathbf{H}^T_{\mathsf{B}}\boldsymbol{\Lambda}_{\mathsf{R}}\mathbf{H}^*_{\mathsf{B}}\right|\right\}, \\ & \overleftrightarrow{\mathrm{R}}_{\mathsf{BA}} \leq \frac{1}{2}\min\left\{\overleftrightarrow{\mathrm{I}}_{\mathsf{BR}}, \log_2\left|\mathbf{I}_{N_{\mathsf{A}}} + \frac{1}{\sigma^2_{n_{\mathsf{A}}}}\mathbf{H}^T_{\mathsf{A}}\boldsymbol{\Lambda}_{\mathsf{R}}\mathbf{H}^*_{\mathsf{A}}\right|\right\}, \\ & \overleftrightarrow{\mathrm{R}}_{\mathsf{AB}} + \overleftrightarrow{\mathrm{R}}_{\mathsf{BA}} \leq \frac{1}{2}\overleftrightarrow{\mathrm{I}}_{\mathsf{AB},\mathrm{MAC}}, \ \ \mathrm{Tr}(\boldsymbol{\Lambda}_{\mathsf{R}}) \leq P_{\mathsf{R}}. \end{array} \right. \quad (7.15)$$

With appropriate modeling, $\mathcal{P}^{\text{mm}}_{\text{two-way}}$ can be efficiently solved through SDP. However, the optimal covariance would "only" result in the maximum possible *equal* rate allocation. In other words, although we maximize the minimum link rate as much as possible, there may be a better covariance choice that can further increase the rate of the non-minimum link rate without decreasing the minimum one. Such a scenario is illustrated in Fig. 7.4. There, the point "max-min" depicts a typical result of $\mathcal{P}^{\text{mm}}_{\text{two-way}}$, where the data transfer from A to B is limited by $\overleftrightarrow{\mathrm{R}}^{(1)}_{\mathsf{AB}}$ for max-min decision. Whereas, assigning

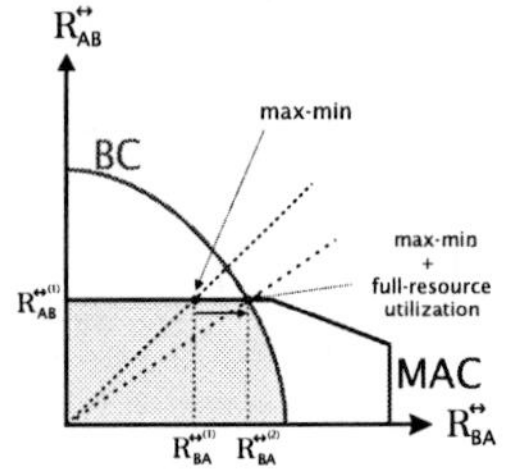

Figure 7.4.: Illustration of alternatives for max-min fairness optimization.

still a rate of $\overset{\leftrightarrow}{\mathrm{R}}_{\mathsf{AB}}^{(1)}$ to this direction, we can improve the rate of the reverse-direction from $\overset{\leftrightarrow}{\mathrm{R}}_{\mathsf{BA}}^{(1)}$ to $\overset{\leftrightarrow}{\mathrm{R}}_{\mathsf{BA}}^{(2)}$ as shown in the figure. Summing up, we can design a better covariance than the one obtained from $\mathcal{P}^{\mathrm{mm}}_{\mathsf{two-way}}$ for efficient utilization of the resources while supporting the most max-min fair rates. Note that such a *tricky* behaviour is in contrast with other similar max-min formulations, and arising from the fact that the capacity region at hand has some horizontal and/or vertical plautos. Putting it differently, in case that we have a convex region without any strict corners, the result from $\mathcal{P}^{\mathrm{mm}}_{\mathsf{two-way}}$ would be the best to achieve.

To this end, we propose the following optimization procedure that is similar to the one employed for sum rate maximization. The optimization steps are given as follows:

- Step 1: We maximize the minimum link within the BC capacity region by ignoring the MAC phase constraints, i.e.,

$$\begin{aligned}&\max_{\mathbf{\Lambda}_{\mathsf{R}}\succeq 0}\ \min\left\{\log_2\left|\mathbf{I}_{N_{\mathsf{A}}}+\frac{1}{\sigma^2_{n_{\mathsf{A}}}}\mathbf{H}_{\mathsf{A}}^T\mathbf{\Lambda}_{\mathsf{R}}\mathbf{H}_{\mathsf{A}}^*\right|,\log_2\left|\mathbf{I}_{N_{\mathsf{B}}}+\frac{1}{\sigma^2_{n_{\mathsf{B}}}}\mathbf{H}_{\mathsf{B}}^T\mathbf{\Lambda}_{\mathsf{R}}\mathbf{H}_{\mathsf{B}}^*\right|\right\}\\&\textbf{subject to}\quad \mathrm{Tr}(\mathbf{\Lambda}_{\mathsf{R}})\leq P_{\mathsf{R}}.\end{aligned}$$

  To do so, we model each logarithmic expression with geomean and lower bound them with a common threshold as we did in similar max-min problems in the first part of this thesis (see Section 3.2).

- Step 2: Having obtained the max-min optimal rate pair $(\overset{\leftrightarrow}{\mathrm{R}}_{\mathsf{RA}}^{\star},\overset{\leftrightarrow}{\mathrm{R}}_{\mathsf{RB}}^{\star})$, check the following clauses:

  - If $\overset{\leftrightarrow}{\mathrm{R}}_{\mathsf{RA}}^{\star}\leq\overset{\leftrightarrow}{\mathrm{I}}_{\mathsf{BR}}$ and $\overset{\leftrightarrow}{\mathrm{R}}_{\mathsf{RB}}^{\star}>\overset{\leftrightarrow}{\mathrm{I}}_{\mathsf{AR}}$,

    Set $\overset{\leftrightarrow}{\mathrm{R}}_{\mathsf{AB}}:=\overset{\leftrightarrow}{\mathrm{I}}_{\mathsf{AR}}$ and go to Step 3.

  - else if $\overset{\leftrightarrow}{\mathrm{R}}_{\mathsf{RA}}^{\star}>\overset{\leftrightarrow}{\mathrm{I}}_{\mathsf{BR}}$ and $\overset{\leftrightarrow}{\mathrm{R}}_{\mathsf{RB}}^{\star}\leq\overset{\leftrightarrow}{\mathrm{I}}_{\mathsf{AR}}$,

Set $\overset{\leftrightarrow}{\mathrm{R}}_{\mathsf{BA}} := \overset{\leftrightarrow}{\mathrm{I}}_{\mathsf{BR}}$ and go to Step 4.

- else

Go to Step 5.

- Step 3: Solve

$$\begin{aligned}&\max_{\boldsymbol{\Lambda}_{\mathsf{R}}\succeq 0}\ \log_2\left|\mathbf{I}_{N_{\mathsf{A}}}+\frac{1}{\sigma^2_{n_{\mathsf{A}}}}\mathbf{H}_{\mathsf{A}}^T\boldsymbol{\Lambda}_{\mathsf{R}}\mathbf{H}_{\mathsf{A}}^*\right|\\ &\textbf{subject to}\ \ \log_2\left|\mathbf{I}_{N_{\mathsf{B}}}+\frac{1}{\sigma^2_{n_{\mathsf{B}}}}\mathbf{H}_{\mathsf{B}}^T\boldsymbol{\Lambda}_{\mathsf{R}}\mathbf{H}_{\mathsf{B}}^*\right|\geq\overset{\leftrightarrow}{\mathrm{I}}_{\mathsf{AR}},\ \mathrm{Tr}(\boldsymbol{\Lambda}_{\mathsf{R}})\leq P_{\mathsf{R}},\end{aligned}$$

set $\overset{\leftrightarrow}{\mathrm{R}}_{\mathsf{BA}} := \mathbf{min}\left\{\log_2\left|\mathbf{I}_{N_{\mathsf{A}}}+\frac{1}{\sigma^2_{n_{\mathsf{A}}}}\mathbf{H}_{\mathsf{A}}^T\boldsymbol{\Lambda}_{\mathsf{R}}^{\star}\mathbf{H}_{\mathsf{A}}^*\right|,\overset{\leftrightarrow}{\mathrm{I}}_{\mathsf{BR}}\right\}$ and go to Step 5.

- Step 4: Solve

$$\begin{aligned}&\max_{\boldsymbol{\Lambda}_{\mathsf{R}}\succeq 0}\ \log_2\left|\mathbf{I}_{N_{\mathsf{B}}}+\frac{1}{\sigma^2_{n_{\mathsf{B}}}}\mathbf{H}_{\mathsf{B}}^T\boldsymbol{\Lambda}_{\mathsf{R}}\mathbf{H}_{\mathsf{B}}^*\right|\\ &\textbf{subject to}\ \ \log_2\left|\mathbf{I}_{N_{\mathsf{A}}}+\frac{1}{\sigma^2_{n_{\mathsf{A}}}}\mathbf{H}_{\mathsf{A}}^T\boldsymbol{\Lambda}_{\mathsf{R}}\mathbf{H}_{\mathsf{A}}^*\right|\geq\overset{\leftrightarrow}{\mathrm{I}}_{\mathsf{BR}},\ \mathrm{Tr}(\boldsymbol{\Lambda}_{\mathsf{R}})\leq P_{\mathsf{R}},\end{aligned}$$

set $\overset{\leftrightarrow}{\mathrm{R}}_{\mathsf{AB}} := \mathbf{min}\left\{\log_2\left|\mathbf{I}_{N_{\mathsf{B}}}+\frac{1}{\sigma^2_{n_{\mathsf{B}}}}\mathbf{H}_{\mathsf{B}}^T\boldsymbol{\Lambda}_{\mathsf{R}}^{\star}\mathbf{H}_{\mathsf{B}}^*\right|,\overset{\leftrightarrow}{\mathrm{I}}_{\mathsf{AR}}\right\}$ and go to Step 5.

- Step 5: Having found candidates $(\overset{\leftrightarrow}{\mathrm{R}}_{\mathsf{AB}},\overset{\leftrightarrow}{\mathrm{R}}_{\mathsf{BA}})$, decide on the final link rates after checking the MAC sum bound:

$$(\overset{\leftrightarrow}{\mathrm{R}}{}^{\star}_{\mathsf{AB}},\overset{\leftrightarrow}{\mathrm{R}}{}^{\star}_{\mathsf{BA}}) := \begin{cases}(\overset{\leftrightarrow}{\mathrm{R}}_{\mathsf{AB}},\overset{\leftrightarrow}{\mathrm{R}}_{\mathsf{BA}}), & \text{if}\ \ \overset{\leftrightarrow}{\mathrm{R}}_{\mathsf{AB}}+\overset{\leftrightarrow}{\mathrm{R}}_{\mathsf{BA}}\leq\overset{\leftrightarrow}{\mathrm{I}}_{\mathsf{AB,MAC}}\\ (\overset{\leftrightarrow}{\mathrm{R}}_{\mathsf{AB}}/2,\overset{\leftrightarrow}{\mathrm{R}}_{\mathsf{BA}}/2), & \text{if}\ \ \overset{\leftrightarrow}{\mathrm{R}}_{\mathsf{AB}}+\overset{\leftrightarrow}{\mathrm{R}}_{\mathsf{BA}}>\overset{\leftrightarrow}{\mathrm{I}}_{\mathsf{AB,MAC}}\\ (\overset{\leftrightarrow}{\mathrm{R}}_{\mathsf{AB}},\mathbf{min}\{\overset{\leftrightarrow}{\mathrm{R}}_{\mathsf{BA}},\overset{\leftrightarrow}{\mathrm{I}}_{\mathsf{AB,MAC}}-\overset{\leftrightarrow}{\mathrm{R}}_{\mathsf{AB}}\}), & \text{if}\ \ \overset{\leftrightarrow}{\mathrm{R}}_{\mathsf{AB}}>\overset{\leftrightarrow}{\mathrm{I}}_{\mathsf{AB,MAC}}/2\\ (\mathbf{min}\{\overset{\leftrightarrow}{\mathrm{R}}_{\mathsf{AB}},\overset{\leftrightarrow}{\mathrm{I}}_{\mathsf{AB,MAC}}-\overset{\leftrightarrow}{\mathrm{R}}_{\mathsf{BA}}\},\overset{\leftrightarrow}{\mathrm{R}}_{\mathsf{BA}}), & \text{else.}\end{cases}$$

To summarize, we first find the optimal *max-min fair* rate allocation within the BC phase in Step 1. A rather sub-optimal (but easy) approach would be to directly skip to Step 5 which is just scaling the rates according to the MAC constraints. However, as noted before, we can do better by searching further within the two-way capacity region. Conditioning the limited data transfer direction to have at least MAC supported rate, the Steps 2 decides towards which direction the search should persist. Accordingly either Step 3 or Step 4 is applied. Finally, in Step 5 we cross-check if the sum of the resultant rate pair is smaller or equal to the MAC sum bound. If not, MAC sum rate is distributed to the two directions in the most max-min fair way.

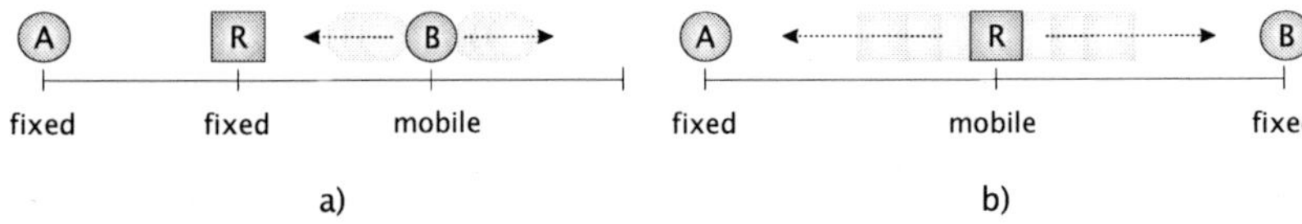

Figure 7.5.: Scenarios for realizing asymmetric channel qualities.

7.2.3. Performance Results

In this section, we present Monte Carlo simulation results. Throughout the simulations, we use the MATLAB based semidefinite tool YALMIP [77] to solve the designed semidefinite problems. The elements of the channel matrices $\mathbf{H}_\mathsf{A}$ and $\mathbf{H}_\mathsf{B}$ are i.i.d. Rayleigh fading coefficients with zero mean and variance $\sigma^2_{H_\mathsf{A}}$ and $\sigma^2_{H_\mathsf{B}}$, respectively. Moreover, these channel matrices are assumed to stay constant over one transmission cycle of the respective protocol. At each phase of both relaying protocols, each transmitting node is assumed to transmit with full average transmit power of 1, i.e., $P \triangleq P_\mathsf{R} = P_\mathsf{A} = P_\mathsf{B} = 1$. Further, we consider the case where all nodes have the same noise variance $\sigma^2_n \triangleq \sigma^2_{n_\mathsf{R}} = \sigma^2_{n_\mathsf{A}} = \sigma^2_{n_\mathsf{B}}$. Hence, the average signal-to-noise ratio is defined as $\text{SNR} = P/\sigma^2_n$. Unless otherwise stated we assume equal time sharing for each phase, i.e., $\alpha_i = \frac{1}{4}\ \forall i \in \{1,2,3,4\}$, for one-way relaying and $\alpha_{\text{MAC}} = \alpha_{\text{BC}} = \frac{1}{2}$, for two-way relaying.

We first study the sum rate performance of both relaying protocols for different cases of with and without CSIT at the relay. As the end-to-end rates are limited by both links A-to-R (R-to-A) and B-to-R (R-to-B), we are interested in a scenario of unbalanced channel qualities. In order to gain complete insight on the impact of each node to the system performance, we consider two different setups to realize this unbalanced channel quality scenario.

- The first setup assumes that terminal A and relay R are fixed, whereas terminal B is mobile (see scenario (a) in Fig. 7.5). In order simulate this, we define the following link based average SNRs: $\text{SNR}_\mathsf{A} = \sigma^2_{H_\mathsf{A}} P/\sigma^2_n$ and $\text{SNR}_\mathsf{B} = \sigma^2_{H_\mathsf{B}} P/\sigma^2_n$. Hence, while fixing $\sigma^2_{H_\mathsf{A}}$ (so as SNR_A), we vary $\sigma^2_{H_\mathsf{B}}$ to change SNR_B.

- In the second setup, we assume that terminals are fixed, but relay R is mobile (see scenario (b) in Fig. 7.5). For such a simulation setup, we set $\sigma^2_{H_\mathsf{A}} = \sigma^2_{H_\mathsf{B}} = 1$ and scale the channel matrices $\mathbf{H}_\mathsf{A}$ and $\mathbf{H}_\mathsf{B}$ respectively with $\frac{1}{p^{\zeta/2}}$ and $\frac{1}{(1-p)^{\zeta/2}}$, where p

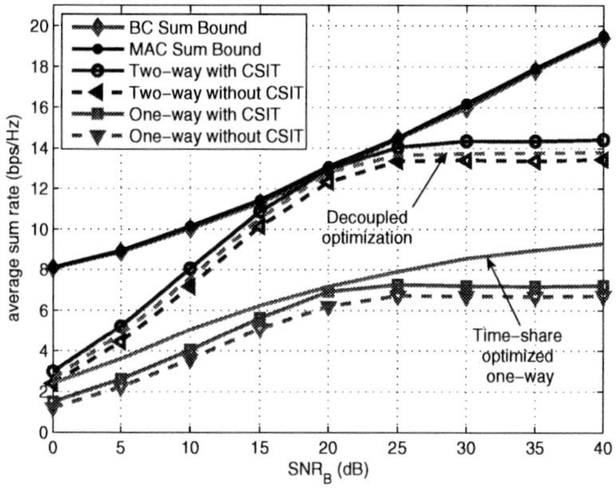

Figure 7.6.: Average sum rate vs. average $\mathrm{SNR_B}$ for $\mathrm{SNR_A} = 20$dB, $N_\mathsf{A} = N_\mathsf{B} = 2$, and $N_\mathsf{R} = 4$.

is the normalized distance between terminal A and relay R, and $\zeta = 3$ is the path loss exponent.

In Fig. 7.6 we consider the first setup and set $\mathrm{SNR_A} = 20$dB, and vary $\mathrm{SNR_B}$ from 0 to 40 dB. The sum rate of two-way relaying is upper bounded by the two maximal sum rates that can be achieved individually from respective phases, i.e., MAC sum bound $\overleftrightarrow{\mathrm{I}}_{\mathsf{AB,MAC}}$ and maximal two-way broadcast sum rate (7.14). For the pre-assumed case of equal time sharing in this figure, the BC sum bound appears to be slightly worse than the MAC bound, and hence, appears to be the bottleneck for the sum rate of two-way relaying. Independent from the employed protocol and the availability of CSIT, the sum rates improve until the two link quality equalize, and saturate afterwards. In essence, as $\mathrm{SNR_A}$ increases, the bottleneck link moves from B to A. In general, two-way relaying almost doubles the sum-rate of the its one-way counterpart, which confirms the claim that two-way relaying is able to avoid the pre-log factor $\frac{1}{2}$. It is observed that CSIT at the relay provides a continuous improvement of almost 1 bps/Hz for both protocols over SNR. Approaching this result from an another perspective, we realize that in fast time-varying channel environments, where adaptation of covariances per channel realization is quite challenging, the absence of CSIT adaptation does not ruin the performance of relaying.

Recall that we aimed for maximizing the sum rate of the two-way protocol optimally

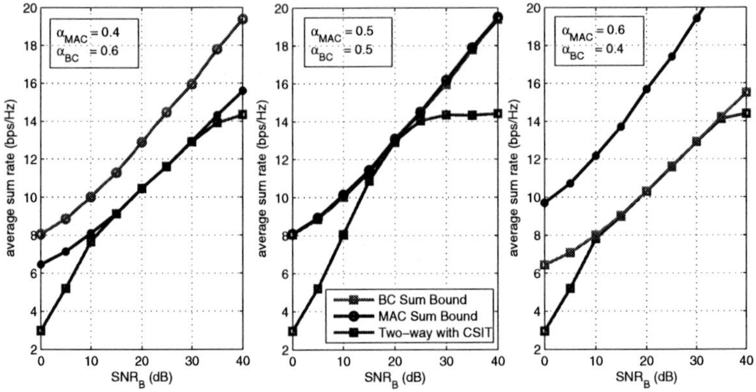

Figure 7.7.: Comparison of different sets of time sharing parameters $(\alpha_{\mathrm{MAC}}, \alpha_{\mathrm{BC}})$ for varying SNR_B, fixed $\mathrm{SNR}_\mathsf{A} = 20\mathrm{dB}$, $N_\mathsf{A} = N_\mathsf{B} = 2$, and $N_\mathsf{R} = 4$.

over two phases instead of decoupling the optimization for each phase. The effort spent for such a two-phase coupled optimal optimization can be appreciated through the dashed curve in the figure, which depicts the sum rate performance of a decoupled optimization. While simulating this curve, we first optimize the broadcast phase such that the sum rate for the broadcast phase individually is maximized. Later, the resultant relay-to-terminal rates are compared with the associated first phase rates, and scaled correspondingly to support each other. We observe that the half of the sum rate gain obtained through CSIT is lost by using such a decoupled approach.

On the other hand, in order to improve the performance of one-way relaying, we optimize the time-sharing parameters α_i over the four signaling phases through (7.1) . We observe that especially at the most unbalanced SNR regimes, the sum rate performance improves drastically, but is still much worse than its two-way counterpart. Although not depicted in the figure, for low and high SNR_B values, the optimization usually results in that the full time resources are allocated to a single uni-direction, i.e., a kind of a multiuser diversity over the reverse data transfer directions.

We focus further on the impact of time sharing in Fig. 7.7, where we plot the MAC and BC bounds and sum rate for two relaying for three different sets of $(\alpha_{\mathrm{MAC}}, \alpha_{\mathrm{BC}})$. For $(\alpha_{\mathrm{MAC}}, \alpha_{\mathrm{BC}}) = (0.4, 0.6)$, the sum rate is strictly bounded by the MAC phase for $15\mathrm{dB} \leq \mathrm{SNR}_\mathsf{B} \leq 30\mathrm{dB}$, whereas for $(\alpha_{\mathrm{MAC}}, \alpha_{\mathrm{BC}}) = (0.6, 0.4)$ the limiting phase happens

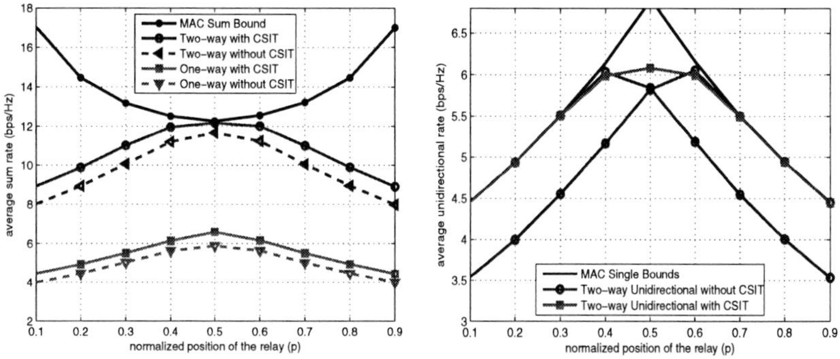

(a) (b)

Figure 7.8.: Impact of relay position with respect to the terminals for equal time-sharing in between phases: (a) Sum rate, (b) Unidirectional rates ($\mathrm{SNR} = 10$dB, $N_{\mathsf{A}} = N_{\mathsf{B}} = 2$, and $N_{\mathsf{R}} = 4$.).

to be the BC. Assigning equal time share to each phase, we equalize the MAC and BC bounds so, that almost maximal sum rate is achieved for all specific $\mathrm{SNR}_{\mathsf{B}}$ choice. Apparently, we observe that equal time sharing is almost optimal "on the average" for an antenna configuration of $(N_{\mathsf{A}}, N_{\mathsf{R}}, N_{\mathsf{B}}) = (2, 4, 2)$. Let us know elaborate on this claim. First, note that different antennas configurations and average SNR values (especially asymmetric numbers of terminal antennas and channel qualities) may lead different time shares to be optimal on the average. We emphasize that these findings are based on *average* performance. Naturally, the optimal set $(\alpha_{\mathrm{MAC}}, \alpha_{\mathrm{BC}})$ per channel realization can be different. In essence, the optimal time share that maximizes sum rate per channel realization has been derived in [145] originating from an idea proposed in [44]. However, note that in fast time-varying fading environment, where the terminals are mobile, adaptation of time-shares to individual channel realizations is an impractical challenging process.

We now shift our focus to the second setup and look for the position that the relay should locate with respect to the terminals such that sum rate is maximized. Fig. 7.8(a) depicts the sum rate performance of both relaying schemes for different positions of R. We observe that regardless of the employed protocol and the availability of CSIT, the sum rate is maximized when the relay is exactly at the middle of two terminals. In

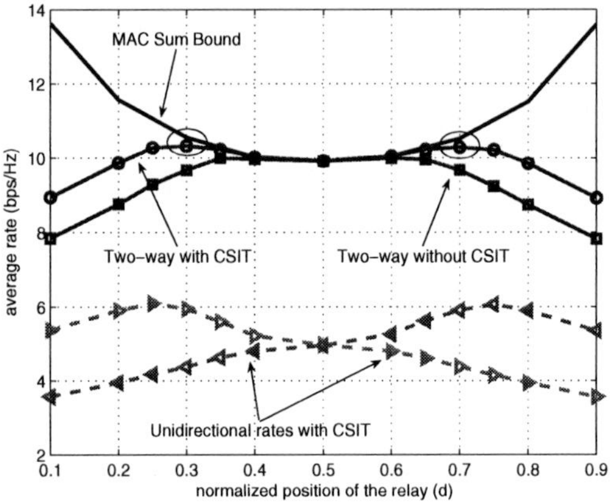

Figure 7.9.: Impact of relay position with respect to the terminals for $(\alpha_{\mathrm{MAC}}, \alpha_{\mathrm{BC}}) = (0.4, 0.6)$ (SNR = 10dB, $N_{\mathrm{A}} = N_{\mathrm{B}} = 2$, and $N_{\mathrm{R}} = 4$.)

Fig. 7.8(b), we show the unidirectional rates $\mathrm{R}_{\mathrm{AB}}^{\leftrightarrow}$ and $\mathrm{R}_{\mathrm{BA}}^{\leftrightarrow}$ of two way relaying, and investigate the effect of CSIT on these. First, observe that precoding with CSIT results in a symmetric unidirectional performance curve with a maximal value at $p = 0.5$. In the absence of CSIT at the relay, the maximal unidirectional rate from the respective source's view is achieved when the relay is a bit closer to the destination terminal than being in the middle. For example, the maximal unidirectional rate from A-to-B, i.e., $\mathrm{R}_{\mathrm{AB}}^{\leftrightarrow}$, for two-way relaying without CSIT, is achieved when the relay is positioned at $p = 0.6$ in the figure. This is mainly because one terminal experiences a stronger channel than the other terminal, and without CSIT the relay can not compensate this difference through precoding. However, when the sum rate of these two unidirectional rates are compared through Fig. 7.8(a), we observe that sum rate equalizes for unbalanced channel qualities leading a maximum at the mid.

In Fig. 7.9, we once again study the impact of time-sharing on the rate performance of two-way relaying. We choose a setup where the MAC is the bottleneck, i.e., $(\alpha_{\mathrm{MAC}}, \alpha_{\mathrm{BC}}) = (0.4, 0.6)$. Consequently, the maximal sum rate is achieved at a relay position of either $p = 0.3$ or 0.7 instead of $p = 0.5$ of the equal-share case. In essence, as α_{MAC} is reduced further, the maximal sum rate achieving relay positions would approach more close to the terminals. On the other hand, note the corresponding unidirectional

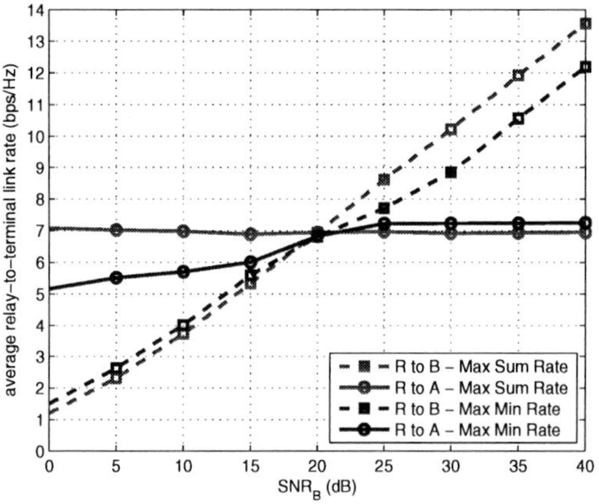

Figure 7.10.: Average relay-to-terminal link rate vs. average $\mathrm{SNR}_{\mathsf{B}}$ for $\mathrm{SNR}_{\mathsf{A}} = 20\mathrm{dB}$, $N_{\mathsf{A}} = N_{\mathsf{B}} = 2$, and $N_{\mathsf{R}} = 4$.

rate behaviour over p, where the symmetry per direction in the equal-share case is replaced by an symmetric curve.

After several sum-rate maximization plots, Fig. 7.10 shows the performance of max-min beamformer in the relay-to-terminal transmission phase. We plot the rate of each link R-to-A and R-to-B individually and compare them for sum rate and max-min rate optimizations. As we fix $\mathrm{SNR}_{\mathsf{A}} = 20\mathrm{dB}$ and vary $\mathrm{SNR}_{\mathsf{B}}$, we expect to observe more response at the the rate through R-to-B than the rate through R-to-A. For sum rate maximization, we see that $\overleftrightarrow{\mathrm{R}}_{\mathsf{RA}}$ changes only marginally, which indicates that the optimal covariance matrix is almost indifferent to noise variance changes at terminal B. Contrary to this, $\overleftrightarrow{\mathrm{R}}_{\mathsf{RB}}$ grows from 1.3 bps/Hz up to 13.6 bps/Hz as $\mathrm{SNR}_{\mathsf{B}}$ varies between 0-to-40 dB. When we aim at max-min fairness instead of sum rate, we improve the performance of the terminal with lower rate w.r.t. to that of the other by 0.3-to-0.4 bps/Hz. In the regime of $0\mathrm{dB} \leq \mathrm{SNR}_{\mathsf{B}} \leq 20\mathrm{dB}$, the beamformer tries to improve $\overleftrightarrow{\mathrm{R}}_{\mathsf{RB}}$, i.e., for the sake of max-min fairness, whereas $\overleftrightarrow{\mathrm{R}}_{\mathsf{RA}}$ is increased in the regime of $20\mathrm{dB} \leq \mathrm{SNR}_{\mathsf{B}} \leq 40\mathrm{dB}$. However, in order to improve one link's rate the other link sacrifices its performance, which is in the nature of fair rate assignment. Note finally that the plotted rates are not yet compared to MAC bounds; hence, would probably reduce to satisfy both phases' bounds (see Section 7.2.2.2).

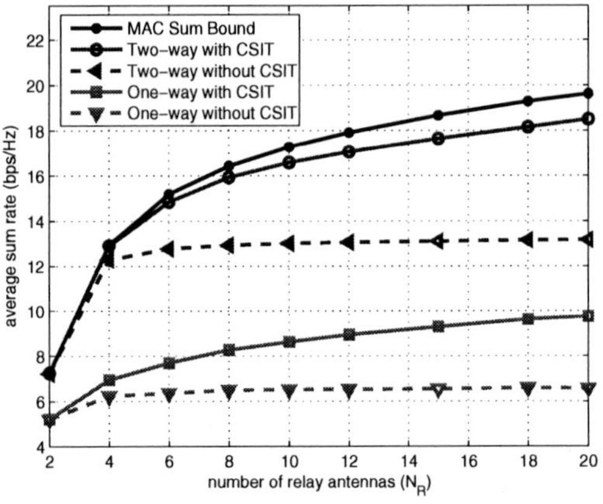

Figure 7.11.: Average sum rate vs. number of relay antennas N_{R} for equal average SNR of $\mathrm{SNR}_{\mathsf{A}} = \mathrm{SNR}_{\mathsf{B}} = 20\mathrm{dB}$, $N_{\mathsf{A}} = N_{\mathsf{B}} = 2$.

Next, we are interested in understanding the impact of number of relay/terminal antennas on the sum rate performance. In Fig. 7.11, we plot average sum rates for various numbers of relay antennas N_{R}, where we fix the number of terminal antennas to 2. The sum-rate differences between the protocols without CSIT and the protocols with CSIT increase with an increasing number of relay antennas N_{R}. In fact, for the protocols without CSIT increasing N_{R} beyond 4-6 antennas, does not anymore bring a considerable gain. Rate doubling feature of two-way relaying with respect to its one-way counterpart is observed over all N_{R} choices. Note the dramatic sum rate improvement for two-way relaying as N_{R} increases from 2 to 4. This can be interpreted as follows. There establishes a multiple access channel at the relay in two-way relaying, where 2-antenna equipped relay needs to perfectly decode information from two terminals each with 2 antennas. In other words, available two spatial degrees of freedom is shared between two terminals (one to each). However, setting $N_{\mathsf{R}} = 4$, there occurs an excess of two spatial degrees of freedom, and hence, provides a surplus array gain. Summing up, the relative gain coming with CSIT at the relay is dependent on the antenna configuration within the network. Hence, CSIT depicts its significance more and more, as the ratio between number of relay antennas and number of node antennas increases.

On the other hand, in Fig. 7.12 we focus on the number of antennas at the terminals side. In Fig. 7.12(a), we simulate an asymmetric scenario, where we vary N_{B} from 1 to 8 while keeping N_{A} and N_{R} fixed at 4 and 8, respectively. Note that for small N_{B}, there is a huge gap between the maximal two-way achievable sum rate and the MAC sum bound, which is due to the fact that both unidirectional links are dominated by the relatively weak B-to-R channel. As the number of relay antennas at both terminals equalize at 4, two-way sum rates enter a saturation stage, where the MAC phase appears to be the performance limiter. The previous spatial degrees of freedom statement also valid in these observations. Consequently, we raise the following natural question: Which antenna configuration is the most efficient in terms of maximizing sum rate? We address this issue with Fig. 7.12(b), where we fix relay antennas $N_{\mathsf{R}} = 8$, assume equal number of terminal antennas $N_{\mathsf{A}} = N_{\mathsf{B}}$ and vary the terminal antennas. For two-way relaying, as long as $N_{\mathsf{A}} + N_{\mathsf{B}} \leq N_{\mathsf{R}}$, we obtain a significant performance improvement for each additional terminal antennas. Although this improvement monotonically continues as terminal antennas increase further, the gain tends to become negligible. Moreover, it is notable that the impact of CSIT at the relay vanishes as spatial degrees per terminal reduces (or equivalently as the number of antennas at the terminals increases). On the other hand for one-way relaying, the absence of a multiple access channel enables a substantial improvement for each additional terminal antenna until $N_{\mathsf{A}} = N_{\mathsf{B}} \leq N_{\mathsf{R}}$. It is simply because there occurs a point-to-point MIMO transmission at each phase in which all available spatial degrees of freedom is assigned to the corresponding terminal.

diyi

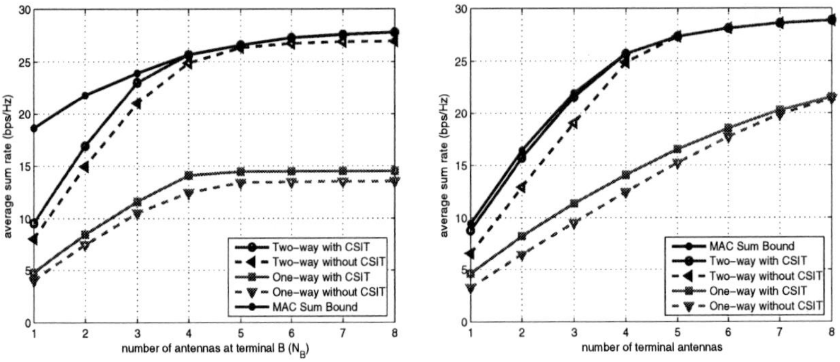

(a) (b)

Figure 7.12.: Impact of number of antennas at the terminals: (a) Fix $N_\mathsf{A} = 4$, $N_\mathsf{R} = 8$ and vary N_B, (b) Fix $N_\mathsf{R} = 8$, and vary $N_\mathsf{A} = N_\mathsf{B}$ ($\mathrm{SNR} = 20\mathrm{dB}$).

7.3. Robust Covariance Optimization for Two-Way Relaying

In the previous section, we studied the achievable data transfer rates for two-way relaying in details, and considered two specific objective functions for rate allocations to individual terminals. Our primary assumption during covariance optimization was that the relay has perfect transit CSI knowledge of both terminals. Moreover, we propose to gain this knowledge in the receive mode of the relay during the MAC phase. In other words, there is no need to feedback any CSI to the relay, which is usual case in general broadcast channel studies [36, 115, 125]. However, there may occur some errors while estimating the channels in the MAC phase or the channels may change in between two phases due to mobility of the users, e.g., a multihop cellular network. That is to say, as the absence of CSIT is a crucial performance limiter, there is also need for considerations on the imperfect CSIT case. An extensive list of references that deal with robust optimization against imperfect CSI has been already provided in Section 4.1.

To this end, in this section we study robust covariance designs in order to protect two-way relaying from negative impacts of imperfect CSI at the relay. Basing our investigations on the know-how developed in Section 4.3, we aim at providing a guideline to extend the optimization in Section 7.2.2 to the imperfect CSIT knowledge case.

We simplify our following considerations by assuming single-antenna terminals, i.e., $N_\mathsf{A} = N_\mathsf{B} = 1$, with the motivation that such a case lets itself for rather manageable mathematical derivations. Correspondingly, the MIMO channel matrices $\mathbf{H}_\mathsf{A}$ and $\mathbf{H}_\mathsf{B}$ boil down to be vector channels $\mathbf{h}_\mathsf{A} \in \mathbb{C}^{N_\mathsf{R}}$ and $\mathbf{h}_\mathsf{B} \in \mathbb{C}^{N_\mathsf{R}}$, respectively.

We assume that each terminal-to-relay channel vector is subject to an independent imperfection source, and is defined within an spherical uncertainty set of $\mathcal{U}_\mathsf{n}$. In other words, all possible realizations of the nominal data $\mathbf{h}_\mathsf{n}$ is assumed to be considered within $\mathcal{U}_\mathsf{n}$. As each channel is independently constrained by the uncertainty, the general uncertainty set is $\mathcal{U} = \mathcal{U}_\mathsf{A} \times \mathcal{U}_\mathsf{B}$. We choose an uncertainty set model that is different than the ones in Section 4.3, and is motivated by the practical view of multiuser downlink communications [25, 106]. That is,

$$\mathcal{U}_\mathsf{n} = \left\{ \mathbf{h}_\mathsf{n} = \hat{\mathbf{h}}_\mathsf{n} + \boldsymbol{\eta}_\mathsf{n} \ : \ \|\boldsymbol{\eta}_\mathsf{n}\| \leq \rho_\mathsf{n} \right\}, \quad \mathsf{n} \in \{\mathsf{A}, \mathsf{B}\},$$

where $\hat{\mathbf{h}}_\mathsf{n}$ is the channel estimate and $\boldsymbol{\eta}_i$ represents the uncertainty around this estimate. Note that this model bounds the size of the uncertainty rather than giving explicit statistics about it, which would be the case for Gaussian noise estimation errors. It has been reported in [25] that such a model is appropriate for systems where CSI is quantized at the receivers and fed back to the transmitter.

Since the designed covariance should consider all possible channel realizations within the defined uncertainty sets, an optimization based on these, focuses on the worst-case performance of two-way relaying. In the following, we first characterize the worst-case impact of channel mismatch on the broadcast phase's capacity region. Later, we incorporate imperfect CSIT with covariance optimization procedures derived in the previous section.

7.3.1. Worst-Case Two-Way Broadcast Capacity Region

In this section, we are interested in to determine the amount of shrinking at the two-way broadcast capacity region due to channel uncertainty. Hence, in the following we derive the *worst-case* two-way broadcast capacity region boundary for given uncertainty sets.

To get insight, let us first elaborate conceptually and start with the capacity region for two-way broadcast channel for a given covariance matrix under defined uncertainty sets. As illustrated in the first sub-figure of Fig. 7.13, we expect the capacity boundary to shrink towards the origin as the size of the uncertainty grows. Following the second

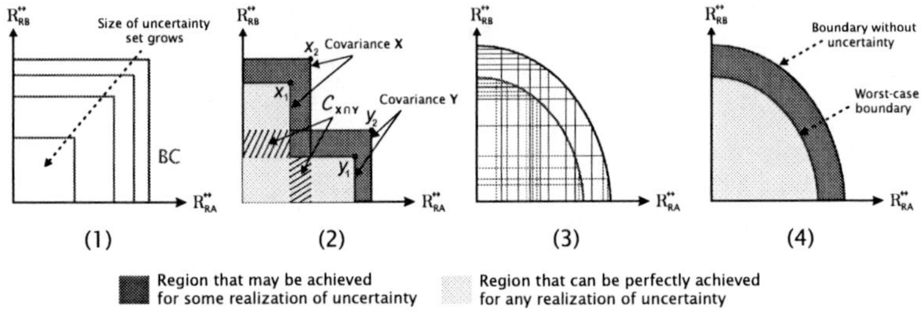

Figure 7.13.: Illustration of shrinking effect on the two-way broadcast capacity region due to channel uncertainty.

sub-figure of Fig. 7.13, assume that we are given a covariance $\mathbf{X}$, which maximally supports a rate pair given by point $x_2 = (\mathrm{R}_{\mathsf{A}x_2}, \mathrm{R}_{\mathsf{B}x_2})$ in the absence of uncertainty and point $x_1 = (\mathrm{R}_{\mathsf{A}x_1}, \mathrm{R}_{\mathsf{B}x_1})$ in the worst-case within given uncertainty regions. Hence, the worst capacity region for given covariance $\mathbf{X}$ and uncertainty set sizes $(\rho_{\mathsf{A}}, \rho_{\mathsf{B}})$ can be defined as

$$\mathcal{C}_{\mathrm{BC}}^{\mathrm{wc}}(\mathbf{X}, \rho_{\mathsf{A}}, \rho_{\mathsf{B}}) = \left\{(\mathrm{R}_{\mathsf{RA}}^{\leftrightarrow}, \mathrm{R}_{\mathsf{RB}}^{\leftrightarrow}) \in \mathbb{R}_+ \ : \ \mathrm{R}_{\mathsf{RA}}^{\leftrightarrow} \leq \mathrm{R}_{\mathsf{A}x_1}, \ \mathrm{R}_{\mathsf{RB}}^{\leftrightarrow} \leq \mathrm{R}_{\mathsf{B}x_1}\right\}.$$

Likewise, the capacity region for given covariance $\mathbf{X}$ without uncertainty is defined as

$$\mathcal{C}_{\mathrm{BC}}(\mathbf{X}) = \left\{(\mathrm{R}_{\mathsf{RA}}^{\leftrightarrow}, \mathrm{R}_{\mathsf{RB}}^{\leftrightarrow}) \in \mathbb{R}_+ \ : \ \mathrm{R}_{\mathsf{RA}}^{\leftrightarrow} \leq \mathrm{R}_{\mathsf{A}x_2}, \ \mathrm{R}_{\mathsf{RB}}^{\leftrightarrow} \leq \mathrm{R}_{\mathsf{B}x_2}\right\}.$$

See the second sub-figure for illustrations of both regions. The amount of loss due to uncertainty can be quantified by the region $\mathcal{C}_{\mathrm{loss}}(\mathbf{X}, \rho_{\mathsf{A}}, \rho_{\mathsf{B}}) \triangleq \mathcal{C}_{\mathrm{BC}}(\mathbf{X}) \backslash \mathcal{C}_{\mathrm{BC}}^{\mathrm{wc}}(\mathbf{X}, \rho_{\mathsf{A}}, \rho_{\mathsf{B}})$. Now, consider an another covariance $\mathbf{Y} \neq \mathbf{X}$. We obtain different achievable regions, which intersect with the ones of covariance $\mathbf{X}$. That is to say, the loss region $\mathcal{C}_{\mathrm{loss}}(\mathbf{X}, \rho_{\mathsf{A}}, \rho_{\mathsf{B}})$ of covariance $\mathbf{X}$ would intersect with the worst-case region $\mathcal{C}_{\mathrm{BC}}^{\mathrm{wc}}(\mathbf{Y}, \rho_{\mathsf{A}}, \rho_{\mathsf{B}})$ of covariance $\mathbf{Y}$, and improve the union worst-case capacity region to be

$$\mathcal{C}_{\mathrm{BC}}^{\mathrm{wc}}(\mathbf{X}, \rho_{\mathsf{A}}, \rho_{\mathsf{B}}) \cup \mathcal{C}_{\mathrm{BC}}^{\mathrm{wc}}(\mathbf{Y}, \rho_{\mathsf{A}}, \rho_{\mathsf{B}}).$$

Hence, taking the union of all possible feasible covariances (see the third sub-figure of Fig. 7.13), we obtain the worst-case two-way broadcast capacity region for given uncertainty set sizes. In the fourth sub-figure of Fig. 7.13, we assume that we have derived the capacity boundary that is achievable for any realization within both uncertainty sets, we specify two non-intersecting regions. For given sizes for uncertainty sets, the inner region plus its boundary consist of all points (rate pairs) that can be achievable independent of the mismatch. Thats why, we refer this union as the *safe region* in the sequel. On the other hand, the region starting (but excluding) from the worst-case boundary until (and including) the boundary of no uncertainty, is called as the *unsafe region*, and the points within it may or may not be achievable depending on the uncertainty realizations $\boldsymbol{\eta}_{\mathsf{n}}$ from the uncertainty sets. Let us highlight once again that although we can not guarantee achievability within this unsafe region, there is a certain probability larger than zero for each of these points that it is achievable. It is possible to quantify this probability of assurance through stochastic optimization methods, but it is out of the scope of this thesis.

After presenting a conceptual explanation of the worst-case capacity, we shift our focus now to theoretical derivations. Recall the mutual information expression between R and A,B given in (6.21). Modifying it according to single-antennas equipped terminals and introducing the definition of uncertain channel knowledge, we re-write (6.21) as

$$\mathrm{R}_{\mathsf{Rn}}^{\leftrightarrow} \leq \mathrm{I}_{\mathsf{Rn}}^{\leftrightarrow} = \log_2\left(1 + \frac{1}{\sigma_{n_{\mathsf{n}}}^2}(\hat{\mathbf{h}}_{\mathsf{n}} + \boldsymbol{\eta}_{\mathsf{n}})^H \boldsymbol{\Lambda}_{\mathsf{R}}(\hat{\mathbf{h}}_{\mathsf{n}} + \boldsymbol{\eta}_{\mathsf{n}})\right), \quad \|\boldsymbol{\eta}_{\mathsf{n}}\| \leq \rho_{\mathsf{n}}, \quad \mathsf{n} \in \{\mathsf{A},\mathsf{B}\}. \tag{7.16}$$

The rate expression is dependent on the realization of $\boldsymbol{\eta}_{\mathsf{n}}$ driven out of $\mathcal{U}_{\mathsf{n}}$. Our aim is now to derive the robust counterpart for (7.16) such that we obtain an expression valid for all realizations of uncertainty.

Robust Counterpart for (7.16) We re-formulate the first inequality in (7.16) so, that we obtain

$$(\hat{\mathbf{h}}_{\mathsf{n}} + \boldsymbol{\eta}_{\mathsf{n}})^H \boldsymbol{\Lambda}_{\mathsf{R}}(\hat{\mathbf{h}}_{\mathsf{n}} + \boldsymbol{\eta}_{\mathsf{n}}) \geq \sigma_{n_{\mathsf{n}}}^2(2^{\mathrm{R}_{\mathsf{Rn}}^{\leftrightarrow}} - 1), \tag{7.17}$$

which can be further expanded in terms of an explicit quadratic of $\boldsymbol{\eta}_{\mathsf{n}}$ as

$$\hat{\mathbf{h}}_{\mathsf{n}}^H \boldsymbol{\Lambda}_{\mathsf{R}}\hat{\mathbf{h}}_{\mathsf{n}} + 2\mathrm{Re}\{\hat{\mathbf{h}}_{\mathsf{n}}^H \boldsymbol{\Lambda}_{\mathsf{R}}\boldsymbol{\eta}_{\mathsf{n}} + \boldsymbol{\eta}_{\mathsf{n}}^H \boldsymbol{\Lambda}_{\mathsf{R}}\boldsymbol{\eta}_{\mathsf{n}}\} \geq \sigma_{n_{\mathsf{n}}}^2(2^{\mathrm{R}_{\mathsf{Rn}}^{\leftrightarrow}} - 1). \tag{7.18}$$

Besides (7.18), we have a second quadratic function of $\boldsymbol{\eta}_{\mathsf{n}}$, i.e., $\boldsymbol{\eta}_{\mathsf{n}}^H\boldsymbol{\eta}_{\mathsf{n}} - \rho_{\mathsf{n}}^2 \geq 0$, which sets the size of uncertainty. The robust counterpart design requires that the non-positivity of the latter uncertainty size related function to lead the non-negativity of (7.18). At this point, we apply the S-lemma to these two quadratic functions, which establishes the aforementioned mathematical relation within these two quadratic functions. That is, we obtain the following SDP formulation for robust counterpart of the rate constraint:

$$\mathcal{WC}_{\mathsf{n}}(\boldsymbol{\Lambda}_{\mathsf{R}}, \mathrm{R}_{\mathsf{Rn}}^{\leftrightarrow}, \rho_{\mathsf{n}}) \triangleq \begin{bmatrix} \hat{\mathbf{h}}_{\mathsf{n}}^H \boldsymbol{\Lambda}_{\mathsf{R}}\hat{\mathbf{h}}_{\mathsf{n}} - \sigma_{n_{\mathsf{n}}}^2(2^{\mathrm{R}_{\mathsf{Rn}}^{\leftrightarrow}} - 1) & \hat{\mathbf{h}}_{\mathsf{n}}^H \boldsymbol{\Lambda}_{\mathsf{R}} \\ \boldsymbol{\Lambda}_{\mathsf{R}}\hat{\mathbf{h}}_{\mathsf{n}} & \boldsymbol{\Lambda}_{\mathsf{R}} \end{bmatrix} - \zeta_{\mathsf{n}} \begin{bmatrix} \rho_{\mathsf{n}}^2 & \mathbf{0} \\ \mathbf{0} & -\mathbf{I} \end{bmatrix} \succeq 0, \tag{7.19}$$

where $\zeta_n \geq 0$. The constraint that $\mathcal{WC}_n(\mathbf{\Lambda}_R, \overleftrightarrow{R}_{Rn}, \rho_n) \succeq 0$ indicates that the pre-set $\overleftrightarrow{R}_{Rn}$ is achievable for all uncertainty realizations under given covariance $\mathbf{\Lambda}_R$ and uncertainty set size ρ_n.

Worst-Case Performance The robust counterpart $\mathcal{WC}_n(\mathbf{\Lambda}_R, \overleftrightarrow{R}_{Rn}, \rho_n)$ is functions of the covariance $\mathbf{\Lambda}_R$, the rate lower bound $\overleftrightarrow{R}_{Rn}$ and the size of uncertainty ρ_n. We are now interested in finding the worst-case broadcast performance of the node n for a given transmit covariance matrix and uncertainty size. In other words, fixing $\mathbf{\Lambda}_R$ and ρ_n, we are interested in to find the maximal $\overleftrightarrow{R}_{Rn}$. The corresponding problem can be immediately formulated as the following SDP:

$$\max_{\overleftrightarrow{R}_{Rn} \geq 0} \overleftrightarrow{R}_{Rn} \text{ subject to } \mathcal{WC}_n(\mathbf{\Lambda}_R, \overleftrightarrow{R}_{Rn}, \rho_n) \succeq 0. \tag{7.20}$$

However, note that (7.20) concerns only with a single chosen node ρ_n, and ignores the performance of the other.

Worst-Case Capacity Region With (7.19) at hand, we express the worst-case two-way broadcast capacity region for given uncertainty set size, similar to (7.21), as

$$\mathcal{C}_{BC}^{wc}(\rho_A, \rho_B) = \bigcup_{\mathbf{\Lambda}_R \succeq 0, \mathrm{Tr}(\mathbf{\Lambda}_R) \leq P_R} \left\{ (\overleftrightarrow{R}_{RA}, \overleftrightarrow{R}_{RB}) \in \mathbb{R}_+ \, : \, \begin{array}{c} \mathcal{WC}_A(\mathbf{\Lambda}_R, \overleftrightarrow{R}_{RA}, \rho_A) \succeq 0, \\ \mathcal{WC}_B(\mathbf{\Lambda}_R, \overleftrightarrow{R}_{RB}, \rho_B) \succeq 0 \end{array} \right\}. \tag{7.21}$$

For the derivation of the boundary of the worst-case capacity, we follow the same optimization procedure described in Section 6.2.2 by only extending the employed optimization problems $\mathcal{P}_{BC,A}$ and $\mathcal{P}_{BC,B}$ therein to a worst-case scenario formulation presented above. Specifically,

- **Initialize:** Solve:

$$\mathcal{P}_{BC,A}^{wc} : \max_{\mathbf{\Lambda}_R \succeq 0, \mu \geq 0} \mu \text{ subject to } \begin{bmatrix} \hat{\mathbf{h}}_A^H \mathbf{\Lambda}_R \hat{\mathbf{h}}_A - \mu & \hat{\mathbf{h}}_A^H \mathbf{\Lambda}_R \\ \mathbf{\Lambda}_R \hat{\mathbf{h}}_A & \mathbf{\Lambda}_R \end{bmatrix} - \zeta_A \begin{bmatrix} \rho_A^2 & \mathbf{0} \\ \mathbf{0} & -\mathbf{I} \end{bmatrix} \succeq 0,$$
$$\mathrm{Tr}(\mathbf{\Lambda}_R) \leq P_R, \ \zeta_A \geq 0$$

  where $\overleftrightarrow{R}_{RA}^{wc\star} = \log_2(1 + \mu^\star / \sigma_{nA}^2)$.

- **Iterate:** Set $\boldsymbol{\tau} := [0, \ \Delta, \ 2\Delta, \ \cdots, \ \overleftrightarrow{R}_{RA}^{wc\star} - \Delta, \ \overleftrightarrow{R}_{RA}^{wc\star}]$, where $\Delta \geq 0$. For each $\tau \in \boldsymbol{\tau}$, solve

$$\mathcal{P}_{BC,B}^{wc} := \max_{\mathbf{\Lambda}_R \succeq 0, \mu \geq 0} \mu \text{ subject to } \begin{bmatrix} \hat{\mathbf{h}}_B^H \mathbf{\Lambda}_R \hat{\mathbf{h}}_B - \mu & \hat{\mathbf{h}}_B^H \mathbf{\Lambda}_R \\ \mathbf{\Lambda}_R \hat{\mathbf{h}}_B & \mathbf{\Lambda}_R \end{bmatrix} - \zeta_B \begin{bmatrix} \rho_B^2 & \mathbf{0} \\ \mathbf{0} & -\mathbf{I} \end{bmatrix} \succeq 0,$$

$$\begin{bmatrix} \hat{\mathbf{h}}_{\mathsf{A}}^H \mathbf{\Lambda}_{\mathsf{R}} \hat{\mathbf{h}}_{\mathsf{A}} - \sigma_{n\mathsf{A}}^2 (2^{\tau} - 1) & \hat{\mathbf{h}}_{\mathsf{A}}^H \mathbf{\Lambda}_{\mathsf{R}} \\ \mathbf{\Lambda}_{\mathsf{R}} \hat{\mathbf{h}}_{\mathsf{A}} & \mathbf{\Lambda}_{\mathsf{R}} \end{bmatrix} - \zeta_{\mathsf{A}} \begin{bmatrix} \rho_{\mathsf{A}}^2 & 0 \\ 0 & -\mathbf{I} \end{bmatrix} \succeq 0$$
$$\mathrm{Tr}(\mathbf{\Lambda}_{\mathsf{R}}) \leq P_{\mathsf{R}}, \zeta_{\mathsf{A}} \geq 0, \zeta_{\mathsf{B}} \geq 0,$$

where $\overset{\leftrightarrow\mathrm{wc}\star}{\mathrm{R}_{\mathsf{RB}}}(\tau) = \log_2(1 + \mu^\star / \sigma_{n\mathsf{B}}^2)$. Then, the boundary point for given τ is

$$(\overset{\leftrightarrow\mathrm{wc}}{\mathrm{R}_{\mathsf{RA}}}, \overset{\leftrightarrow\mathrm{wc}}{\mathrm{R}_{\mathsf{RB}}}) := (\tau, \overset{\leftrightarrow\mathrm{wc}\star}{\mathrm{R}_{\mathsf{RB}}}(\tau)).$$

Fig. 7.14 depicts a typical two-way broadcast capacity region boundaries of one specific channel realization for various uncertainty set sizes. We set $N_{\mathsf{R}} = 3$, $P_{\mathsf{R}} = 1$, $\sigma_{n\mathsf{A}}^2 = \sigma_{n\mathsf{B}}^2 = 10^{-2}$ (corresponds to an average SNR of 20 dB). The vector channels employed in the example are

$$\mathbf{h}_{\mathsf{A}} = \begin{bmatrix} 0.6562 - 0.5113i \\ 0.1225 - 0.4062i \\ -0.4890 - 0.2176i \end{bmatrix}, \quad \mathbf{h}_{\mathsf{B}} = \begin{bmatrix} 0.1890 + 0.9893i \\ 0.9436 - 0.4048i \\ -0.9412 - 0.2451i \end{bmatrix}.$$

Moreover, for the sake of fair comparison in between the impact of different uncertainty set sizes, we take the norm of the channel realization as the basis, and scale the channel estimate and the radius of the uncertainty set correspondingly. That is, if we have the channel realization $\mathbf{h}_{\mathsf{n}}$ and a desired uncertainty sphere radius of ρ_{n}, then the estimate and the radius become

$$\hat{\mathbf{h}}_{\mathsf{n}} = \sqrt{\frac{1}{1+\rho_{\mathsf{n}}^2}}\mathbf{h}_{\mathsf{n}} \quad \text{and} \quad \alpha_{\mathsf{n}} = \sqrt{\frac{\rho_{\mathsf{n}}^2 \mathbf{h}_{\mathsf{n}}^H \mathbf{h}_{\mathsf{n}}}{1+\rho_{\mathsf{n}}^2}}, \tag{7.22}$$

so that on the average the norm of $(\hat{\mathbf{h}}_{\mathsf{n}} + \boldsymbol{\eta}_{\mathsf{n}})$ is fixed and equal to the norm of $\mathbf{h}_{\mathsf{n}}$.

As we projected similarly in Fig. 7.13, we observed that the two-way broadcast capacity regions shrinks towards the origin as the uncertainty size ρ grows. Although it may not attribute to the general performance, this specific channel realization performance suggests that the impact of uncertainty accelerates as ρ increases, e.g., the relative fractional capacity loss from $\rho = 0.2$ to $\rho = 0.3$ is larger than that of from $\rho = 0.1$ to $\rho = 0.2$.

7.3.2. Worst-Case Robust Covariance Optimization

Having introduced the robust counterpart derivations in the previous section, the extension of sum rate and minimum link rate maximization optimizations to robust forms

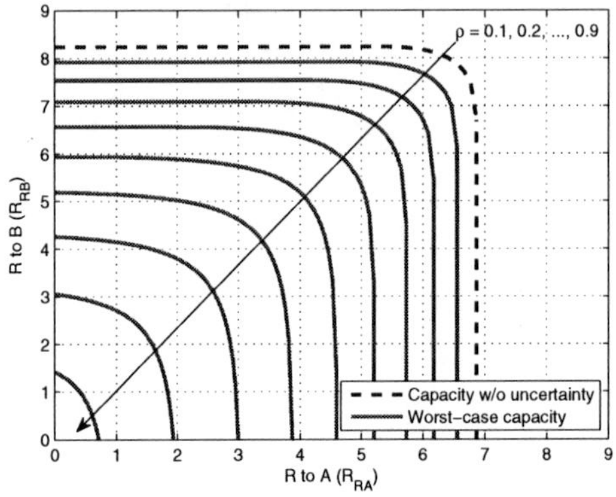

Figure 7.14.: Worst-case capacity boundary of two-way broadcast channel.

are rather trivial. In essence, we directly follow the steps of both optimization procedures given in Sections 7.2.2.1 and 7.2.2.2 by appropriately changing the employed SDP formulations. Instead of repeating the optimization procedures explicitly, we give a guideline for extending the SDPs.

Throughout these optimizations, we face with three kind of SDPs: Either we maximize the minimum of two rate expressions, or we maximize one rate expression while lower bounding the other, or we maximize the sum of two rate expressions. The SDP of the robust counterpart for the former one is

$$\max_{\mathbf{\Lambda}_\mathsf{R}\succeq 0, \mu\geq 0} \mu \textbf{ subject to } \begin{bmatrix} \hat{\mathbf{h}}_\mathsf{B}^H \mathbf{\Lambda}_\mathsf{R} \hat{\mathbf{h}}_\mathsf{B} - \mu & \hat{\mathbf{h}}_\mathsf{B}^H \mathbf{\Lambda}_\mathsf{R} \\ \mathbf{\Lambda}_\mathsf{R} \hat{\mathbf{h}}_\mathsf{B} & \mathbf{\Lambda}_\mathsf{R} \end{bmatrix} - \zeta_\mathsf{B} \begin{bmatrix} \rho_\mathsf{B}^2 & \mathbf{0} \\ \mathbf{0} & -\mathbf{I} \end{bmatrix} \succeq 0,$$
$$\begin{bmatrix} \hat{\mathbf{h}}_\mathsf{A}^H \mathbf{\Lambda}_\mathsf{R} \hat{\mathbf{h}}_\mathsf{A} - \mu & \hat{\mathbf{h}}_\mathsf{A}^H \mathbf{\Lambda}_\mathsf{R} \\ \mathbf{\Lambda}_\mathsf{R} \hat{\mathbf{h}}_\mathsf{A} & \mathbf{\Lambda}_\mathsf{R} \end{bmatrix} - \zeta_\mathsf{A} \begin{bmatrix} \rho_\mathsf{A}^2 & \mathbf{0} \\ \mathbf{0} & -\mathbf{I} \end{bmatrix} \succeq 0$$
$$\mathrm{Tr}(\mathbf{\Lambda}_\mathsf{R}) \leq P_\mathsf{R}, \zeta_\mathsf{A} \geq 0, \zeta_\mathsf{B} \geq 0.$$

Whereas, SDP of the robust counterpart for the second type optimization is written as

$$\max_{\mathbf{\Lambda}_\mathsf{R}\succeq 0, \mu\geq 0} \mu \textbf{ subject to } \begin{bmatrix} \hat{\mathbf{h}}_\mathsf{n}^H \mathbf{\Lambda}_\mathsf{R} \hat{\mathbf{h}}_\mathsf{n} - \mu & \hat{\mathbf{h}}_\mathsf{n}^H \mathbf{\Lambda}_\mathsf{R} \\ \mathbf{\Lambda}_\mathsf{R} \hat{\mathbf{h}}_\mathsf{n} & \mathbf{\Lambda}_\mathsf{R} \end{bmatrix} - \zeta_\mathsf{n} \begin{bmatrix} \rho_\mathsf{n}^2 & \mathbf{0} \\ \mathbf{0} & -\mathbf{I} \end{bmatrix} \succeq 0,$$
$$\begin{bmatrix} \hat{\mathbf{h}}_\mathsf{m}^H \mathbf{\Lambda}_\mathsf{R} \hat{\mathbf{h}}_\mathsf{m} - \sigma_{n_\mathsf{m}}^2 (2^\tau - 1) & \hat{\mathbf{h}}_\mathsf{m}^H \mathbf{\Lambda}_\mathsf{R} \\ \mathbf{\Lambda}_\mathsf{R} \hat{\mathbf{h}}_\mathsf{m} & \mathbf{\Lambda}_\mathsf{R} \end{bmatrix} - \zeta_\mathsf{m} \begin{bmatrix} \rho_\mathsf{m}^2 & \mathbf{0} \\ \mathbf{0} & -\mathbf{I} \end{bmatrix} \succeq 0$$
$$\mathrm{Tr}(\mathbf{\Lambda}_\mathsf{R}) \leq P_\mathsf{R}, \zeta_\mathsf{n} \geq 0, \zeta_\mathsf{m} \geq 0,$$

for $\mathsf{m}, \mathsf{n} \in \{\mathsf{A}, \mathsf{B}\}$, $\mathsf{m} \neq \mathsf{n}$, where τ is the lower bound rate for the terminal m. Note that in the above presented SDPs, we benefit from the monotonic equivalence of rate and SNR. Hence, after solving them, we need to make a conversion of $\log_2(1 + \mu^\star/\sigma^2_{n_\mathsf{n}})$.

Lastly, the robust counterpart for sum of two rate expressions is formulated as

$$\max_{\mathbf{\Lambda}_\mathsf{R} \succeq 0, \mu_\mathsf{A} \geq 0, \mu_\mathsf{B} \geq 0} \quad \mu_\mathsf{A} + \mu_\mathsf{B} \quad \textbf{subject to} \quad \mathrm{Tr}(\mathbf{\Lambda}_\mathsf{R}) \leq P_\mathsf{R}, \zeta_\mathsf{A} \geq 0, \zeta_\mathsf{B} \geq 0,$$

$$\begin{bmatrix} \hat{\mathbf{h}}_\mathsf{B}^H \mathbf{\Lambda}_\mathsf{R} \hat{\mathbf{h}}_\mathsf{B} - \sigma^2_{n_\mathsf{B}}(2^{\mu_\mathsf{B}} - 1) & \hat{\mathbf{h}}_\mathsf{B}^H \mathbf{\Lambda}_\mathsf{R} \\ \mathbf{\Lambda}_\mathsf{R} \hat{\mathbf{h}}_\mathsf{B} & \mathbf{\Lambda}_\mathsf{R} \end{bmatrix} - \zeta_\mathsf{B} \begin{bmatrix} \rho_\mathsf{B}^2 & 0 \\ 0 & -\mathbf{I} \end{bmatrix} \succeq 0,$$

$$\begin{bmatrix} \hat{\mathbf{h}}_\mathsf{A}^H \mathbf{\Lambda}_\mathsf{R} \hat{\mathbf{h}}_\mathsf{A} - \sigma^2_{n_\mathsf{A}}(2^{\mu_\mathsf{A}} - 1) & \hat{\mathbf{h}}_\mathsf{A}^H \mathbf{\Lambda}_\mathsf{R} \\ \mathbf{\Lambda}_\mathsf{R} \hat{\mathbf{h}}_\mathsf{A} & \mathbf{\Lambda}_\mathsf{R} \end{bmatrix} - \zeta_\mathsf{A} \begin{bmatrix} \rho_\mathsf{A}^2 & 0 \\ 0 & -\mathbf{I} \end{bmatrix} \succeq 0,$$

which is not directly SDP applicable due to the nonlinearity involved with the optimization variables μ_A and μ_B. We did not come across with such a nonlinearity problem in the previous two formulations, because there either both rate expressions are incorporated by the same optimization variable (μ) and hence, we could use monotonic equivalence; or only one rate expression is constrained by such variable.

As we can not solve the above mentioned sum rate problem directly, we take an iterative approach similar to the one used in capacity boundary computation in the previous section. That is, we first find the maximal achievable worst-case rate $\mathrm{R}_{\mathsf{wc-max}}$ in one chosen direction, say R-to-A. Later, for each point on the line between zero and $\mathrm{R}_{\mathsf{wc-max}}$, we lower bound the rate of the R-to-A link, and search for the maximal worst-case rate of the R-to-B link. While iterating, we decide on the maximal sum rate at the iteration that incrementing the lower bound of R-to-A link does not anymore increase the resultant sum rate. Such a stopping criterion arises from the fact that capacity regions is convex and hence, there is only one global maxima.

To sum up, in order to design robust covariance for the sake of maximal sum or minimum link rate, we follow the procedures presented in the previous section, and substitute the corresponding robust counterparts derived in this section into the appropriate steps therein.

7.3.2.1. Performance Results

Simulation Setup In the following, we study the performance of robust covariance optimization through Monte Carlo simulations. The elements of the channel vectors $\mathbf{h}_\mathsf{A}$ and $\mathbf{h}_\mathsf{B}$ are i.i.d. Rayleigh fading coefficients with zero mean and unit variance. Moreover, these channel matrices are assumed to stay constant over one transmission cycle of the respective protocol. We further assume $P \triangleq P_{\mathsf{R}} = P_{\mathsf{A}} = P_{\mathsf{B}} = 1$ and $\sigma_n^2 \triangleq \sigma_{n_{\mathsf{R}}}^2 = \sigma_{n_{\mathsf{A}}}^2 = \sigma_{n_{\mathsf{B}}}^2 = 10^{-2}$. According to the chosen ρ, the channel vectors and the uncertainty sphere are scaled as explained in the previous section.

Robust Algorithm and its Benchmarks We specifically focus on worst-case robust max-min covariance optimization. Hence, the optimization procedure in Section 7.2.2.2 is employed by changing the corresponding SDPs with their robust counterparts given above.

As benchmarks, we compute the worst-case performances of the optimal covariance computed through (7.15) by neglecting the possibility of uncertainty, and through the identity covariance of the no-CSIT case. The worst-case max-min performance of the two-way relaying system for given a covariance $\mathbf{\Lambda}$ and channel vectors $\hat{\mathbf{h}}_{\mathsf{A}}, \hat{\mathbf{h}}_{\mathsf{B}}$, is computed as follows. We first compute the worst-case broadcast phase rates per each link through

$$\mathcal{P}_{\mathrm{wc}}(\mathbf{\Lambda}, \mathbf{h}_{\mathsf{n}}) \; : \max_{\mu \geq 0} \quad \mu \; \textbf{subject to} \begin{bmatrix} \mathbf{h}_{\mathsf{n}}^H \mathbf{\Lambda}_{\mathsf{R}} \mathbf{h}_{\mathsf{n}} - \mu & \mathbf{h}_{\mathsf{n}}^H \mathbf{\Lambda}_{\mathsf{R}} \\ \mathbf{\Lambda}_{\mathsf{R}} \mathbf{h}_{\mathsf{n}} & \mathbf{\Lambda}_{\mathsf{R}} \end{bmatrix} - \zeta_{\mathsf{n}} \begin{bmatrix} \rho_{\mathsf{n}}^2 & 0 \\ 0 & -\mathbf{I} \end{bmatrix} \succeq 0, \zeta_{\mathsf{n}} \geq 0$$

for $\mathsf{n} \in \{\mathsf{A}, \mathsf{B}\}$, which is a standard SDP problem. Hence, we compute the worst-case broadcast rate of terminal n for the given $\mathbf{\Lambda}$ as $\mathrm{R}_{\mathsf{Rn}}^{\leftrightarrow\mathrm{wc}}(\mathbf{\Lambda}) = \log_2(1 + \mu^\star / \sigma_{n_{\mathsf{n}}}^2)$. Next, we check these rates with the MAC single bounds, and then reach the final worst-case max-min two-way unidirectional rates through applying Step 5 of the procedure in Section 7.2.2.2.

Simulation Results We first study the average minimum worst-case performance of the two reference non-robust approaches and the proposed robust design. The minimum is taken over the two unidirectional links (terminals), whereas the average is computed over 1000 channel realizations. Fig. 7.15 depicts the performance comparison for a fixed time-share of $(\alpha_{\mathrm{MAC}}, \alpha_{\mathrm{BC}}) = (0.55, 0.45)$. The robust design outperforms the non-robust design substantially for all possible uncertainty sizes. The largest gain is achieved at $\rho = 0.7$ where more than 1 bps/Hz rate improvement is attained with the robust approach. Note that our objective function through the robust design was to maximize the minimum worst-case performance over two-phase unidirectional rates. However, in order to highlight the difference of the minimum, i.e., $\mathbf{min}\{\mathrm{R}_{\mathsf{AB}}^{\leftrightarrow\mathrm{wc}}, \mathrm{R}_{\mathsf{BA}}^{\leftrightarrow\mathrm{wc}}\}$, and the average, i.e., $\frac{1}{2}(\mathrm{R}_{\mathsf{AB}}^{\leftrightarrow\mathrm{wc}} + \mathrm{R}_{\mathsf{BA}}^{\leftrightarrow\mathrm{wc}})$, performances, we also plot latter one. The robust design is still effective over the average performance although its primary objective was

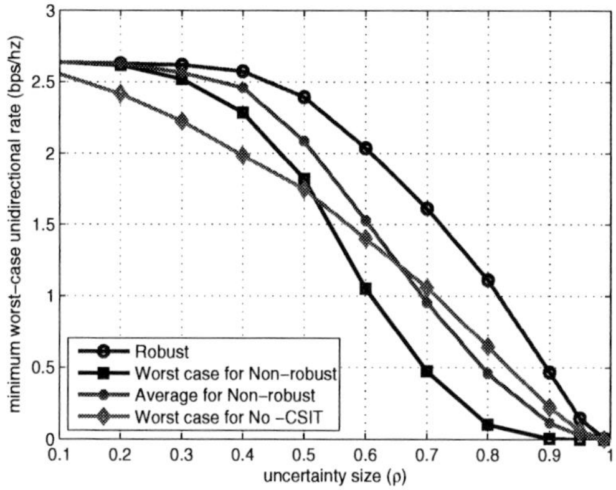

Figure 7.15.: Average minimum worst-case unidirectional rate vs. uncertainty size ρ for an average SNR of $\mathrm{SNR} = 20\mathrm{dB}$, $N_\mathsf{R} = 4$ relay antennas, and $(\alpha_{\mathrm{MAC}}, \alpha_{\mathrm{BC}}) = (0.55, 0.45)$.)

not such. On the other hand, another curve depicts the worst-case performance of feeding spatially white inputs to the antennas (no CSIT). Comparing it with the non-robust covariance design, we report that CSIT without robust design is useless when $\rho > 0.5$. Finally, let us focus on the two-extreme cases $\rho < 0.1$ and $\rho \to \infty$. As ρ gets larger, the rates go to zero independently from the covariance choice, because the uncertainty at each link nulls the estimate. Besides, observing that both robust and non-robust approaches perform the same for $\rho < 0.1$, it can be thought at the first glance that uncertainty is ineffective at that regime. However, the fact is that the performance there is anyhow limited by the MAC phase through pre-assigned time-shares. We further investigate the impact of time sharing with the following figure.

In Fig. 7.16(a), we compare the robust and non-robust covariance designs for different time-share sets. In general, the advantage of robust approach over the non-robust depicts itself as long as MAC rates do not limit the overall two-phase performance. Fig. 7.16(b) shows the impact of time-sharing more explicitly. Although the robust design results in better rates for the BC phase, the overall two-phase results are dominated by the MAC phase, and hence there occurred the flat plateaus in the figure. In essence, we can benefit from the figure in order to choose the maximal unidirectional rate resulting time-share set for given uncertainty size ρ. For example, larger α_{MAC} is preferable for

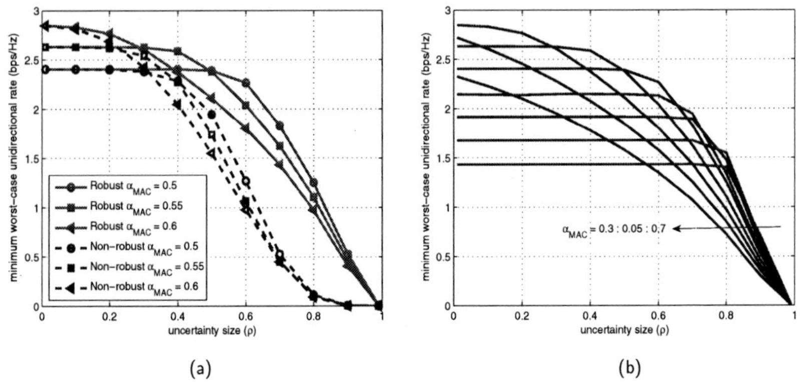

Figure 7.16.: Impact of time sharing between signaling phases: (a) Comparison of robust and non-robust optimization designs, (b) Dependence of robust performance to the choice of $(\alpha_{\mathrm{MAC}}, \alpha_{\mathrm{BC}})$ $(\mathrm{SNR} = 20\mathrm{dB}, N_{\mathsf{R}} = 4)$.

larger ρ, and vice-versa for smaller α_{MAC}.

7.4. Concluding Remarks

In this chapter, we have addressed the availability of transmit CSI at the transmitting nodes for both one- and two-way relaying protocols. For one-way relaying, we characterized the capacity region of unidirectional data rates over four phases in terms of time-sharing parameters. Moreover, for two-way relaying, we respectively characterized the capacity regions of MAC phase, BC phase and the overall two-way protocol over two phases. Then, we considered two objective functions, i.e., maximum sum rate and maximum minimum unidirectional link rate, and proposed optimal solutions through efficient SDP modeling.

In general, we showed that two-way relaying achieves a quite substantial improvement in spectral efficiency compared to its one-way counterpart independently from the CSIT assumption. Further, we reported that the impact of CSIT at the relay highly correlates with the ratio between number of relay antennas and number of terminal antennas. That is to say, the higher the ratio is, the more substantial gain is achieved through precoding.

Our motivation to assume transmit CSI at the relay in two-way protocol, was incorporating the requisite channel estimation in first phase, channel reciprocity and slow-fading environment. However, besides the commonly encountered channel estimation errors, the available channel estimates may be outdated in the second phase due to mobility of the users in between two phases. In order to provide reliability to the network, we have considered the imperfect transmit CSI case in a worst-case framework. We conclude that robust designs are compulsory for reliable data support in the presence of uncertainty or mobility. More interestingly, we have observed that the non-robust design employing CSIT may result in a worst-case performance much worse than employing no CSIT. This highlights the importance of *uncertainty awareness* while exploiting CSIT.

Chapter 8

Multiuser MIMO Decode-and-Forward Relay Networks

After establishing a general understanding on MIMO DF relaying protocols, we extend this framework to the simultaneous communication of multiple S-D pairs through a single DF relay. While employing zero-forcing based spatial separation in the downlink phase of one-way relaying, we propose a novel two-level separation for its two-way counterpart. That is, we manage the interference in between the members of each pair through network coding, whereas different pairs are still separated spatially. Further, we propose an SDP based sum rate maximization algorithm, which is optimal over two signaling phases to exploit the CSIT at the relay.

8.1. Introduction

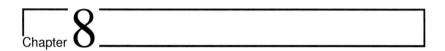

Distributed spatial multiplexing has been extensively studied in Part I through coherent amplify-and-forward relaying. In this chapter, we focus on achieving distributed spatial multiplexing through decode-and-forward relaying. Specifically, we are interested in providing a reliable physical-medium for simultaneous MIMO communications of multiple S-D pairs. Such a multiuser structure can be adapted in a both multihop local area and cellular networks. For instance, considering a WLAN scenario, the access point equipped with multiple antennas, i.e., following the IEEE 802.11n standard, can act as the relay for the multiple (uni- or) bidirectional communications of wireless devices in an office environment. Alternatively in a cellular network, a base station, which has already been approved to have multiple antennas in 4G, can provide (uni- or) bidirectional data services for inter- or intra-cell communications of several mobile stations pairs. Another cellular scenario can be that, there is a relay station placed at the cell edge to serve neighbouring cells simultaneously. That is, it can spatially multiplex several pairs of base and mobile stations within different cells. Please see Fig. 8.1 for illustrative graphics of the aforementioned cooperative multiuser scenarios.

State-of-the-Art In contrary with its AF counterpart [1, 30, 133, 134], there is hardly any research on multiuser MIMO DF relaying except that there is the capacity scaling results [23, 80, 124]. However, although not MIMO, there have several proposals to enable communications in between multiple pair of users through DF relaying [31, 33, 79, 104, 105]. In [105], two interfering S-D pairs have been assumed to be assisted by a single relay. They provided coding strategies that combine DPC, beamforming and interference mitigation techniques at the relay and gave an achievable rate region for Gaussian interference relay channel. It has been claimed that proposed strategies can be carried over a DF relaying scenario. A similar two-pair scenario is considered in [79] for a special purpose of multicast traffic from each source to both destinations. In [33], a multicast scenario is divided into two phases, where in the first phase source broadcasts a common message to all intended terminals. The terminals, which are able to decode the common message perfectly in the first phase, act as relays in the second phase and concurrently forward re-encoded common message to the others using distributed space time coding. In [104], a twisted setup of multiple terminals has been considered, where each node has a packet to send to all the others with the assistance of a single relay node. They analyzed three schemes each respectively based on time-division, network coding and superposition encoding. On the other hand, a typical multiuser two-way relay network has been studied in [31], where multiple pairs of terminals communicate with their pre-assigned partners via a single relay node. Inter-terminal interference within a pair is mitigated through self-interference cancellation, whereas inter-pair separation is realized through CDMA. A joint demodulate-and-XOR forward relaying scheme is proposed where all terminals transmit to the relay concurrently in the first phase, and the relay jointly demodulates and generates an estimate of the XORed symbol for each pair to broadcast back in the second phase.

In this chapter, we specifically focus on a multiuser MIMO network setup, which is composed of multiple S-D terminal pairs and a single half-duplex MIMO relay node. Communication takes place bidirectionally between the pairs, where either four-phase one-way relaying protocol or two-phase two-way relaying protocol is employed.

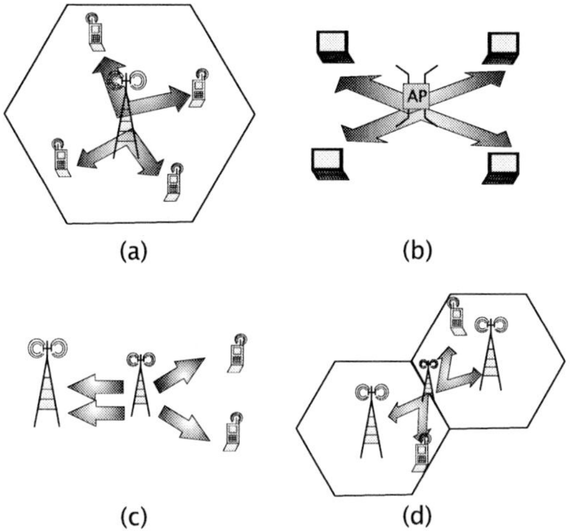

Figure 8.1.: Graphical illustration of multiuser DF relaying: (a) Cellular scenario, (b) WLAN scenario, (c) Cellular/MIMO/WLAN/Ad-hoc range extension scenario for one-to-many multiuser communications, (d) Relay on the cell-edge serves the pairs in different cells, which can not communicate due to topological conditions.

Contribution of this Chapter Our contribution in this chapter can be summarized in three parts. First, we extend the MIMO one- and two-way relaying signal models and the corresponding capacity regions to the multiple S-D pairs case. Independent of the employed protocol, we observe a multiple access channel at each transmission towards the relay. For one-way relaying, any transmission from the relay to the terminals is described by the conventional MIMO broadcast channel. However, there establishes a *modified* broadcast channel for the second phase of two-way relaying, where some terminals, i.e., members of each pair, intend for the same message. Our second contribution is addressing this issue, where we propose a two-level terminal separation for the broadcast phase of two-way relaying: bit-level and spatial-level. Note that in [31], our spatial-level separation was replaced with spread spectrum based separation due to the absence of multiple antennas.

The optimization of such a multiuser scenario may seem trivial at first glance since both MAC and BC phases have been individually well-studied in the literature. However, as previously mentioned for the single-pair case, the two phases are coupled in terms of overall end-to-end transfer rates, i.e., the capacity region of the respective protocol is represented by the intersection of the capacity regions of the two phases. Hence, an optimal optimization approach should take this coupling into account, which unfortunately does not boil down to checking some clauses as it was for the single-pair case in Sections 7.2.2.1 and 7.2.2.2. To this end, we focus on this optimization problem and propose a modular iterative sum rate maximization algorithm which is independent from the employed relaying protocol and the number of terminals. Further, it can be immediately extended to max-min type optimizations or to the introduction of QoS constraints.

Outline of this Chapter The rest of this chapter is organized as follows. We introduce the principles of multiuser one- and two-way relaying protocols in Section 8.2, where we also introduced our two-level terminal separation for two-way relaying. In Section 8.3, we individually treat each transmission phase, i.e., terminals-to-relay and relay-to-terminals. There, we design the structure of precoders at the relay, derive the achievable rate expressions and characterize the corresponding capacity regions. The proposed sum rate maximization algorithm is presented in Section 8.4, where we first elaborate on a simplified case in Section 8.4.1.1, and then generalize it in Section 8.4.1.2. In Section 8.5 we provide Monte Carlo simulations results and correspondingly discuss the performance of both multiuser two-relaying protocols. We finally present some conclusive remarks in Section 8.6.

8.2. Multiuser Protocols

We consider a multiuser relaying scenario, where K terminal pairs communicate via a single DF relay, i.e., there are $2K+1$ nodes in total. We define two terminal sets $\mathcal{A}$ and $\mathcal{B}$ such that each terminal in $\mathcal{A}$ wants to communicate with one specific terminal in $\mathcal{B}$. We assume that the pairings are non-intersecting and there is one-to-one terminal mapping from set $\mathcal{A}$ to set $\mathcal{B}$, and vice-versa. In the sequel, we denote the ith pair with P_i and its members with A_i and B_i. Following the considerations from the previous chapters, we assume that there is no direct-link between sets $\mathcal{A}$ and $\mathcal{B}$.

The members of ith terminal pair have N_{A_i} and N_{B_i} antennas, respectively; whereas the relay is equipped with N_{R} antennas both in transmit and receive modes. We denote the frequency flat fading MIMO channels from A_i to R and B_i to R with $\mathbf{H}_{\mathsf{A}_i} \in \mathbb{C}^{N_{\mathsf{R}} \times \mathsf{N}_{\mathsf{A}_i}}$ and $\mathbf{H}_{\mathsf{B}_i} \in \mathbb{C}^{N_{\mathsf{R}} \times \mathsf{N}_{\mathsf{B}_i}}$, respectively. All other assumptions regarding the characteristics of

MIMO channels for the single pair case in the previous chapters are also valid here and in the sequel.

The pair P_i wants to exchange the bit sequence tuple $(\mathbf{x}_{\mathsf{A}_i}, \mathbf{x}_{\mathsf{B}_i})$, where the elements of each are either 1 or 0. The corresponding coded transmit signal vectors are denoted respectively by $\mathbf{s}_{\mathsf{A}_i} \in \mathbb{C}^{N_{\mathsf{A}_i}}$ and $\mathbf{s}_{\mathsf{B}_i} \in \mathbb{C}^{N_{\mathsf{B}_i}}$, with covariances of $\mathbf{\Lambda}_{\mathsf{A}_i} \triangleq \mathbb{E}\{\mathbf{s}_{\mathsf{A}_i}\mathbf{s}_{\mathsf{A}_i}^H\}$ and $\mathbf{\Lambda}_{\mathsf{B}_i} \triangleq \mathbb{E}\{\mathbf{s}_{\mathsf{B}_i}\mathbf{s}_{\mathsf{B}_i}^H\}$.

We extend both relaying protocols (one- and two-way relaying) to the multiuser setup. Although in theory each pair can choose the relaying protocol to employ individually and independently from the others, for the sake of simplicity throughout the presentation, we will assume that all pairs together employ either one- or two-way relaying protocol.

8.2.1. Multiuser One-Way Relaying

The multiuser one-way protocol is accomplished in four time-slots (channel-uses):

1. a concurrent transmission from all terminals in $\mathcal{A}$ to R,
2. a broadcast transmission from R to $\mathcal{B}$,
3. a concurrent transmission from all terminals in $\mathcal{B}$ to R,
4. and a broadcast transmission from R to $\mathcal{A}$.

The data transfer model is symmetric for both information flow directions. Hence, in the following we focus only on the transmission from $\mathcal{A}$ to $\mathcal{B}$ from which the model for the reverse direction can be trivially deduced.

In the first time-slot, all K terminals in $\mathcal{A}$ concurrently transmit to R through the same physical channel (see Fig. 8.2(a)). The received signal at the relay is

$$\mathbf{r}_{\mathsf{R}} = \sum_{i=1}^{K} \mathbf{H}_{\mathsf{A}_i}\mathbf{s}_{\mathsf{A}_i} + \mathbf{n}_{\mathsf{R}}, \tag{8.1}$$

where each transmit signal is subject to the power constraint of $\mathrm{Tr}(\mathbf{\Lambda}_{\mathsf{A}_i}) \leq P_{\mathsf{A}_i}$. Hence, we observe a multiple access channel at the relay. Assuming that the relay has at least K antennas, it efficiently decodes all information from $\mathcal{A}$ using an appropriate MAC decoding scheme, e.g., successive interference cancellation [119], and perfectly obtain the bit sequence tuple $(\mathbf{x}_{\mathsf{A}_1}, \ldots, \mathbf{x}_{\mathsf{A}_K})$.

In the second time-slot, the relay needs to re-encode the decoded information from the previous time-slot, and then broadcast them to $\mathcal{B}$. Specifically, the relay encodes

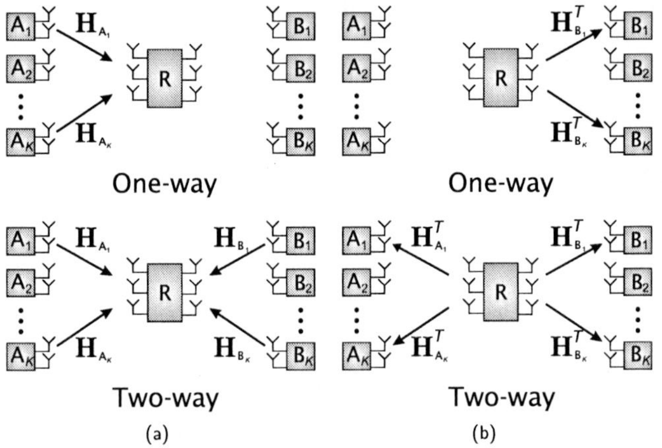

Figure 8.2.: Network models and transmission protocols for multiuser one- and two-way relaying: (a) MAC phase, (b) BC phase.

the ith information sequence, intended for B_i, to a complex signal vector $\mathbf{x}_{\mathsf{A}_i} \rightarrow \mathbf{s}_{\mathsf{R}_i}$, and then pre-codes with the matrix $\vec{\mathbf{W}}_i$. The relay transmit signal $\mathbf{s}_{\mathsf{R}} \in \mathbb{C}^{N_{\mathsf{R}}}$ is obtained by superposing these precoded vectors:

$$\mathbf{s}_{\mathsf{R}} = \sum_{i=1}^{K} \vec{\mathbf{W}}_i \mathbf{s}_{\mathsf{R}_i}.$$

We observe a conventional broadcast channel through the transmission from R to $\mathcal{B}$, where the relay spatially separates the terminals in $\mathcal{B}$ and sends private information to each terminal [49, 125]. Referring to the following sections for specific structures of the precoders, here we only note that we impose an average power constraint of P_{R}.

Next, the received signal at the ith terminal in $\mathcal{B}$ is

$$\mathbf{r}_{\mathsf{B}_i} = \underbrace{\mathbf{H}_{\mathsf{B}_i}^T \vec{\mathbf{W}}_i \mathbf{s}_{\mathsf{R}_i}}_{\text{intended signal from } \mathsf{A}_i} + \underbrace{\sum_{j=1, j\neq i}^{K} \mathbf{H}_{\mathsf{B}_i}^T \vec{\mathbf{W}}_j \mathbf{s}_{\mathsf{R}_j}}_{\text{interference}} + \mathbf{n}_{\mathsf{B}_i}, \tag{8.2}$$

where $\mathbf{n}_{\mathsf{B}_i} \sim \mathcal{CN}(\mathbf{0}, \sigma_{n_{\mathsf{B}}}^2 \mathbf{I}_{N_{\mathsf{B}_i}})$ represents the AWGN effect at the terminal (see Fig. 8.2(b)).

8.2.2. Multiuser Two-Way Relaying

The multiuser two-way protocol starts with the MAC phase, where all terminals within both $\mathcal{A}$ and $\mathcal{B}$ simultaneously transmit to the relay in the first time-slot (see Fig. 8.2(a)). Hence, the received signal at the relay is

$$\mathbf{r}_\mathsf{R} = \sum_{i=1}^{K} (\mathbf{H}_{\mathsf{A}_i}\mathbf{s}_{\mathsf{A}_i} + \mathbf{H}_{\mathsf{B}_i}\mathbf{s}_{\mathsf{B}_i}) + \mathbf{n}_\mathsf{R}, \tag{8.3}$$

where we assume per-terminal transmit power constraint. Through the observed MAC, the relay is assumed to efficiently decode all information from the terminals and perfectly obtain the bit sequence tuple $(\mathbf{x}_{\mathsf{A}_1}, \mathbf{x}_{\mathsf{B}_1}, \dots, \mathbf{x}_{\mathsf{A}_K}, \mathbf{x}_{\mathsf{B}_K})$.

In the second broadcast phase, the relay re-encodes the decoded information from the previous time-slot, and forwards them back to both $\mathcal{A}$ and $\mathcal{B}$ (see Fig. 8.2(b)). Through this downlink transmission, while broadcasting $2K$ independent information, the relay needs to separate $2K$ terminals and manage the inter-user interference. In fact, this can be simply accomplished through space, but it is well-know that the efficiency of such a spatial separation is constrained in practice by the number of transmit antennas as the number of terminals increases [36, 115]. However, we note that it is always possible to use non-linear (but impractical) interference management schemes like dirty paper coding [34, 125], which are independent from transmit antenna numbers. Hence, any scheme that can reduce the need for spatial degrees of freedom while ensuring practical implementation, is at most welcome. At this point, we incorporate XOR precoding with spatial separation and propose a novel *two-levels of separation* for multiuser two-way broadcast channel: *bit-level* and *spatial-level.* To summarize:

- **Bit-level Separation:** We couple the two bit sequences $(\mathbf{x}_{\mathsf{A}_i}, \mathbf{x}_{\mathsf{B}_i})$ belonging to the pair P_i, and combine them on bit-level through XOR-addition, i.e.,

  $$\mathbf{x}_{\mathsf{R}_i} = \mathbf{x}_{\mathsf{A}_i} \oplus \mathbf{x}_{\mathsf{B}_i}.$$

  Further, we encode the resultant bit sequence to a complex signal vector $\mathbf{x}_{\mathsf{R}_i} \to \mathbf{s}_{\mathsf{R}_i}$. Thus, as the terminals decode $\mathbf{s}_{\mathsf{R}_i}$ at the receive sides, they can remove the back-propagated self-interference so that $\mathbf{x}_{\mathsf{A}_i}$ and $\mathbf{x}_{\mathsf{B}_i}$ do not anymore cause interference to each other.

- **Spatial-level Separation:** At the end of bit-level combination, instead of $2K$ information to separate spatially, we have a K-tuple of $(\mathbf{s}_{\mathsf{R}_1}, \dots, \mathbf{s}_{\mathsf{R}_K})$. In order to further separate these K transmit signals we employ spatial methods and linearly pre-code each with a matrix of $\breve{\mathbf{W}}_i$. The final relay transmit signal is obtained by superposing all:

$$\mathbf{s}_{\mathsf{R}} = \sum_{i=1}^{K} \breve{\mathbf{W}}_i \mathbf{s}_{\mathsf{R}_i}.$$

The structural details on spatial separation, i.e., how to design $\breve{\mathbf{W}}_i$, will be given in the following section.

Finally, the received signals at the terminals of pair P_i are

$$\begin{aligned} \mathbf{r}_{\mathsf{A}_i} &= \underbrace{\mathbf{H}_{\mathsf{A}_i}^T \breve{\mathbf{W}}_i \mathbf{s}_{\mathsf{R}_i}}_{\text{intended signal from } \mathsf{B}_i} + \underbrace{\sum_{j=1, j\neq i}^{K} \mathbf{H}_{\mathsf{A}_i}^T \breve{\mathbf{W}}_j \mathbf{s}_{\mathsf{R}_j}}_{\text{interference}} + \mathbf{n}_{\mathsf{A}_i}, \\ \mathbf{r}_{\mathsf{B}_i} &= \underbrace{\mathbf{H}_{\mathsf{B}_i}^T \breve{\mathbf{W}}_i \mathbf{s}_{\mathsf{R}_i}}_{\text{intended signal from } \mathsf{A}_i} + \underbrace{\sum_{j=1, j\neq i}^{K} \mathbf{H}_{\mathsf{B}_i}^T \breve{\mathbf{W}}_j \mathbf{s}_{\mathsf{R}_j}}_{\text{interference}} + \mathbf{n}_{\mathsf{B}_i}, \end{aligned} \tag{8.4}$$

where $\mathbf{n}_{\mathsf{n}_i} \sim \mathcal{CN}(\mathbf{0}, \sigma_{n_\mathsf{n}}^2 \mathbf{I}_{N_{\mathsf{n}_i}})$ represents the AWGN affect at terminal n, $\mathsf{n} \in \{\mathsf{A}, \mathsf{B}\}$. After each terminal decodes the intended signal, the self-interference cancellation is done by applying a simple XOR operation, i.e., $\mathbf{x}_{\mathsf{R}_i} \oplus \mathbf{x}_{\mathsf{A}_i} = \mathbf{x}_{\mathsf{B}_i}$, $\mathbf{x}_{\mathsf{R}_i} \oplus \mathbf{x}_{\mathsf{B}_i} = \mathbf{x}_{\mathsf{A}_i}$.

8.3. Achievable Rates and Precoding at the Relay

8.3.1. Multiple Access Phase

In both relaying protocols multiple users (K for one-way, $2K$ for two-way protocol) are trying to access to the relay. Upon receiving the received signal $\mathbf{r}_{\mathsf{R}}$ as given in (8.1) for one-way and in (8.3) for two-way, the relay decodes all incoming information perfectly assuming that it uses an appropriate MAC decoding scheme with sufficient number of antennas. The terminals' transmit signals are assumed to be drawn from independent Gaussian codebooks, and correspondingly, the achievable rates in the first phase are described by the MIMO MAC capacity region [49]. Assuming that there is no transmit CSI at the terminals, the MIMO MAC capacity region is defined as the following L-dimensional polyhedron

$$\mathcal{C}_{\text{MAC}}^{\text{mu}} = \left\{ (\mathrm{R}_1, \ldots, \mathrm{R}_L) \in \mathbb{R}_+ \ : \ \textstyle\sum_{i\in\mathcal{S}} \mathrm{R}_i \leq \log_2 \left| \mathbf{I} + \textstyle\sum_{i\in\mathcal{S}} \frac{P_i}{N_i \sigma_{n_\mathsf{R}}^2} \mathbf{H}_i \mathbf{H}_i^H \right| \right\} \tag{8.5}$$

where

$$\begin{cases} L := K \\ \mathcal{S} := \mathcal{A} \\ (\mathrm{R}_1, \ldots, \mathrm{R}_L) := (\mathrm{R}_{\mathsf{A}_1\mathsf{R}}^{\rightarrow}, \ldots, \mathrm{R}_{\mathsf{A}_K\mathsf{R}}^{\rightarrow}) \\ (\mathbf{H}_1, \ldots, \mathbf{H}_L) := (\mathrm{H}_{\mathsf{A}_1}, \ldots, \mathrm{H}_{\mathsf{A}_K}) \\ (N_1, \ldots, N_L) := (N_{\mathsf{A}_1}, \ldots, N_{\mathsf{A}_K}) \\ (P_1, \ldots, P_L) := (P_{\mathsf{A}_1}, \ldots, P_{\mathsf{A}_K}) \end{cases} \quad \text{for one-way relaying,} \tag{8.6}$$

$$\begin{cases} L := 2K \\ \mathcal{S} := \mathcal{A} \cup \mathcal{B} \\ (\mathrm{R}_1, \ldots, \mathrm{R}_L) := (\mathrm{R}_{\mathsf{A}_1\mathsf{R}}^{\leftrightarrow}, \mathrm{R}_{\mathsf{B}_1\mathsf{R}}^{\leftrightarrow}, \ldots, \mathrm{R}_{\mathsf{B}_K\mathsf{R}}^{\leftrightarrow}, \mathrm{R}_{\mathsf{B}_K\mathsf{R}}^{\leftrightarrow}) \\ (\mathbf{H}_1, \ldots, \mathbf{H}_L) := (\mathrm{H}_{\mathsf{A}_1}, \mathrm{H}_{\mathsf{B}_1}, \ldots, \mathrm{H}_{\mathsf{B}_K}, \mathrm{H}_{\mathsf{A}_K}) \\ (N_1, \ldots, N_L) := (N_{\mathsf{A}_1}, N_{\mathsf{B}_1}, \ldots, N_{\mathsf{A}_K}, N_{\mathsf{B}_K}) \\ (P_1, \ldots, P_L) := (P_{\mathsf{A}_1}, P_{\mathsf{B}_1}, \ldots, P_{\mathsf{A}_K}, P_{\mathsf{B}_K}) \end{cases} \quad \text{for two-way relaying.} \tag{8.7}$$

The capacity region (8.5) is constituted of $2^L - 1$ inequalities. Because of no CSIT assumption, we use diagonal covariances with equal weights for the elements on the diagonal. Hence, the relay can assign any rate-tuple to the terminals as long as it is described as a point within $\mathcal{C}_{\mathrm{MAC}}^{\mathrm{mu}}$. The capacity can be extended to the case of having CSIT knowledge at the transmitting terminals by taking union over all transmit covariances satisfying power constraints [49].

8.3.2. Broadcast Phase

Assuming perfect transmit CSI at the relay, the achievable downlink rates from R to the corresponding terminals are totally dependent on the efficiency of inter-user interference management, which is determined by the design of the precoding matrices, i.e., $\overleftrightarrow{\mathbf{W}}_i$, $\vec{\mathbf{W}}_i$, and the transmit signal covariances, i.e., $\mathbf{\Lambda}_i$, (consequently transmit power allocation). For both relaying protocols, there are K information stream that should be separated spatially. For one-way relaying, there established a conventional *MIMO broadcast channel* with K destinations, whereas for two-way relaying, it is a *modified MIMO broadcast channel* with $2K$ destinations, where the members of each pair P_i is demanding the same information.

As investigated intensively in the literature, there have been proposed several MIMO BC transmission schemes such as general beamforming (BF) [142], zero-forcing beamforming (ZF-BF) [36, 115, 142], dirty paper coding (DPC) [34, 125, 142], and zero-forcing dirty-paper coding (ZF-DPC) [29, 36]. In general BF, beamforming directions are chosen such that different information streams are separated by letting only a *controlled* amount of interference at the destinations, whereas an additional zero-interference condition is enforced in ZF-BF, i.e., through channel inversion. For the sake of feasibility of channel inversion, the transmitter requires to have a certain minimum number of antennas, which depends on the number of destinations and receive antennas. Basing on interference pre-subtraction, DPC employs the fact that the interference is ineffective on channel capacity as long as non-causal interference knowledge is available in advance at the transmitter. Although DPC is the optimal in terms of maximizing the sum rate of the conventional MIMO broadcast channel, it is difficult to be implemented in practical systems due to its nonlinearity and high computational complexity of successive encodings/decodings. In order to reduce complexity, DPC is incorporated with zero-forcing, i.e., ZF-DPC, so that only partial inter-user interference is cancelled through ZF pre-coders, and the rest is handled through DPC.

We focus on low-complexity solutions of zero-forcing based schemes. The reason to exclude DPC from our considerations is its computational burden. We emphasize here that it is a difficult non-convex problem to optimize even for the conventional MIMO BC channel [125], whereas, as it will be shown in the next section, our problem deals with a modified version of BC channel and is constrained with non-convex constraints arising from the MAC phase. Nevertheless, it has been shown in [142] that ZF-BF can achieve the sum capacity of DPC at the asymptotic limit of number of destinations.

In the following, we adapt the block diagonalization based ZF-BF [115] to our relaying scenarios, and treat both protocols individually by depicting the differences in between.

8.3.2.1. Multiuser One-Way Relaying

Aiming at zero-interference in between the terminals in $\mathcal{B}$, the relay chooses the ith precoding matrix $\vec{\mathbf{W}}_i$ such that it fulfills the zero-forcing condition $\mathbf{H}_{\mathsf{B}_j}^T \vec{\mathbf{W}}_i = 0$ for all $j \neq i, j = \{1, \ldots, K\}$. In other words, $\vec{\mathbf{W}}_i$ is forced to lie in the nullspace of

$$\vec{\mathbf{G}}_i \in \mathbb{C}^{\sum_{j \neq i} N_{\mathsf{B}j} \times N_{\mathsf{R}}} \triangleq \left[\mathbf{H}_{\mathsf{B}_1} \;\cdots\; \mathbf{H}_{\mathsf{B}_{i-1}} \; \mathbf{H}_{\mathsf{B}_{i+1}} \;\cdots\; \mathbf{H}_{\mathsf{B}_K}\right]^T. \tag{8.8}$$

As the condition for a nonempty nullspace is $N_{\mathsf{R}} > \operatorname{rank}(\vec{\mathbf{G}}_i)$, the relay needs to transmit with at least $N_{\mathsf{R}} > \mathbf{max}_i\left(\operatorname{rank}(\vec{\mathbf{G}}_i)\right)$ antennas to transmit interference-free data concurrently to all terminals. Choosing the ith ZF sub-precoder $\vec{\mathbf{W}}_i^{\mathrm{zf}}$ from the corresponding nullspace $\vec{\mathbf{V}}_i \triangleq \operatorname{null}(\vec{\mathbf{G}}_i)$, which can be computed through SVD of $\vec{\mathbf{G}}_i$, the interference terms in (8.2) are removed. Hence, the achievable data transfer rate from R to B_i is written as

$$\mathrm{R}_{\mathsf{RB}_i}^{\rightarrow} \leq \mathrm{I}_{\mathsf{RB}_i}^{\rightarrow} = \log_2 \left| \mathbf{I}_{N_{\mathsf{B}_i}} + \frac{1}{\sigma_{n_\mathsf{B}}^2} \mathbf{H}_{\mathsf{B}_i}^T \vec{\mathbf{V}}_i \mathbf{\Lambda}_{\mathsf{R}_i} \vec{\mathbf{V}}_i^H \mathbf{H}_{\mathsf{B}_i}^\star \right|, \tag{8.9}$$

where $\vec{\mathbf{V}}_i \in \mathbb{C}^{N_\mathsf{R} \times N_\mathsf{R} - \mathrm{rank}(\vec{\mathbf{G}}_i)}$, and $\mathbf{\Lambda}_{\mathsf{R}_i} \triangleq \mathbb{E}\{\mathbf{s}_{\mathsf{R}_i}\mathbf{s}_{\mathsf{R}_i}^H\}$ is the ith sub-covariance. Correspondingly, note the power constraint imposed on the relay becomes

$$\sum_{i=1}^{K} \mathrm{Tr}(\vec{\mathbf{V}}_i \mathbf{\Lambda}_{\mathsf{R}_i} \vec{\mathbf{V}}_i^H) = \sum_{i=1}^{K} \mathrm{Tr}(\mathbf{\Lambda}_{\mathsf{R}_i}) \leq P_\mathsf{R}, \tag{8.10}$$

where the equality follows from the orthonormality of $\vec{\mathbf{V}}_i$. Assuming a fixed individual power constraint P_{R_i} for $\mathbf{\Lambda}_{\mathsf{R}_i}$, such that $\sum_{i=1}^{K} P_{\mathsf{R}_i} = P_\mathsf{R}$, the optimal $\mathbf{\Lambda}_{\mathsf{R}_i}$ that leads to a maximum data transmission rate, can be found by applying water-filling on the singular value matrix $\mathbf{\Sigma}_i$. This is simply obtained through SVD of $\mathbf{H}_{\mathsf{B}_i}^T \vec{\mathbf{V}}_i$ [115]. Hence, computing the optimal $\mathbf{\Lambda}_{\mathsf{R}_i}^\star$ according to a unit power constraint, i.e. $P_{\mathsf{R}_i} = 1$, the supported rate (8.9) becomes

$$\mathrm{R}_{\mathsf{RB}_i}^{\rightarrow} \leq \mathrm{I}_{\mathsf{RB}_i}^{\rightarrow} = \log_2 \left| \mathbf{I}_{N_{\mathsf{B}_i}} + \frac{P_{\mathsf{R}_i}}{\sigma_{n_\mathsf{B}}^2} \mathbf{\Sigma}_i^2 \mathbf{\Lambda}_{\mathsf{R}_i}^\star \right|, \tag{8.11}$$

which is now only a function of allocated power P_{R_i}. Henceforth, the members of the rate tuple $(\mathrm{R}_{\mathsf{RB}_1}^{\rightarrow}, \ldots, \mathrm{R}_{\mathsf{RB}_K}^{\rightarrow})$ are coupled only through the power constraint. Hence, the capacity region of the broadcast phase for multiuser one-way relaying is given by

$$\mathcal{C}_{\mathrm{BC}}^{\mathrm{mu-one}} = \bigcup_{P_{\mathsf{R}_i} \geq 0, \sum_{i=1}^{K} P_{\mathsf{R}_i} \leq P_\mathsf{R}} \left\{ (\mathrm{R}_{\mathsf{RB}_1}^{\rightarrow}, \ldots, \mathrm{R}_{\mathsf{RB}_K}^{\rightarrow}) \in \mathbb{R}_+ \ : \ \mathrm{R}_{\mathsf{RB}_i}^{\rightarrow} \leq \mathrm{I}_{\mathsf{RB}_i}^{\rightarrow}, k \in \{1, \ldots, K\} \right\} \tag{8.12}$$

where $\mathrm{I}_{\mathsf{RB}_i}^{\rightarrow}$ is defined in (8.11). Note that the region (8.12) stands for the information flow from $\mathcal{A}$ to $\mathcal{B}$. As above stated, similar rate expressions and capacity regions can also be derived for the reverse direction of information flow.

8.3.2.2. Multiuser Two-Way Relaying

The relay has K information streams to broadcast, each of which is intended for two separate terminals, i.e., the corresponding A_i and B_i. Hence, the zero-forcing condition differs from the one for one-way relaying such that it should now null $2(K-1)$ terminal directions. Specifically, the pre-coder $\overleftrightarrow{\mathbf{W}}_i$ serving the pair P_i, should fulfill the zero-forcing condition

$$\mathbf{H}_{\mathsf{B}_j}^T \overleftrightarrow{\mathbf{W}}_i = 0, \mathbf{H}_{\mathsf{A}_j}^T \overleftrightarrow{\mathbf{W}}_i = 0, \quad \text{for all } j \neq i, j = \{1, \ldots, K\}. \tag{8.13}$$

Consequently, $\overleftrightarrow{\mathbf{W}}_i$ is forced to lie in

$$\overleftrightarrow{\mathbf{G}}_i \in \mathbb{C}^{\sum_{j\neq i}(N_{\mathsf{A}_j}+N_{\mathsf{B}_j})\times N_{\mathsf{R}}} \triangleq \left[\mathbf{H}_{\mathsf{A}_1}\ \mathbf{H}_{\mathsf{B}_1} \cdots \mathbf{H}_{\mathsf{A}_{i-1}}\ \mathbf{H}_{\mathsf{B}_{i-1}}\ \mathbf{H}_{\mathsf{A}_{i+1}}\ \mathbf{H}_{\mathsf{B}_{i+1}} \cdots \mathbf{H}_{\mathsf{A}_K}\ \mathbf{H}_{\mathsf{B}_K}\right]^T . \tag{8.14}$$

For perfect zero-forcing, the requirement on the number of relay transmit antennas becomes $N_{\mathsf{R}} > \mathbf{max}_i\left(\text{rank}(\overleftrightarrow{\mathbf{G}}_i)\right)$, $i \in \{1, \ldots, K\}$. The multi-pair interference terms in (8.4) are simply canceled by substituting appropriate precoder designs $\overleftrightarrow{\mathbf{W}}_i^{\text{zf}}$ out of the corresponding nullspaces $\overleftrightarrow{\mathbf{V}}_i \triangleq \text{null}(\overleftrightarrow{\mathbf{G}}_i)$. Consequently, the terminals decode the (multi-pair) interference-free intended signal and then apply self-interference cancellation. The achievable data transfer rate from R to the terminal n in P_i is then written as

$$\overleftrightarrow{\mathrm{R}}_{\mathsf{Rn}_i} \leq \overleftrightarrow{\mathrm{I}}_{\mathsf{Rn}_i} = \log_2 \left| \mathbf{I}_{N_{\mathsf{n}_i}} + \frac{1}{\sigma_{n_{\mathsf{n}}}^2} \mathbf{H}_{\mathsf{n}_i}^T \overleftrightarrow{\mathbf{V}}_i \mathbf{\Lambda}_{\mathsf{R}_i} \overleftrightarrow{\mathbf{V}}_i^H \mathbf{H}_{\mathsf{n}_i}^* \right|, \quad \mathsf{n} \in \{\mathsf{A}, \mathsf{B}\}. \tag{8.15}$$

By definition, the nullspace $\overleftrightarrow{\mathbf{V}}_i$ is $N_{\mathsf{R}} \times N_{\mathsf{R}} - \text{rank}(\overleftrightarrow{\mathbf{G}}_i)$ dimensional and it correspondingly defines the dimensions of the sub-precoder $\mathbf{\Lambda}_{\mathsf{R}_i}$. However, the maximal number of independent substream that can be exchanged in between the ith pair is not only limited by the excess degrees of freedom at the transmitter, i.e., $N_{\mathsf{R}} - \text{rank}(\overleftrightarrow{\mathbf{G}}_i)$, but also with the number of terminal antennas.

Observe that the achievable rates of the members of the ith pair P_i are coupled through both precoder designs and the covariance matrix. Hence, we can not directly apply a water-filling solution here in order to find the optimal maximum rates valid for both terminals. Because the beamformer needs to direct to two different directions in contrary to the one-way protocol, we should further shape the beam with the covariance design. However, as the channels of the two terminals become more and more correlated, e.g., get closer to each other, the relay can approach to use a single beam which can be then optimized by conventional broadcast channel methods. This issue has also been reported in [138] as an affect of the philosophy of network coding on the signal processing at the relay.

Finally, the capacity region of the broadcast phase for multiuser two-way relaying is given by the following union over all feasible sub-covariances

$$\mathcal{C}_{\text{BC}}^{\text{mu-two}} = \bigcup_{\substack{\mathbf{\Lambda}_{\mathsf{R}_1} \succeq 0, \ldots, \mathbf{\Lambda}_{\mathsf{R}_K} \succeq 0, \\ \sum_{i=1}^K \text{Tr}(\mathbf{\Lambda}_{\mathsf{R}_i}) \leq P_{\mathsf{R}}}} \left\{ \begin{array}{c} (\overleftrightarrow{\mathrm{R}}_{\mathsf{RA}_1}, \overleftrightarrow{\mathrm{R}}_{\mathsf{RB}_1}, \ldots, \overleftrightarrow{\mathrm{R}}_{\mathsf{RA}_K}, \overleftrightarrow{\mathrm{R}}_{\mathsf{RB}_K}) \in \mathbb{R}_+ : \\ \overleftrightarrow{\mathrm{R}}_{\mathsf{RA}_i} \leq \overleftrightarrow{\mathrm{I}}_{\mathsf{RA}_i}, \overleftrightarrow{\mathrm{R}}_{\mathsf{RB}_i} \leq \overleftrightarrow{\mathrm{I}}_{\mathsf{RB}_i}, k \in \{1, \ldots, K\} \end{array} \right\} \tag{8.16}$$

where $\overleftrightarrow{\mathrm{I}}_{\mathsf{Rn}_i}$ is defined in (8.15).

Summing up, the optimization of multiuser one-way relaying protocol is simplified to be a problem of real power allocation to each of K information stream under fixed average power constraint. Whereas, for the multiuser two-way relaying protocol, we need to design K semidefinite covariances under fixed average power constraint. The corresponding optimization issues will be addressed in Section 8.4.

Remark 8.3.1: Although we have focused on a pure zero-forcing scheme, the broadcast phase of the above mentioned relaying protocols can be immediately extended to hybrid schemes like successive zero-forcing DPC [36]. So that, we can further exploit the tradeoff between complexity and performance. Such an adaptation of the protocols and also the optimization algorithms that will be given below, are trivial in the sense that the whole formulation structure of rate expressions (8.9) and (8.15) stay the same except that zero-forcing conditions and the motivations for the covariances change. For further details, please refer to [44], where we have studied successive ZF-DPC in a similar multiuser two-way relaying protocol for cellular communications.

8.3.3. The Overall Two-Phase Capacity Region

Assuming that the time sharing parameters for the two phases are denoted by α_{MAC} and α_{BC}, the capacity region for the end-to-end information rate tuples

$$\vec{\mathcal{R}}(K) \triangleq (\vec{\mathrm{R}}_{\mathsf{A}_1\mathsf{B}_1}, \vec{\mathrm{R}}_{\mathsf{A}_2\mathsf{B}_2}, \ldots, \vec{\mathrm{R}}_{\mathsf{A}_K\mathsf{B}_K}) \tag{8.17}$$

for one-way relaying, and

$$\overleftrightarrow{\mathcal{R}}(K) \triangleq (\overleftrightarrow{\mathrm{R}}_{\mathsf{A}_1\mathsf{B}_1}, \overleftrightarrow{\mathrm{R}}_{\mathsf{B}_1\mathsf{A}_1}, \overleftrightarrow{\mathrm{R}}_{\mathsf{A}_2\mathsf{B}_2}, \overleftrightarrow{\mathrm{R}}_{\mathsf{B}_2\mathsf{A}_2} \ldots, \overleftrightarrow{\mathrm{R}}_{\mathsf{A}_K\mathsf{B}_K}, \overleftrightarrow{\mathrm{R}}_{\mathsf{B}_K\mathsf{A}_K}) \tag{8.18}$$

for two-way relaying, are given respectively by the intersection of the corresponding MAC and BC phase capacity regions, i.e.,

$$\begin{aligned}
\mathcal{C}^{\mathrm{mu}}_{\mathrm{one-way}}(\alpha_{\mathrm{MAC}}, \alpha_{\mathrm{BC}}) &= \mathcal{C}^{\mathrm{mu-one}}_{\mathrm{MAC}}(\alpha_{\mathrm{MAC}}) \cap \mathcal{C}^{\mathrm{mu-one}}_{\mathrm{BC}}(\alpha_{\mathrm{BC}}),\\
\mathcal{C}^{\mathrm{mu}}_{\mathrm{two-way}}(\alpha_{\mathrm{MAC}}, \alpha_{\mathrm{BC}}) &= \mathcal{C}^{\mathrm{mu-two}}_{\mathrm{MAC}}(\alpha_{\mathrm{MAC}}) \cap \mathcal{C}^{\mathrm{mu-two}}_{\mathrm{BC}}(\alpha_{\mathrm{BC}}).
\end{aligned}$$

The capacity regions are respectively $K-$ and $2K-$dimensional, and both are convex due to the fact that the intersecting regions are already convex by definition. Note that unless otherwise is stated, we assume that $\alpha_{\mathrm{MAC}} = \alpha_{\mathrm{BC}} = 0.5$ for simplicity in the sequel.

8.4. Optimization for Multiuser One- and Two-Way Relaying

So far we have characterized the achievable rates only for given covariances (or transmit powers). In this section we design the covariances such that we achieve specific points within the capacity region that correspond to chosen figures of merit. Although we focus primarily on sum rate maximization throughout this section, in a last part we briefly discuss possible extensions to max-min type optimizations and QoS requirements on the achievable rates.

As suggested for the single pair case, a suboptimal but trivial choice of optimizing achievable rates is to decouple multiple access and broadcast phases. The maximal rates for the first phase is readily available and can be chosen accordingly out of $\mathcal{C}_{\mathrm{MAC}}$. The broadcast phase can be optimized through the well-known schemes and optimization techniques (max-sum or max-min) proposed in the literature for the general MIMO broadcast channel [36, 115, 125, 142]. Next, in order to find the overall two-phase rates, the resultant broadcast phase rates are cross checked with the related MAC phase rates, i.e., the rate tuple point in the BC capacity region is linearly scaled to be placed in the MAC capacity region, or vice versa. Obviously, such a decoupling approach is suboptimal as the resources can not be fully utilized, i.e., the bottleneck phase can be different for each terminal/link. Thus, although more complex, we aim to find the overall rates optimally over two phases avoiding decoupling, which differentiates the optimization procedure from that for conventional broadcast channel.

8.4.1. Sum Rate Maximization

We define the sum rate for one- and two-way relaying respectively as

$$\mathrm{R}_{\mathrm{sum}}^{\mathrm{mu-one}} = \sum_{i=1}^{K} (\mathrm{R}^{\rightarrow}_{\mathsf{A}_i\mathsf{B}_i} + \mathrm{R}^{\rightarrow}_{\mathsf{B}_i\mathsf{A}_i}), \quad \mathrm{R}_{\mathrm{sum}}^{\mathrm{mu-two}} = \sum_{i=1}^{K} (\mathrm{R}^{\leftrightarrow}_{\mathsf{A}_i\mathsf{B}_i} + \mathrm{R}^{\leftrightarrow}_{\mathsf{B}_i\mathsf{A}_i}). \tag{8.19}$$

For one-way relaying, the sum rates for different information flow directions are decoupled. Hence, as the corresponding optimization can be performed independently and identically, in the sequel we only focus on the case of transmission from A to B.

The general sum rate maximization problems for one- and two-way are formulated respectively as

$$\mathcal{P}^{\text{mu-sum}}_{\text{one-way}} : \max_{P_{\mathsf{R}_i} \geq 0,\, \vec{\mathrm{R}}_{\mathsf{A}_i\mathsf{B}_i} \geq 0} \sum_{i=1}^{K} \vec{\mathrm{R}}_{\mathsf{A}_i\mathsf{B}_i} \quad \textbf{subject to} \quad \vec{\mathcal{R}}(K) \in \mathcal{C}^{\text{mu-one}}_{\text{MAC}} \cap \mathcal{C}^{\text{mu-one}}_{\text{BC}},$$
$$\sum_{i=1}^{K} P_{\mathsf{R}_i} \leq P_{\mathsf{R}},\; i \in \{1, \ldots, K\},$$

$$\mathcal{P}^{\text{mu-sum}}_{\text{two-way}} : \max_{\boldsymbol{\Lambda}_{\mathsf{R}_i} \succeq 0,\, \vec{\mathrm{R}}_{\mathsf{A}_i\mathsf{B}_i} \geq 0,\, \vec{\mathrm{R}}_{\mathsf{B}_i\mathsf{A}_i} \geq 0} \sum_{i=1}^{K} \vec{\mathrm{R}}_{\mathsf{A}_i\mathsf{B}_i} \quad \textbf{subject to} \quad \overleftrightarrow{\mathcal{R}}(K) \in \mathcal{C}^{\text{mu-two}}_{\text{MAC}} \cap \mathcal{C}^{\text{mu-two}}_{\text{BC}},$$
$$\sum_{i=1}^{K} \mathrm{Tr}(\boldsymbol{\Lambda}_{\mathsf{R}_i}) \leq P_{\mathsf{R}},\; i \in \{1, \ldots, K\},$$

where the rate tuples $\vec{\mathcal{R}}(K)$ and $\overleftrightarrow{\mathcal{R}}(K)$ are defined in (8.17) and (8.18), respectively.

Both capacity regions we are searching in are convex by definition, and hence, there is one global maxima for each sum rate maximization. However, there maybe multiple power allocations/covariance designs that may achieve it. Unfortunately, with the above given formulations, none of the maximizations can be solved through a trivial water-filling solution or a simple SDP. Both compromise several logdet consisting upper bound constraints on the variables building the objective function, which are well-known for not being trivially solvable by the available numerical SDP tools (see the above discussions for the single pair case).

In the following, we propose an iterative algorithm which exploits the geometry of the intersection of MAC and BC capacity regions, and is independent of the number of terminals and different relaying protocols. For the sake of brevity, the subsequent considerations will be on the basis of two-way relaying, where we drop further details for one-way relaying. However, as it will be reported in the related sections, the derivations for two-way protocol are immediately extendable to the one-way protocol by just changing the corresponding constraints and optimization variables (i.e., real powers instead of semidefinite covariances), which affects neither the solution method nor the structure of the algorithm.

We define a generalized expression $\tilde{K}$ for denoting the number of terminals for simultaneous information flow, e.g., $\tilde{K} = K$ and $\tilde{K} = 2K$ for one- and two-way protocols respectively. Our target is to search for the sum-rate optimal rate tuple $\mathcal{R}(\tilde{K})$ (either $\vec{\mathcal{R}}(K)$ or $\overleftrightarrow{\mathcal{R}}(K)$) inside a space with $\tilde{K}$ dimensions, which is defined by the constraints in the corresponding MAC and BC phases' capacity regions. At first, we elaborate on an intuitive and simple example for the $\tilde{K} = 2$ case. Consequently, building on this framework, the general sum rate maximization algorithm for arbitrary $\tilde{K}$ is presented.

8.4.1.1. 2-Dimensional Case

Let us choose a random rate couple $(\mathrm{R}_a, \mathrm{R}_b)$ out of the rate tuple $\mathcal{R}(\tilde{K})$ such that $\{a, b \in \{1, \ldots, \tilde{K}\}, a \neq b,\}$, and discard all the others. The respective couple is defined in a 2-dimensional convex rate region (see the shaded area in Fig. 8.3(a) for an illustration of the rate region). Note that we follow the rate ordering defined in (8.17) and (8.18), e.g., for one-way relaying, if $(a, b) = (1, 2)$, this corresponds to $(\mathrm{R}_{\overrightarrow{\mathsf{A}_1\mathsf{B}_1}}, \mathrm{R}_{\overrightarrow{\mathsf{A}_2\mathsf{B}_2}})$; and for two-way relaying if $(a, b) = (1, 3)$, this corresponds to $(\mathrm{R}_{\overrightarrow{\mathsf{A}_1\mathsf{B}_1}}, \mathrm{R}_{\overrightarrow{\mathsf{A}_2\mathsf{B}_2}})$ or if $(a, b) = (1, 4)$, this corresponds to $(\mathrm{R}_{\overrightarrow{\mathsf{A}_1\mathsf{B}_1}}, \mathrm{R}_{\overrightarrow{\mathsf{B}_2\mathsf{A}_2}})$.

We want to maximize the sum rate of this chosen couple. The relation between any specific realization $(\hat{\mathrm{R}}_a, \hat{\mathrm{R}}_b)$ of this couple can be expressed through an angle $\phi_{a,b} \in [0, \pi/2]$ as $\hat{\mathrm{R}}_b = \hat{\mathrm{R}}_a \tan(\phi_{a,b})$. Hence, the sum rate of this couple for a given direction $\phi_{a,b}$ is expressed as

$$\hat{\mathrm{R}}_a + \hat{\mathrm{R}}_b = \hat{\mathrm{R}}_a\Big(1 + \tan(\phi_{a,b})\Big).$$

For two way relaying, the sum rate of $(\mathrm{R}_a, \mathrm{R}_b)$ is optimized over either one or two covariances. The mapping from the notation (a, b) to covariance index (i, j) is simply $(i, j) = (\lceil a/2 \rceil, \lceil b/2 \rceil)$. Hence, if the mapping results in that $i = j$, which means that (a, b) represents two rates belonging to a single pair P_i but in reverse directions, e.g., $(\mathrm{R}_{\overrightarrow{\mathsf{A}_i\mathsf{B}_i}}, \mathrm{R}_{\overrightarrow{\mathsf{B}_i\mathsf{A}_i}})$, then we deal with a single covariance of $\mathbf{\Lambda}_{\mathsf{R}_i}$. Otherwise, we have two covariances $\mathbf{\Lambda}_{\mathsf{R}_i}$ and $\mathbf{\Lambda}_{\mathsf{R}_j}$ to optimize over. For one-way relaying, the corresponding mapping is one-to-one: $(i, j) = (a, b)$.

The sum rate maximization of $(\mathrm{R}_a, \mathrm{R}_b)$ is computed through the following quasi-convex problem for one- and two-way relaying respectively:

$$\mathcal{P}^{\text{one}}_{\text{sum}}(\phi_{a,b}) : \max_{\hat{\mathrm{R}}_a \geq 0,\, P_{\mathsf{R}_i} \geq 0, P_{\mathsf{R}_j} \geq 0} \hat{\mathrm{R}}_a \quad \textbf{subject to} \quad \Big(\hat{\mathrm{R}}_a, \hat{\mathrm{R}}_a \tan(\phi_{a,b})\Big) \in \mathcal{C}^{(a,b)}_{\text{MAC}} \cap \mathcal{C}^{(a,b)}_{\text{BC}},$$
$$P_{\mathsf{R}_i} + P_{\mathsf{R}_j} \leq P_{\mathsf{R}},$$

$$\mathcal{P}^{\text{two}}_{\text{sum}}(\phi_{a,b}) : \max_{\hat{\mathrm{R}}_a \geq 0,\, \mathbf{\Lambda}_{\mathsf{R}_i} \succeq 0,\, \mathbf{\Lambda}_{\mathsf{R}_j} \succeq 0} \hat{\mathrm{R}}_a \quad \textbf{subject to} \quad \Big(\hat{\mathrm{R}}_a, \hat{\mathrm{R}}_a \tan(\phi_{a,b})\Big) \in \mathcal{C}^{(a,b)}_{\text{MAC}} \cap \mathcal{C}^{(a,b)}_{\text{BC}},$$
$$\mathrm{Tr}(\mathbf{\Lambda}_{\mathsf{R}_i}) + \mathrm{Tr}(\mathbf{\Lambda}_{\mathsf{R}_j}) \leq P_{\mathsf{R}}.$$

For the corresponding protocol, $\mathcal{C}^{(a,b)}_{\text{MAC}}$ and $\mathcal{C}^{(a,b)}_{\text{BC}}$ represent the convex MAC and BC region constraints associated only with the rate couple $(\mathrm{R}_a, \mathrm{R}_b)$, respectively. Such

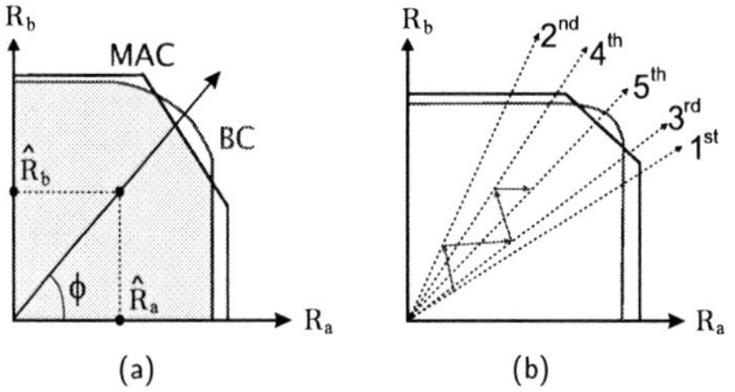

Figure 8.3.: Illustrations for sum rate optimization algorithm: (a) The achievable rate region for any rate couple $(\mathrm{R}_a, \mathrm{R}_b)$, (b) Examplary iterations for search direction $\phi_{a,b}$.

a constraining is nothing but allocating all available resources to the chosen couple. Moreover, note that we formulated $\mathcal{P}_{\text{sum}}^{\text{two}}(\phi_{a,b})$ as if there are two covariances for the sake of generality; however, one of them should be removed if the choice of rate couple requires to do so as aforementioned.

Modeling the constraints of BC phase with semidefinite cones, the problem $\mathcal{P}_{\text{sum}}(\phi_{a,b})$ can be efficiently and optimally solved by a bisection method combined with SDP feasibility checks [26, 77]. In order to get more insight on the individual structure of the optimization problems and the corresponding optimization, we present an example towards the end of this section.

As soon as the optimal $\hat{\mathrm{R}}_a^\star$ is computed, we can find the optimal rate couple

$$(\hat{\mathrm{R}}_a^\star, \hat{\mathrm{R}}_a^\star \tan(\phi_{a,b}))$$

that maximizes the sum rate on the direction of $\phi_{a,b}$. Note that the change of optimization coordinates from cartesian to polar gives us the opportunity to use the efficient bisection algorithm and reduce the computational complexity significantly as $\tilde{K}$ increases. Moreover, as it will be explained in the sequel, it provides the flexibility to reduce the optimization search space.

Next, since the achievable rate region is convex (independently from the chosen pair or the employed relaying protocol), we can search over the optimal $\phi_{a,b}$ that maximizes $\hat{\mathrm{R}}_a+\hat{\mathrm{R}}_b$, using an unconstrained minimization method, e.g., the gradient descent method. In summary, for each iteration of the descent algorithm, i.e. for each chosen $\phi_{a,b}$, we solve $\mathcal{P}_{\text{sum}}(\phi_{a,b})$, and iterate until iterating $\phi_{a,b}$ further does not induce significant change in sum rate (see Fig. 8.3(b) for an illustration).

- **Example:** Say we are considering two-way relaying and choose $(a,b) = (1,4)$, which corresponds to the rate couple $(\overset{\leftrightarrow}{\mathrm{R}}_{\mathsf{A}_1\mathsf{B}_1}, \overset{\leftrightarrow}{\mathrm{R}}_{\mathsf{B}_2\mathsf{A}_2})$. Further, assume that we are searching along the direction determined by an angle of $\phi_{1,4}$, i.e., $\overset{\leftrightarrow}{\mathrm{R}}_{\mathsf{B}_2\mathsf{A}_2} = \overset{\leftrightarrow}{\mathrm{R}}_{\mathsf{A}_1\mathsf{B}_1} \tan(\phi_{1,4})$. In order to maximize

$$\overset{\leftrightarrow}{\mathrm{R}}_{\mathsf{A}_1\mathsf{B}_1} + \overset{\leftrightarrow}{\mathrm{R}}_{\mathsf{B}_2\mathsf{A}_2} = \overset{\leftrightarrow}{\mathrm{R}}_{\mathsf{A}_1\mathsf{B}_1}\Big(1+\tan(\phi_{1,4})\Big),$$

we need to optimize over the covariances $\mathbf{\Lambda}_{\mathsf{R}_1}$ and $\mathbf{\Lambda}_{\mathsf{R}_2}$. The formulation of the sum rate maximization $\mathcal{P}^{\text{two}}_{\text{sum}}(\phi_{1,4})$ is written as

$$\begin{aligned}
\mathcal{P}^{\text{two}}_{\text{sum}}(\phi_{1,4}) \; : \quad & \max_{\overset{\leftrightarrow}{\mathrm{R}}_{\mathsf{A}_1\mathsf{B}_1}\geq 0,\; \mathbf{\Lambda}_{\mathsf{R}_1}\succeq 0,\; \mathbf{\Lambda}_{\mathsf{R}_2}\succeq 0} \quad \overset{\leftrightarrow}{\mathrm{R}}_{\mathsf{A}_1\mathsf{B}_1} \\
\textbf{subject to} \quad & \mathrm{Tr}(\mathbf{\Lambda}_{\mathsf{R}_1}) + \mathrm{Tr}(\mathbf{\Lambda}_{\mathsf{R}_2}) \leq P_{\mathsf{R}}, \\
\begin{array}{c}\text{MAC}\\ \text{constraints}\end{array} & \left\{\begin{array}{r}
\overset{\leftrightarrow}{\mathrm{R}}_{\mathsf{A}_1\mathsf{B}_1} \leq \log_2\left|\mathbf{I} + \frac{P_{\mathsf{A}_1}}{N_{\mathsf{A}_1}\sigma^2_{n_{\mathsf{R}}}}\mathbf{H}_{\mathsf{A}_1}\mathbf{H}^H_{\mathsf{A}_1}\right|, \\
\overset{\leftrightarrow}{\mathrm{R}}_{\mathsf{A}_1\mathsf{B}_1}\tan(\phi_{a,b}) \leq \log_2\left|\mathbf{I} + \frac{P_{\mathsf{B}_2}}{N_{\mathsf{B}_2}\sigma^2_{n_{\mathsf{R}}}}\mathbf{H}_{\mathsf{B}_2}\mathbf{H}^H_{\mathsf{B}_2}\right|, \\
\overset{\leftrightarrow}{\mathrm{R}}_{\mathsf{A}_1\mathsf{B}_1}\Big(1+\tan(\phi_{1,4})\Big) \leq \log_2\left|\mathbf{I} + \frac{P_{\mathsf{A}_1}}{N_{\mathsf{A}_1}\sigma^2_{n_{\mathsf{R}}}}\mathbf{H}_{\mathsf{A}_1}\mathbf{H}^H_{\mathsf{A}_1} + \frac{P_{\mathsf{B}_2}}{N_{\mathsf{B}_2}\sigma^2_{n_{\mathsf{R}}}}\mathbf{H}_{\mathsf{B}_2}\mathbf{H}^H_{\mathsf{B}_2}\right|.
\end{array}\right. \\
\begin{array}{c}\text{BC}\\ \text{constraints}\end{array} & \left\{\begin{array}{r}
\overset{\leftrightarrow}{\mathrm{R}}_{\mathsf{A}_1\mathsf{B}_1} \leq \log_2\left|\mathbf{I} + \frac{1}{\sigma^2_{n_{\mathsf{B}}}}\mathbf{H}^T_{\mathsf{B}_1}\mathbf{\Lambda}_{\mathsf{R}_1}\mathbf{H}^*_{\mathsf{B}_1}\right|, \\
\overset{\leftrightarrow}{\mathrm{R}}_{\mathsf{A}_1\mathsf{B}_1}\tan(\phi_{a,b}) \leq \log_2\left|\mathbf{I} + \frac{1}{\sigma^2_{n_{\mathsf{A}}}}\mathbf{H}^T_{\mathsf{A}_2}\mathbf{\Lambda}_{\mathsf{R}_2}\mathbf{H}^*_{\mathsf{A}_2}\right|.
\end{array}\right.
\end{aligned} \tag{8.20}$$

The first three rate constraints are related with the chosen rates (dimensions) through the MAC phase, whereas the last two constraints are describing the BC phase capacity associated with the chosen rate couple. As there is nothing to optimize with the MAC rate expressions, they affect as real-valued upper bound constraints, which can be trivially checked to satisfy or not. Moreover, the last two BC phase related rate expression can be immediately modeled as SDP constraints (see Sections 7.2.2.1 and 7.2.2.2 for implementation details). In other words, for a given $\overset{\leftrightarrow}{\mathrm{R}}_{\mathsf{A}_1\mathsf{B}_1}$, the problem turns out to be a simple SDP feasibility check. Hence, incorporating the bisection method over $\overset{\leftrightarrow}{\mathrm{R}}_{\mathsf{A}_1\mathsf{B}_1}$, and solving an SDP feasibility problem at each iteration, the optimal value of $\mathcal{P}^{\text{two}}_{\text{sum}}(\phi_{1,4})$ can be efficiently found [26, 77].

8.4.1.2. The General Case

Previously, we disregarded the constraints enforced by the relation between the chosen couple $(\mathrm{R}_a, \mathrm{R}_b)$ and the others within the index set $\{1,\ldots,\tilde{K}\}\backslash\{a,b\}$, and allocate

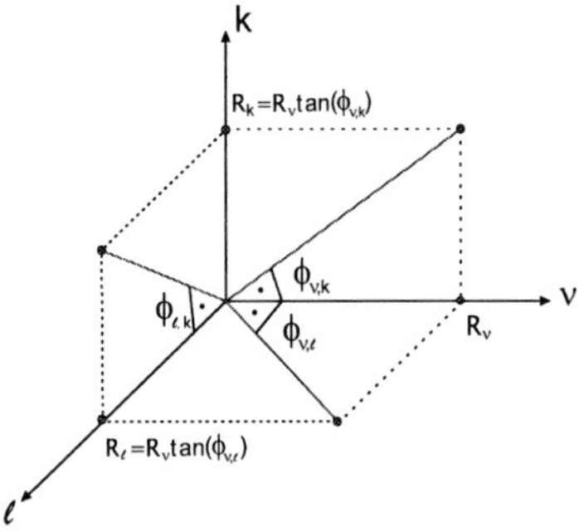

Figure 8.4.: Illustration of the equivalence in (8.22): The tangent of the angle between the dimensions ℓ and k is written as $\tan(\phi_{\ell,k}) = \frac{\mathrm{R}_\nu \tan(\phi_{\nu,k})}{\mathrm{R}_\nu \tan(\phi_{\nu,\ell})} = \frac{\tan(\phi_{\nu,k})}{\tan(\phi_{\nu,\ell})}$.

resources for the sake of only $(\mathrm{R}_a, \mathrm{R}_b)$. Since each couple $(\mathrm{R}_a, \mathrm{R}_b)$ out of available $\tilde{K}$ can be associated through an angle (direction) in between, it may be conjectured that we need $\tilde{K}!/(2(\tilde{K}-2)!)$ angles to represent all relations between all $\tilde{K}$ dimensions. Whereas, in essence, we need only $\tilde{K}-1$ angles. This statement can be immediately proven as follows: Set one dimension fixed, say the νth, and associate each of all the rest $\tilde{K}-1$ dimensions with the νth dimension through a corresponding angle. Thus, defining

$$\boldsymbol{\phi} = [\phi_{\nu,1}, \ldots, \phi_{\nu,\nu-1}, \phi_{\nu,\nu+1}, \ldots, \phi_{\nu,\tilde{K}}],$$

the rates for all dimensions can be expressed as

$$\mathcal{R}_\nu(\boldsymbol{\phi}) \triangleq \Big(\mathrm{R}_\nu \tan(\phi_{\nu,1}), \ldots, \mathrm{R}_\nu \tan(\phi_{\nu,\nu-1}), \mathrm{R}_\nu, \mathrm{R}_\nu \tan(\phi_{\nu,\nu+1}), \ldots, \mathrm{R}_\nu \tan(\phi_{\nu,\tilde{K}})\Big), \tag{8.21}$$

for $\nu \in \{1, \ldots, \tilde{K}\}$. With the knowledge of the vector $\boldsymbol{\phi}$, one can derive all other angles through the relation (see Fig. 8.4)

$$\tan(\phi_{\ell,k}) = \frac{\tan(\phi_{\nu,k})}{\tan(\phi_{\nu,\ell})}, \quad \forall k, \ell \in \{1, \ldots, \nu-1, \nu+1, \ldots \tilde{K}\}, k \neq \ell. \tag{8.22}$$

Hence, for a given vector $\boldsymbol{\phi}$ and ν, the sum rate is given by

$$\sum_{i=1}^{\tilde{K}} \mathrm{R}_i = \mathrm{R}_\nu \Big(1 + \sum_{i=1, i\neq\nu}^{\tilde{K}} \tan(\phi_{\nu,i})\Big). \tag{8.23}$$

This sum rate can be maximized through the following problems for one- and two-way relaying respectively:

$$\mathcal{P}_{\text{sum}}^{\text{one}}(\boldsymbol{\phi}) \; : \max_{\text{R}_\nu \geq 0,\, P_{\mathsf{R}_1} \geq 0, \ldots, P_{\mathsf{R}_K} \geq 0} \text{R}_\nu \quad \textbf{subject to } \mathcal{R}_\nu(\boldsymbol{\phi}) \in \mathcal{C}_{\text{MAC}}^{\text{mu-one}} \cap \mathcal{C}_{\text{BC}}^{\text{mu-one}}, \quad \sum_{i=1}^{K} P_{\mathsf{R}_i} \leq P_{\mathsf{R}},$$

$$\mathcal{P}_{\text{sum}}^{\text{two}}(\boldsymbol{\phi}) \; : \max_{\text{R}_\nu \geq 0,\, \boldsymbol{\Lambda}_{\mathsf{R}_1} \succeq 0, \ldots, \boldsymbol{\Lambda}_{\mathsf{R}_K} \succeq 0} \text{R}_\nu \quad \textbf{subject to } \mathcal{R}_\nu(\boldsymbol{\phi}) \in \mathcal{C}_{\text{MAC}}^{\text{mu-two}} \cap \mathcal{C}_{\text{BC}}^{\text{mu-two}}, \quad \sum_{i=1}^{K} \text{Tr}(\boldsymbol{\Lambda}_{\mathsf{R}_i}) \leq P_{\mathsf{R}},$$

which can be efficiently solved with a bisection method combined with SDP feasibility checks [26, 77]. Notice that $\mathcal{R}_\nu(\boldsymbol{\phi})$ defines itself through (8.21) according to the employed relaying protocol. Since the $\tilde{K}$ dimensioned achievable rate region is convex by definition, we can further search over $\boldsymbol{\phi}$, whose elements $\phi_i \in [0, \pi/2]$, $i = 1, \ldots, \tilde{K} - 1$, using an unconstrained minimization method.

The overall sum rate maximization algorithm is summarized with **Algorithm 5**. While implementing the bisection part, $\text{R}_\nu^{\max}$ is chosen large enough according to the operation average SNR value, and ϵ is a small positive number indicating the precision of the bisection algorithm. Moreover, the search direction for ϕ is found through a numerical first derivative computation, i.e., $(f(x + \varepsilon) - f(x))/\varepsilon$, $0 < \varepsilon \ll 1$. The iterations for the descent algorithm continues until the improvement at the sum rate with a new iteration becomes smaller than ϵ, where $0 < \epsilon \ll 1$. The convergence and the optimality of the algorithm are ensured through the related conditions of the used methods and convexity of the problem [26].

Remark 8.4.1: The search space for each ϕ_i in $\boldsymbol{\phi}$ can be reduced through some *pre*-optimizations, so that we consider $\phi_i \in [\phi_i^{\min}, \phi_i^{\max}]$ instead of $[0, \pi/2]$. Since each information flow's MAC single bound is readily known through (8.5), the feasibility of the corresponding maximal rate through the BC phase can be *pre*-checked to satisfy the given MAC rate or not. In other words, we check if MAC and BC intersect for single terminal bounds. If they intersect, calculating the angle between this rate and its corresponding couple's rate, this portion can be excluded from the optimization direction space. In Fig. 8.5, we illustrate such a search space reduction: Following the figure, we compute the maximal rates $\text{R}_{\mathsf{BR}}^\star$ and $\text{R}_{\mathsf{RA}}^\star$ for the directions B-to-R and

Algorithm 5 Sum Rate Maximization Algorithm for Multiuser DF Relaying

initiate $\phi_0 \in [0, 2\pi]$ and set ν .
repeat Solve $\mathcal{P}^{\text{one}}_{\text{sum}}(\phi)$ or $\mathcal{P}^{\text{two}}_{\text{sum}}(\phi)$
  initiate $\epsilon > 0, \mathrm{R}^{\max}_{\nu} > 0, \mathrm{R}^{\min}_{\nu} = 0.$
  repeat
    Step 1. Set. $\mathrm{R}_{\nu} \Leftarrow (\mathrm{R}^{\max}_{\nu} + \mathrm{R}^{\min}_{\nu})/2.$
    Step 2. Check the feasibility of either

$$\mathcal{R}_{\nu}(\phi) \in \mathcal{C}^{\text{mu-one}}_{\text{MAC}} \cap \mathcal{C}^{\text{mu-one}}_{\text{BC}}, \; P_{\mathsf{R}_1} \geq 0, \ldots, P_{\mathsf{R}_K} \geq 0, \sum_{i=1}^{K} P_{\mathsf{R}_i} \leq P_{\mathsf{R}}$$

    or

$$\mathcal{R}_{\nu}(\phi) \in \mathcal{C}^{\text{mu-two}}_{\text{MAC}} \cap \mathcal{C}^{\text{mu-two}}_{\text{BC}}, \; \mathbf{\Lambda}_{\mathsf{R}_1} \succeq 0, \ldots, \mathbf{\Lambda}_{\mathsf{R}_K} \succeq 0, \sum_{i=1}^{K} \mathrm{Tr}(\mathbf{\Lambda}_{\mathsf{R}_i}) \leq P_{\mathsf{R}}.$$

    Step 3. **if** feasible
        Set. $\mathrm{R}^{\min}_{\nu} \Leftarrow \mathrm{R}_{\nu}.$
      else
        Set. $\mathrm{R}^{\max}_{\nu} \Leftarrow \mathrm{R}_{\nu}.$
    Step 4. Check and **quit** if $\mathrm{R}^{\max}_{\nu} - \mathrm{R}^{\min}_{\nu} < \epsilon.$
  compute the new search direction for $\phi : \nabla\phi$.
  search for the optimal step size: Δ.
  update for the optimal step size: $\phi \Leftarrow \phi + \Delta \cdot \nabla\phi$.
until $\epsilon-$optimal sum rate is achieved.

R-to-A, respectively. We realize that $\mathrm{R}^{\star}_{\mathsf{RA}}$ is not supportable by the MAC phase. i.e., $\mathrm{R}^{\star}_{\mathsf{RA}} > \mathrm{R}^{\star}_{\mathsf{BR}}$, and hence, the rate tuple maximizing the sum rate can not consist any point larger than $\mathrm{R}^{\star}_{\mathsf{BR}}$. Next, we constraint the R-to-A direction to have at most a rate of $\mathrm{R}^{\star}_{\mathsf{BR}}$, and maximize the R-to-B direction to find $\mathrm{R}^{\star}_{\mathsf{AB}}$ as shown in the figure. Hence, calculating the angle of $\phi^{\star} = \tan^{-1}(\mathrm{R}^{\star}_{\mathsf{AB}}/\mathrm{R}^{\star}_{\mathsf{BR}})$, we update our search space such that we exclude this portion and reduce to $\phi_i \in [0, \pi/2 - \phi^{\star}]$.

8.4.2. Extensions to Max-min Fairness and QoS Assurance

In this section, we extend the iterative algorithm proposed for sum rate maximization to employ other objective functions like maximizing the minimum rate. Further, exploiting

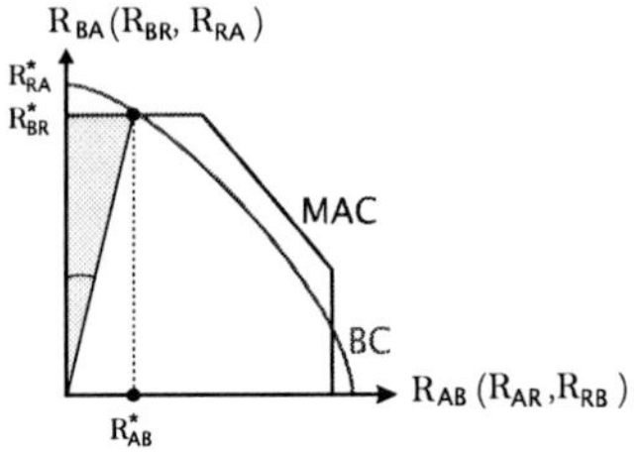

Figure 8.5.: Illustration for reduction of search direction space.

the modularity of the algorithm, we introduce QoS constraints to the problem while maximizing sum- or minimum-rate.

Maxmin Fairness We want to design the transmit covariances (or the real power allocation) such that max-min fairness is introduced between the terminals over the whole network. Specifically, for one-way relaying we allocate powers to individual streams such that $\mathbf{min}_i(\overrightarrow{\mathrm{R}}_{\mathsf{A}_i\mathsf{B}_i})$ is maximized. Similarly, for two-way relaying the covariance matrices are chosen such that either max-min fairness between all terminals, i.e., $\mathbf{min}_i\mathbf{min}\{\overleftrightarrow{\mathrm{R}}_{\mathsf{A}_i\mathsf{B}_i}, \overleftrightarrow{\mathrm{R}}_{\mathsf{B}_i\mathsf{A}_i}\}$, or between all terminal pairs, i.e., $\mathbf{min}_i(\overleftrightarrow{\mathrm{R}}_{\mathsf{A}_i\mathsf{B}_i} + \overleftrightarrow{\mathrm{R}}_{\mathsf{B}_i\mathsf{A}_i})$ is maximized. For both protocols, the problem formulation will be same as $\mathcal{P}^{\mathrm{mu-sum}}_{\mathrm{one-way}}$ or $\mathcal{P}^{\mathrm{mu-sum}}_{\mathrm{two-way}}$, except that the objective functions are changed with the corresponding aforementioned ones.

We briefly outline the max-min-fairness maximization algorithm basing on the implementation structure developed in **Algorithm** 5. We change the objective function from sum to max-min rate and perform **Algorithm** 5 once over the whole terminal set $\mathcal{N} \triangleq \{1, \ldots, \tilde{K}\}$. Consequently, say that the ith dimension has a resultant rate $\mathrm{R}_i^{\mathrm{mm}}$, which is minimum of all dimensions (i can be any one dimension). We create a constraint set $\mathcal{C}_{\mathrm{mm}}$ and add the constraint $\mathrm{R}_i \geq \mathrm{R}_i^{\mathrm{mm}}$ to it. Moreover, we update the set of nodes $\mathcal{N} \Leftarrow \mathcal{N}\backslash\{i\}$ for max-min optimization. In the next iteration, we add $\mathcal{C}_{\mathrm{mm}}$ to the constraint set of **Algorithm** 5, and maximize again the minimum of the rates in the updated set $\mathcal{N}$. In other words, while assigning a rate of at least $\mathrm{R}_i^{\mathrm{mm}}$ to the ith terminal's information flow, we are assigning the largest max-min rates to the rest of the nodes in $\mathcal{N}$. Continuing with this fashion until $\mathcal{N} = \emptyset$, and reducing the number elements of $\mathcal{N}$ with each iteration, we allocate resources such that the most max-min fair rates are allocated to the terminals.

QoS Assurance In order to provide desired transmission rates to all terminals, QoS assurance can be introduced to the system. In other words, we assure that each terminal is supported with the least transmission rate it requires. Collecting all QoS related constraints on linear combinations of R_i in a constraint set $\mathcal{C}_{\mathrm{QoS}}$, we can add the constraint $\{\mathrm{R}_i, \cdots, \mathrm{R}_{\tilde{K}}\} \in \mathcal{C}_{\mathrm{QoS}}$ to both protocols' optimization problems. The QoS constraints will be taken into account during the SDP feasibility checks.

8.5. Performance Results

Simulation Setup Here, we study the sum rate performance of multiuser DF relaying protocols through Monte Carlo simulations. There are $K = 2$ terminal pairs, where each terminal is equipped with two antennas. The elements of the channel matrices $\mathbf{H}_{\mathsf{A}_i}$ and $\mathbf{H}_{\mathsf{B}_i}$ are i.i.d. Rayleigh fading coefficients with zero mean and variances $\sigma^2_{H_\mathsf{A}}$ and $\sigma^2_{H_\mathsf{B}}$, respectively. Moreover, these channel matrices are assumed to stay constant over one transmission cycle of the respective protocol. We further assume $P \triangleq P_{\mathsf{A}_i} = P_{\mathsf{B}_i} = 1$, $P_\mathsf{R} = KP$, $\sigma^2_n \triangleq \sigma^2_{n\mathsf{R}} = \sigma^2_{n\mathsf{A}} = \sigma^2_{n\mathsf{B}}$. Hence, the average SNR at the terminal set $\mathcal{A}$ and $\mathcal{B}$ are defined as $\mathrm{SNR}_\mathsf{A} = \sigma^2_{H_\mathsf{A}} P/\sigma^2_n$ and $\mathrm{SNR}_\mathsf{B} = \sigma^2_{H_\mathsf{B}} P/\sigma^2_n$, respectively. Unless otherwise stated we assume equal time sharing for each phase in both protocols, i.e., $\alpha_{\mathrm{MAC}} = \alpha_{\mathrm{BC}} = 0.5$.

Benchmark Systems We consider two different benchmark perspectives: multiuser diversity and decoupled optimization. In the former one, we drop the constraint from both relaying protocols that the relay serves all terminal pairs simultaneously. Hence, the relay operates in a time-division multiplex mode and serves each pair one by one. Moreover, we employ multiuser diversity (MD) such that the relay chooses the best pair to serve depending on the instantaneous channel conditions, and allocates all available resources (time) to this corresponding one.

As a second benchmark, we drop the constraint that the covariance optimization is coupled over the two phases. Hence, we can optimize the terminal-to-relay and relay-to-terminal rates independently from each other. To this end, we maximize the sum rate in the second BC phase through a water-filling technique [36, 115], and compare the consequent individual relay-to-terminal rates with the corresponding MAC phase rates through (8.5). If a chosen BC rate is not supported by the associated MAC phase, we scale down all BC rates correspondingly. In other words, we find the sum rate

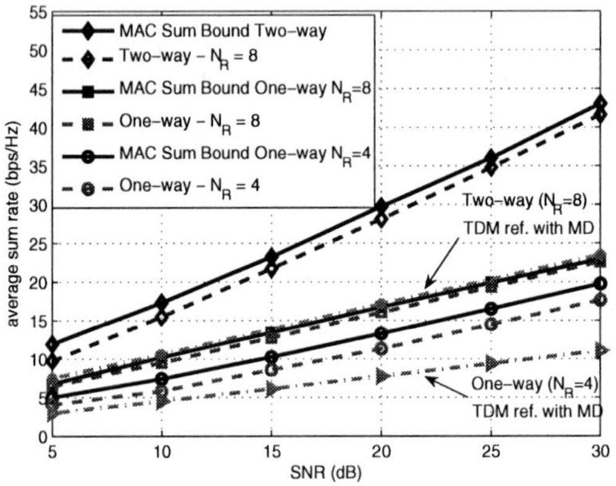

Figure 8.6.: Average sum rate vs. average SNR for $K = 2$ terminal-pairs, $N_{\mathsf{A}_i} = N_{\mathsf{B}_i} = 2, \forall i$ terminal antennas ($\mathrm{SNR} = \mathrm{SNR}_{\mathsf{A}} = \mathrm{SNR}_{\mathsf{B}}$).

maximizing point in the BC capacity region, and scale it appropriately such that it fits in the MAC capacity region.

Simulation Results Fig. 8.6 shows the sum rate performance of the two multiuser relaying protocols against increasing average SNR, e.g., $\mathrm{SNR} = \mathrm{SNR}_{\mathsf{A}} = \mathrm{SNR}_{\mathsf{B}}$. We choose the number of relay antennas N_{R} to be as much as the sum of transmitting terminal antennas in the MAC phase, i.e., $N_{\mathsf{R}} = 8$ for two-way and $N_{\mathsf{R}} = 4$ for one-way relaying. In general, it is depicted that two-way relaying offers a significant sum rate improvement over one-way relaying and recovers the spectral efficiency loss observed with its one-way counterpart. For fair comparison in between two multiuser protocols, the one-way results are also shown for $N_{\mathsf{R}} = 8$ relay antennas; but even this choice can not recover the loss with respect to two-way relaying. On the other hand, the significant advantage of using multiuser precoding is confirmed by comparing the proposed multiuser protocols performance with the TDM based references. Moreover, it is interesting to notice that multiuser one-way relaying with $N_{\mathsf{R}} = 8$ can only perform as good as the TDM based reference scenario for two-way relaying does.

Sum rate performances of both multiuser protocols are ultimately upper bounded by the MAC sum bound given in (8.5). Note that there is an unnegligible performance gap between the MAC upper bound and the actual sum rate for both protocols. These gaps

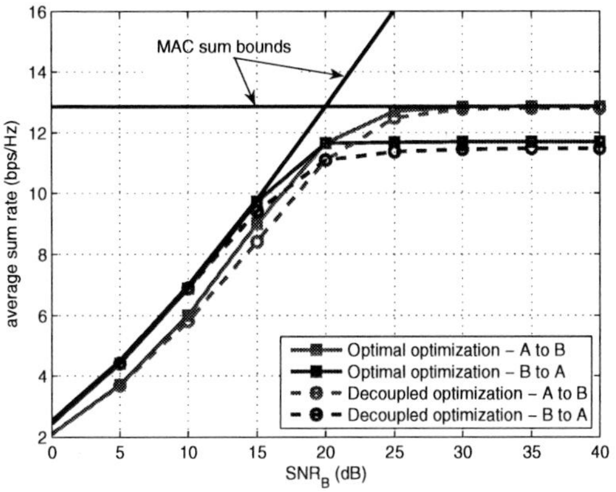

Figure 8.7.: Average sum rate performance of one-way relaying for unbalanced links for $K = 2$ terminal-pairs, $N_{\mathsf{R}} = 4$ relay antennas, $N_{\mathsf{A}_i} = N_{\mathsf{B}_i} = 2, \forall i$ terminal antennas.

can be further reduced by using higher complexity precoder designs, e.g., successive ZF-DPC [36], DPC [125], instead of the block diagonalization based ZF scheme. As above mentioned, adaptation of successive ZF-DPC to the presented multiuser protocols and the corresponding extension of the proposed sum rate maximization algorithm, are pretty straightforward. In [44], the performance improvement gained by employing using ZF-DPC instead of ZF-BF, has been depicted, where the gap between the MAC sum bound and the achieved sum rate was shrinked considerably.

In Figures 8.7 and 8.8, we investigate the impact of unbalanced link quality for both protocols, where we fix the average SNR at one side while varying it on the other side.

Specifically, in Fig. 8.7 we consider the one-way relaying protocol for a setup of $K = 2$ terminal-pairs, $N_{\mathsf{R}} = 4$ relay antennas, $N_{\mathsf{A}_i} = N_{\mathsf{B}_i} = 2, \forall i$ terminal antennas. We plot the achieved sum rate for both information flow directions, i.e., $\mathcal{A}$-to-$\mathcal{B}$ and $\mathcal{B}$-to-$\mathcal{A}$. For both cases, we fix $\mathrm{SNR}_{\mathsf{A}}$ to 20 dB, and vary $\mathrm{SNR}_{\mathsf{B}}$ from 0 to 40 dB. Let us first elaborate on the direction from $\mathcal{A}$ to $\mathcal{B}$. Note that as $\mathrm{SNR}_{\mathsf{A}}$ is fixed, the MAC sum bound (vertical line) stays the same over $\mathrm{SNR}_{\mathsf{B}}$. Hence, the increase in $\mathrm{SNR}_{\mathsf{B}}$ results in higher BC phase rates, which correspondingly improves the sum rate until $\mathrm{SNR}_{\mathsf{B}} = 20 - 25$dB. Beyond these values, the MAC phase ($\mathcal{A}$-to-R link) turns to be the bottleneck for the information flow $\mathcal{A}$-to-$\mathcal{B}$ and the sum rate saturates, i.e., the sum

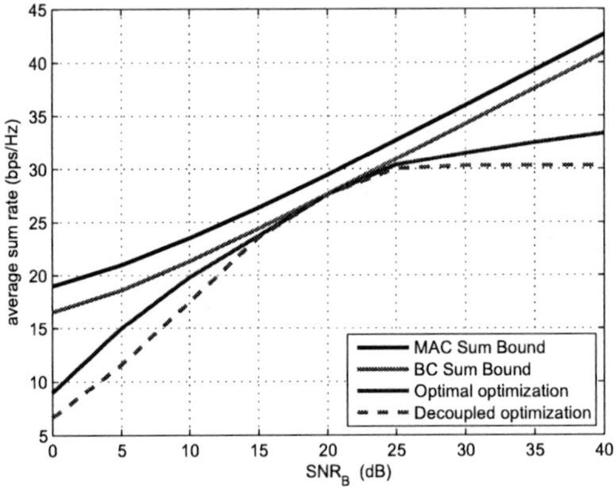

Figure 8.8.: Average sum rate performance of two-way relaying for unbalanced links for $K = 2$ terminal-pairs, $N_{\mathsf{R}} = 8$ relay antennas, $N_{\mathsf{A}_i} = N_{\mathsf{B}_i} = 2, \forall i$ terminal antennas.

rate is limited by the MAC phase. On the other hand, for the direction from $\mathcal{B}$ to $\mathcal{A}$, the MAC sum bound (solid curve) improves with SNR_{B}, whereas now the capacity of the BC phase stays constant. Hence, as SNR_{B} increases, the sum rate improves until a point that the BC phase happens to be the bottleneck. In accordance, we note that the sum rate of the $\mathcal{A}$-to-$\mathcal{B}$ direction is smaller than that of the $\mathcal{B}$-to-$\mathcal{A}$ direction for $\text{SNR}_{\mathsf{B}} < 20\text{dB}$, and higher for $\text{SNR}_{\mathsf{B}} > 20\text{dB}$. In order to appreciate the impact of high-complexity but high-performance coupled optimization, in Fig. 8.7 we also plot the sum rate that would be achieved through a decoupled optimization (dashed curves). Besides the extreme unbalanced cases of $\text{SNR}_{\mathsf{B}} \to 0$ or 40 dB, the two-phase coupled optimization provides sum rate gains up to 0.8-1 bps/Hz.

We shift our focus to the two-way protocol in Fig. 8.8, and consider a multiuser setup of $K = 2$ terminal-pairs, $N_{\mathsf{R}} = 8$ relay antennas, $N_{\mathsf{A}_i} = N_{\mathsf{B}_i} = 2, \forall i$ terminal antennas. In general we observe that for equal time sharing between the phases, the BC phase appears as the fundamental bottleneck when compared with the MAC sum bound. Note that, as mentioned before, the BC performance can be further improved by employing higher-complexity beamformer designs instead of the lower-complexity block-diagonalization we consider here. The sum rate performance of the two-way protocol improves with SNR_{B} until it saturates at around 25 dB. This saturation occurs since

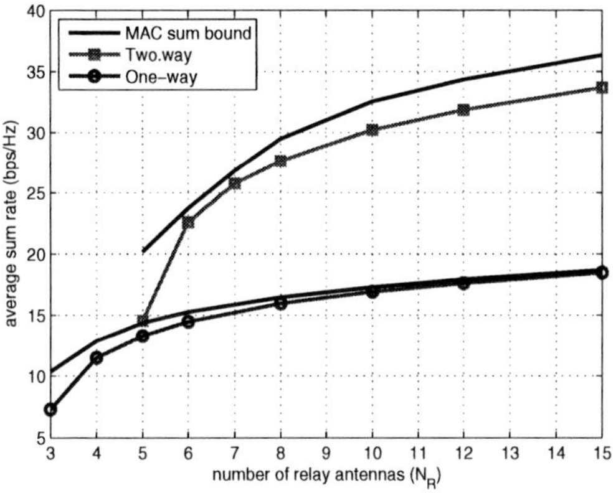

Figure 8.9.: Average sum rate performance vs. number of relay antennas N_{R} for $K = 2$ terminal-pairs and $N_{\mathsf{A}_i} = N_{\mathsf{B}_i} = 2, \forall i$ terminal antennas ($\mathrm{SNR}_{\mathsf{A}} = \mathrm{SNR}_{\mathsf{B}} = 20\mathrm{dB}$).

the fixed $\mathrm{SNR}_{\mathsf{A}}$ limits both uplink of $\mathcal{A}$ to R in the first phase and downlink of R to $\mathcal{A}$ in the second phase. Comparing the sum rates of two-phase coupled and decoupled optimizations, we observe that especially in the extreme unbalanced SNR regimes, the coupled optimization offers substantial gains. Note that this behaviour is in contrast with that of the one-way protocol, where the coupled optimization depicted its strength rather at the balanced SNR regime. This can be explained as follows that in one-way protocol there is always a single bottleneck phase as $\mathrm{SNR}_{\mathsf{B}}$ varies (either MAC or BC, but not together). However, for the two-way protocol, both MAC and BC phases are affected similarly by $\mathrm{SNR}_{\mathsf{B}}$. Hence, at the balanced SNR regime, the performance of both phases of two-way relaying are anyhow good, which results in that the resultant two-way rates do not suffer.

Finally in Fig. 8.9, we study the impact of relay antennas on the sum rate performance. We again consider $K = 2$ terminal-pairs and fix the number of terminal antennas to $N_{\mathsf{A}_i} = N_{\mathsf{B}_i} = 2, \forall i$. Correspondingly, the curves for one- and two-way relaying starts from $N_{\mathsf{R}} = 3$ and $N_{\mathsf{R}} = 5$ antennas, respectively, since these values are the least required relay antenna numbers for efficient zero-forcing beamforming in the second phase. Observe that as N_{R} is increased by one, there occurs a rapid sum rate jump in both protocols, i.e., $N_{\mathsf{R}} = 3$ to 4 for one-way and $N_{\mathsf{R}} = 5$ to 6 for two-way.

This is due that there is an excess degree of freedom to employ at the relay for decoding in the first phase and beamforming in the second phase. Although both protocols are ultimately limited by the MAC sum bound, the sum rate of the one-way protocol approaches to this upper bound quickly at $N_\mathsf{R} = 10$. Whereas, the sum rate of the two-way protocol seems to be rather limited by the BC phase, and hence does not reach the MAC upper bound. Needless to say, the two-way relaying almost doubles the sum rate performance of its one-way counterpart for all N_R choices except of the $N_\mathsf{R} = 5$ case, which corresponds to the minimum antenna configuration for ZF in two-way relaying.

8.6. Concluding Remarks

In this chapter, we have extended one- and two-way MIMO DF relaying schemes to the framework of multiuser interference channel. The key enabler for the provided distributed spatial multiplexing was the use of multiple antennas at the relay node. Multiple antennas at the relay in combination with self-interference cancellation have been efficiently utilized for spatial and bit-level separation of terminal pairs. However, the practical limitations on the maximal number of antennas that can be supported efficiently within such a relay, i.e., AP or relay/base station, will definitely be the capacity-limiter for our proposal.

We have proposed an efficient optimization method based on iterative SDP optimization that maximizes the sum rate over the whole network. The algorithm is quite modular in the sense that it is independent of the network parameters and of the employed relaying protocol as long as the constraints can be modeled in SDP structure. Hence, it can be employed in some other similar problems consisting capacity region intersections.

The proposed DF relaying structure can be of interest to several field of applications in different multiuser network setups. for instance, a simple variant of the considered setup can be to relay the bidirectional uplink-downlink communications in a cellular network. That is, instead of multiple S-D pairs, we can consider a single base station, which is broadcasting private information to multiple mobile stations and concurrently receiving information from these. We have addressed such a scenario in [44], where, in general, theoretical foundations are in line with that of this chapter.

Chapter 9

Conclusions

In order to expand the capacity of current wireless networks, such that large node densities can be further accommodated within, we need to fully utilize the three fundamental resource dimensions: time, frequency, and space. However, each of these dimensions is limited respectively with the stringent bandwidth regulations, low delay-tolerant data applications and practical restriction to employ multiple antennas at mobile nodes. The so called "*relay dimension*" incorporates the pros and cons of all above and provides a versatile distributed solution that can be employed within a broad range of network applications. The introduction of relays gives rise to the implementation of heterogeneous networks, where different sets of terminal nodes with various complexities, processing capabilities or transmission range, can coexist and collaborate in the same medium.

9.1. Concluding Remarks

In this thesis, we have primarily focused on new design perspectives to enable interference free multi-user/pair communications and identified cooperative relaying as an efficient means to achieve this in a *distributed* manner. In two parts, we have addressed both AF and DF relaying schemes, whose application scenarios can be centered at multihop ad-hoc and cellular/local area networks, respectively. Let us now elaborate on our findings in the view of the sufficiency and achievability related design questions that we have committed to answer in Section 1.2.

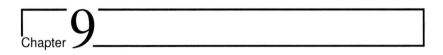

Part 1 In the first part, we have explored how to orthogonalize multiple source-destination terminal pairs in space through a set of distributed AF relay nodes, which

comes with two key requirements. First, there is a certain need for *minimum* number of AF relays to assist the communication in a half-duplex scheme without private benefit. Second, each relay must share its local channel knowledge with the others so that the complex relay gains are jointly chosen to result in multiple interference-free point-to-point links. Such a distributed spatial multiplexing achievement through coherent AF relaying is particularly appealing due to low-cost signal processing applied in linear AF relays, which are transparent to coding and adaptive modulation techniques employed in between each S-D pair. In Chapter 3, we have further realized that additional relays (spatial degrees of freedom) on top of the minimum set required for interference cancellation, opens a new gateway for coherent AF relaying to achieve further distributed MIMO gains of array and diversity. These gains have been exploited through efficient distributed beamforming designs, where we have considered several different options such as maximizing the sum rate (array gain) or maximizing the minimum link rate (array/outage gain and fairness). Each of these have been efficiently optimized both with and without zero-forcing constraints, through either gradient descent based methods or semidefinite programming based algorithms. Therein, by either enforcing the interference ZF constraint or not, we trade-off between optimization computation complexity and rate performance. Further, we proposed two schemes to convert the available excess relay diversity to full effective diversity gain at the destinations: either select the best relay set out of all available or apply max-min type gain optimization through all available relays.

All in all, we have shown that efficient MIMO gains can be, in fact, achievable through a completely distributed network without a need for a central processing unit. The viability of the provided spectral efficiency is under threat of the least required number of single-antenna relays, which grows quadratically with number of pairs N, and the corresponding overhead of local CSI dissemination. Nevertheless, considering the simple and low-cost structure of AF relays, the former threat may not be that compelling to deal with. Alternatively, we can introduce some MIMO relays within the network, where an M_r antenna equipped relays removes the need for M_r^2 single-antenna relays.

In order to cope with the heavy load of CSI dissemination, we have proposed distributed calculation of the relay gain factors in Section 3.3. Doing so, we are intended to attain exactly the same performance that would be achieved by a global CSI requiring solution, but with a bounded CSI dissemination overhead independent from the number of relays. That is, we do not trade-off performance with complexity, but instead we are devising a novel scheme which better utilizes the available resources (in terms of channel-use). The proposed distributed scheme relies on gradient descent based gain allocation, where we have proved that each relay can compute the associated element of the gradient independently from the others by only employing *local* CSI plus *limited feedback* from the destinations. The inherent iterative nature of the new distributed gradient approach eases the adaptation of the network to the propagation channel variations or modified relay sets. Further, it provides a scalable channel occupation cost for relay gain decision, which is determined according to the desired performance levels.

A proper transmission scheme should always incorporate potential system imperfections within the design objectives. Essentially, coherence in AF relaying requires that the channel estimates at the relays and the anticipated global LO phase reference are error-free. Consequently, in Chapter 4 we have studied robust relay gain optimization in the presence of data uncertainty from a worst-case perspective. The LO phase noise, which is triggered by re-transmit processing time of relays and channel estimation instants, has been identified to be an unavoidable uncertainty source. We proposed a computationally tractable worst-case max-min beamforming approach, which employs a probabilistic Gaussian bounding on each relay's LO phase offset. We have shown that especially in high uncertainty regime the robust problem provides substantial worst-case SINR improvement over its non-robust counterpart. Afterwards in the same Chapter 4, we shifted our focus to CSI uncertainty at the relays. We modeled each source-to-relay and relay-to-source channel vectors within independent spherical uncertainty sets. The generalized problem involves both worst-case SINR and power optimizations, and the corresponding SINR expressions consist of multiplicative uncertainty sets. That's why, the problem did not let itself to a computationally tractable mathematical modeling. Instead, we considered two simplified cases, where either only S-R or R-D links are assumed to be imperfect. We proposed SDP based robust counterparts for both cases, where the minimum (over S-D pairs) worst-case SINR has been considerably improved to provide a reliable and fair data link to the individual pairs.

In Chapter 5, we aimed at linking the developed know-how to the big picture of large relay networks, where some relays may not be able establish a direct contact with the others due to topological conditions, or the complete dissemination of local CSI may not be cost-effective. Addressing this issue, we proposed two new relaying protocols within which the relays are grouped in small clusters. The first protocol assumes that there are multiple relay clusters, each of which manages the multiuser interference on its own totally independent from the others. Although the coherence in between the clusters is ruined, the need for *network-wide* global CSI shrinks to be *cluster-wide* global

CSI. Further, applying time-and-cluster specific relay phase rotations, we guarantee an effective diversity order at least as much as the number of clusters. In the second protocol, we introduced hierarchy to the network and assigned different missions to the relays, which require different amounts of CSI knowledge. A big portion of the relays in the network is assumed to employ only their own local CSI and contribute to array gain, with which we further reduced the CSI requirement from being global in clusters to be relay-wise local, i.e., no CSI dissemination. The interference management is assigned to a single smaller relay cluster, where we still assume global CSI within it. It has been shown that the hierarchical relaying protocol augmented with relay selection for clusters, achieves almost the same sum rate performance and exactly the same diversity gains with that of a network-wide global CSI requiring protocol. Note that the distributed gradient scheme can be as well adapted within the global cluster to further reduce the required CSI dissemination overhead.

To conclude, we have conveyed the theoretical basis on the distributed orthogonalization through coherent AF relays to a practically implementable protocol. The CSI dissemination overhead of the required (almost) N_{R}^2 distributed antennas in large relay networks was the fact that hinders us to raise the headline "*from theory to practice: distributed orthogonalization*". However, with the aforementioned relaying protocols in combination with distributed gradient methods, we can release the scaling of this CSI dissemination overhead from being dependent on the number of relays, and reduce it to be linearly proportional only with number of S-D pairs N.

Part 2 Starting from Chapter 6, we have assumed that the relays are more complex and have further decoding and re-encoding abilities. Without relying on T/F/OF/C-DMA techniques, in order for DF relays to efficiently decode the incoming information streams from multiple sources and separate the independent transmit signals to multiple destinations, employment of multiple antennas at relay(s) emerges as a compulsory requirement. Henceforth, we considered a MIMO network, where each node equipped with multiple antennas at both transmit and receive modes. In Section 6, we have extended the single-pair single-relay one- and two-way DF protocols to a MIMO setup. We further proposed a *per-symbol zero-padding* based signal combination approach for XOR precoding based two-way relaying, which enables the relay to support unbalanced rates in the downlink to the destinations.

In Section 7, we studied transmit covariance optimization with CSI at the transmitting nodes for both relaying protocols. We first charaterized the capacity regions of one- and two-way relaying. The case of one-way protocol was rather trivial due to established point-to-point MIMO links. However, for two-way relaying, we provided efficient optimization procedures to optimally determine the covariance matrices that form the boundary points of the MAC and BC phases' achievable rate regions. We have employed an iterative semidefinite optimization for both. Next, we devised two optimization methods based on SDP techniques, each of which optimizes the relay transmit covariance such that either the sum or the minimum of unidirectional link rates is maximized. Both optimizations are two-phase optimal in the sense that the optimal rate tuple within the intersection of MAC and BC rate regions, is searched for. We have observed that the constructive impact of transmit CSI at the relay, emphasizes itself more and more as the fraction of relay antennas to terminal antennas grows. In a last part, we addressed the imperfectness of transmit CSI at the relay for the special case of two-way relaying. Assuming that the estimate of each terminal-to-relay link is defined within a spherical uncertainty set, we first characterized the worst-case capacity of the broadcast phase. In accordance, we provided a robust counterpart guideline to extend the transmit beamforming approaches derived for perfect CSIT to the case of imperfect CSIT. There observed two important behaviours over the size of the uncertainty. First, as expected, the robust max-min beamforming proposal provides substantial worst-case unidirectional link rate improvement. And second, more interestingly, we noted that employing CSIT without being aware of uncertainty may lead much worse worst-case performance when compared with identical covariance matrix of the no CSIT case.

In order for exploiting distributed spatial multiplexing in a multi-pair communications via DF relaying, we have considered the corresponding extension of MIMO one- and two-way relaying in Section 8. The fundamental question to answer there was how to separate the transmit signals of different users/pairs in the downlink from the relay. For one-way protocol we employed block diagonalization based zero-forcing methods, whereas for two-way protocol we proposed a two level of separation through a combination of XOR precoding and spatial separation. We further designed a modular iterative algorithm for maximizing the sum rate of all unidirectional rates, where the achieved feasible solution is optimal over the two phase. The algorithm is mainly based on SDP feasibility checks and exploits the geometry of the multi-dimensional multiuser capacity region. The corresponding sum rate performances depicted two main results: 1) Multiuser two-way relaying proposes significant spectral efficiency over its one-way counterpart, 2) In terms of maximal achievable sum rate, performing a simultaneous multi-pair communications is superior to serving a single pair per transmission in a sequential order or through multiuser diversity.

The MIMO DF relaying (one- or two-way) serves as a potent candidate for adaptation of cellular and local are networks to the next generation communications systems. Several application scenarios for the multiuser case have been already illustrated in Fig. 8.1. In addition to the provided range extension through both one- and two-way protocols, which is anyhow required due to prospected high carrier frequencies, spectral efficiency can be considerably enhanced with two-way relaying.

As a general conclusion to this thesis we would like to highlight that independent of the employed relaying protocol and forwarding scheme, the *quality* and the *quantity* of channel knowledge per relay node, come into prominence as performance determiners in efficient resource optimization.

9.2. Outlook

In spite of our extensive analysis of multiuser AF and DF relaying schemes, there still exist many interesting questions to be answered as future work that may extend the results obtained within this thesis. In the following, we list these open topics in the order of corresponding chapters.

- We have introduced the out-of-cluster interference mitigation in Section 2 and presented the sufficiency conditions for complete interference management. However, in the presence of excess relays, we have not optimized the relay gain vector such that relay transmit power is minimized or sum rate is maximized. Although we have associated the corresponding mathematical formulations with Remark 5.3.1, it would be interesting to further investigate the optimized performance to provide a fair comparison between the absence and presence of interference.

- In the performance evaluations of Part I, we have commonly restricted ourselves to non-cooperating single antenna relays. Although the derived optimization problems are readily applicable for partial relay cooperation or MIMO relays, it is not that clear if there is a certain cooperation pattern or MIMO relay configuration (how many relays with how many antennas) that leads higher sum or outage rates.

- We have reported in Section 3.3 that the convergence speed of gradient based gain allocations, independent from being distributed or not, are highly dependent on the starting point. That is, there is a still need for further investigation on designing better starting vector decisions, which would improve the spectral efficiency through employing less channel uses with the accelerated convergence. Distributed gradient based gain allocation can be extended for achieving further gains such as diversity or minimum transmit power, or may be applied in some other similar relaying networks. For instance, multihop (also called multi-stage) multiuser relaying scenarios may benefit from the distributed gradient based gain decisions in order to reduce the CSI information flow in between different relay stages.

- On the other hand, we have observed in Chapter 4 that deriving computationally tractable robust counterparts for uncertain data sets is not a trivial task to achieve in general. During derivations, we had to impose some relaxations to the constraints or approximations on to the defined uncertainty sets, which results in a reduced set of feasible solutions in reality. Hence, further special attention should be paid especially on CSI uncertainty, for deriving tractable and effective robust solutions. We have already noted that CSI uncertainty is inherently coupled with LO phase uncertainty through the RF chain processing during the channel estimation. That is to say, a joint robust phase and channel uncertainty problem can be investigated, though it may be complex, in order to get more insight on system imperfections in coherent relaying. Note lastly that although we have here focused on a generalized multiuser communications scenario in Section 4.2, the phase noise is also effective on the optimal power allocation for the conventional multiple relay channel with a single S-D pair. Hence, the developed understanding on robust complex gain optimization can be similarly adapted in the robust real power allocation against phase noise.

- In Chapter 8, we have mentioned that the maximum number of pairs that can be efficiently supported with MIMO DF relaying is limited by the maximum number of antennas that can be implemented within the relay in reality. An approach to solve this problem could be to introduce multiple MIMO DF relays into the network. In such a multi-relay setup, however, the transmission schemes both towards and from the relay should be re-defined. Because then, the source transmit signals would be broadcasted to multiple relays through the uplink, and correspondingly the relay transmit signals should be combined appropriately at each destination terminal in the downlink. Further, the robust covariance optimization proposed for MISO downlink channels can be extended to the case of MIMO channels.

Appendix A

A.1. Convex Optimization

We present here a brief overview on convex optimization, where we summarize the concept convexity and convex optimization problems in combination with implementations issues. The content and general structure follow closely with the book of Boyd *et al.* [26] and some other state-of-art references [9–14, 76, 78]. Specifically, we first introduce the concept of convex sets, which is followed by the definition of convex function and convex optimization problem. Next, we provide a summary on commonly used convex problem definitions. After elaborating on the optimality of convex solutions, we present numerical techniques to solve convex optimization problems. Finally in a last part, we shortly touch on the robust convex optimization methodology proposed by [9–14].

A.1.1. The Concept of Convexity

Some important definitions are listed as follows.

Affine set: Let C be a set in $\mathbb{R}^n$, $n \in \mathbb{R}_+$. If the line between any distinct points in C lies in C, then C is an affine set. In other words, if for any x_1, $x_2 \in C$, $\theta \in \mathbb{R}$, we have $\theta x_1 + (1-\theta)x_2 \in C$, this means that C contains the linear combination of any two points in C conditioned on that the sum of the coefficients is one.

Convex set: If the line segment between an two points in C lies in C, then C is a convex set. In other words, for any x_1, $x_2 \in C$ and $0 \leq \theta \leq 1$, we have should have $\theta x_1 + (1-\theta)x_2 \in C$.

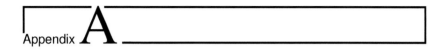

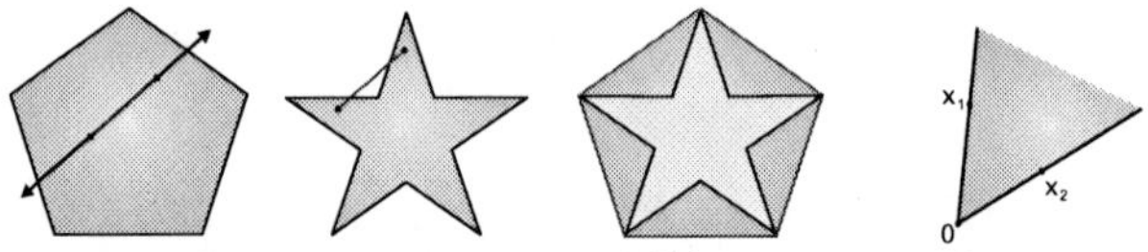

Figure A.1.: Examples for convex/affine/non-convex sets and convex hull of a set.

Convex hull: The convex hull of a set C, denoted by **conv** C, is the set of all convex combinations of all points in C, and as the name implies it is convex. See Fig.A.1 for illustration of the previously defined three definitions.

Cone: If for every $x \in C$ and $\theta \geq 0$, we have $\theta x \in C$, then C is a cone. Moreover, if it is also convex, it is called convex cone (nonnegative orthant), which means that for any x_1, $x_2 \in C$ and $\theta_1, \theta_2 \geq 0$, we have $\theta_1 x_1 + \theta_2 x_2 \in C$. Fig.A.1 presents an example for a convex cone, which is a two-dimensional pie-slice with apex 0 and edges passing through x_1 and x_2.

Ball: A ball (also called as Euclidean ball) in $\mathbb{R}^n$ is defined as

$$B(\mathbf{x}_c, r) = \{\mathbf{x} | (\mathbf{x} - \mathbf{x}_c)^T (\mathbf{x} - \mathbf{x}_c) \leq r^2\} \quad \text{or} \quad B(\mathbf{x}_c, r) = \{\mathbf{x}_c + r\mathbf{u} | \ \|\mathbf{u}\| \leq 1\},$$

where $r > 0$ represents the radius, $\mathbf{x}_c$ is the center. Note that ball forms a convex set.

Ellipsoid: An ellipsoid has the form

$$E(\mathbf{x}_c, r) = \{\mathbf{x} | (\mathbf{x} - \mathbf{x}_c)^T \mathbf{P}^{-1} (\mathbf{x} - \mathbf{x}_c) \leq 1\} \quad \text{or} \quad E(\mathbf{x}_c, r) = \{\mathbf{x}_c + \mathbf{P}\mathbf{u} | \ \|\mathbf{u}\| \leq 1\},$$

where $\mathbf{P}$ is a symmetric and positive definite matrix which defines how far the ellipsoid extends in every direction from $\mathbf{x}_c$. Moreover, the square root of eigenvalues of $\mathbf{P}$ are the lengths of semi-axes of E.

Second order cone: The second order cone is the norm cone for the Euclidean norm

$$C = \{(\mathbf{x}, t) \in \mathbb{R}^{n+1} | \ \|\mathbf{x}\|_2 \leq t\}.$$

Positive semidefinite cone: The set of $n \times n$ symmetric positive semidefinite matrices is denoted by

$$\mathbf{S}^n_+ = \{\mathbf{X} \in \mathbf{S}^n | \mathbf{X} \succeq 0\}. \tag{A.1}$$

If $\theta_1 \mathbf{A} + \theta_2 \mathbf{B} \in \mathbf{S}^n_+$ for $\theta_1, \theta_2 \geq 0$, $\mathbf{A}, \mathbf{B} \in \mathbf{S}^n_+$, then the set $\mathbf{S}^n_+$ is a convex cone. As an example, the positive semidefinite cone in $\mathbf{S}^2$ is

$$\mathbf{X} = \begin{bmatrix} x & y \\ y & z \end{bmatrix} \in \mathbf{S}^2_+ \Leftrightarrow x \geq 0,\ z \geq 0,\ xz \geq y^2. \tag{A.2}$$

A.1.2. Convex Functions and Optimization Problems

Convex Functions Consider a function $f : \mathbb{R}^n \to \mathbb{R}$. If for all vectors $\mathbf{x}$, $\mathbf{y}$ within the domain (set of all inputs) of f and $0 \leq \theta \leq 1$, we have

$$f(\theta \mathbf{x} + (1-\theta)\mathbf{y}) \leq \theta f(\mathbf{x}) + (1-\theta) f(\mathbf{y}), \tag{A.3}$$

then the function f is called to be *convex*. Assuming that $n = 1$, this can be geometrically translated as that the line segment between any x and y on the f lies above the graph of f, where the graph is composed of all $(x, f(x))$. Note that if f is convex, then $-f$ is concave. Some typical convex (concave) functions can be listed as: $|x|$, e^x, x^2, $\log x$, $x \log x$, $\mathbf{a}^T\mathbf{x} + b$, $\|\mathbf{A}\mathbf{x}\|^2$, $\max\{x_1, \ldots, x_n\}$, $\log\{e^{x_1} + \cdots + e^{x_n}\}$, $\log\det$, geometric mean, and etc. On the other hand, there are several operations that preserve convexity such as nonnegative weighted summation, scaling with a positive scalar, pointwise maximization, i.e., $\max\{f_1(x), f_2(x)\}$.

The $\alpha-$sublevel set of a convex function $f : \mathbb{R}^n \to \mathbb{R}$ is defined as $S_\alpha = \{\mathbf{x} \in \mathbf{dom} f | f(\mathbf{x}) \leq \alpha\}$, where $\alpha \in \mathbb{R}$, and **dom** denotes the domain of f. Then, the function f is called *quasiconvex* if its domain and all its sublevels sets are convex. If $-f$ is quasiconvex, then we call f as *quasiconcave*. If a function is both quasiconvex and quasiconcave, then it is called *quasilinear*. A quasilinear function's domain and all level set that $\{\mathbf{x} | f(\mathbf{x}) = \alpha\}$ are convex. Convex functions have convex sublevels, and hence they are quasiconvex but the converse is not necessarily true. Some typical quasiconvex (quasiconcave) functions are listed as: $\log x$ (quasilinear), ceiling function (quasilinear), length of a vector (quasiconvex), linear-fractional function $\frac{\mathbf{a}^T\mathbf{x}+b}{\mathbf{c}^T\mathbf{x}+d}$ (quasilinear), distance ratio function $\frac{\|\mathbf{x}-\mathbf{a}\|}{\|\mathbf{x}-\mathbf{b}\|}$ (quasiconvex), and etc.

An example for quasiconvex function on $\mathbb{R}$ is shown in Fig. A.2. There, the $\alpha-$sublevel set S_α is in $[a, b]$ and convex, whereas S_β is in $[-\infty, c]$.

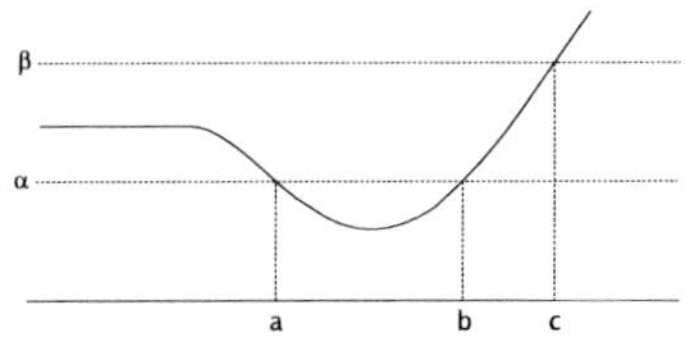

Figure A.2.: A quasiconvex function on $\mathbb{R}$ [26].

Convex Optimization Problems We consider the following general optimization problem notation

$$\min_{\mathbf{x}} \; f_0(\mathbf{x}) \quad \textbf{subject to} \; f_i(\mathbf{x}) \leq 0, \;\; i = 1, \ldots, m$$
$$h_i(\mathbf{x}) = 0, \;\; i = 1, \ldots, p \tag{A.4}$$

where $\mathbf{x} \in \mathbb{R}^n$ is called the optimization variable, $f_0 : \mathbb{R}^n \rightarrow \mathbb{R}$ is the objective (cost) function, $f_i(\mathbf{x}) \leq 0$ are the inequality constraints and $h_i(\mathbf{x}) = 0$ are equality constraints.

The optimal value p^* of (A.4) is defined as

$$p^* = \inf\{f_0(\mathbf{x}) | f_i(\mathbf{x}) \leq 0, i = 1, \ldots, m, h_i(\mathbf{x}) = 0, \;\; i = 1, \ldots, p\}. \tag{A.5}$$

If the problem is *infeasible*, then $p^* = \infty$. If there are some feasible points $\mathbf{x}_k$ which leads the objective to approach $-\infty$ as $k \rightarrow \infty$, then $p^* = -\infty$, and the problem is called to be *unbounded below*. If $f_0(\mathbf{x}^\star) = p^\star$ is achieved at a feasible $\mathbf{x}^\star$, then the $\mathbf{x}^\star$ is an *optimal* solution. A feasible $\mathbf{x}^\star$ is referred as *globally optimal* if $f_0(\mathbf{x}^\star) \leq f_0(\mathbf{x})$ for all feasible $\mathbf{x}$. On the other hand, if $f(\mathbf{x}^\diamond) \leq f(\mathbf{x})$ for all feasible $\mathbf{x}$ satisfying $\|\mathbf{x} - \mathbf{x}^\diamond\| \leq \epsilon$ for some $\epsilon > 0$, then $\mathbf{x}^\diamond$ is said to be *locally optimal*. Assume that a feasible $\mathbf{x}$ results in $f_0(\mathbf{x}^\star) \leq p^\star + \epsilon$ for $\epsilon > 0$, then the $\mathbf{x}$ is called $\epsilon-$*(sub)optimal* solution. We note that the maximization version of the minimization problem (A.4) can be immediately obtained by just changing the objective to be $-f_0(\mathbf{x})$. Hence, all above definitions for minimization can be trivially extended for maximization.

A *convex optimization problem* is defined as

$$\min_{\mathbf{x}} \; f_0(\mathbf{x}) \quad \textbf{subject to} \; f_i(\mathbf{x}) \leq 0, \;\; i = 1, \ldots, m$$
$$\mathbf{a}_i^T \mathbf{x} = b_i, \;\; i = 1, \ldots, p, \tag{A.6}$$

$b_i \in \mathbb{R}$, $\mathbf{a}_i \in \mathbb{R}^n$. This is a special case of the general problem in (A.4), where

1. the objective function $f_0(\mathbf{x})$ must be convex,

2. the inequality constraint functions $f_i(\mathbf{x})$ must be convex,

3. and the equality constraint functions $\mathbf{a}_i^T\mathbf{x} = b_i$ must be affine.

By definition, all locally optimal solutions of the convex problem (A.6) are also globally optimal, i.e., there is one single optimal solution. Assuming that f_0 is differentiable, the optimality criterion for a given feasible $\mathbf{x}$ is that

$$\nabla f_0(\mathbf{x})^T(\mathbf{y} - \mathbf{x}) \geq 0 \text{ for all feasible } \mathbf{y}, \tag{A.7}$$

where $\nabla f_0(\mathbf{x})$ denotes the gradient of f_0 at $\mathbf{x}$.

The quasiconvex optimization problem definition can be carried over from (A.6) by replacing the convexity constraint on $f_i, i = 0, \ldots, m$, with quasiconvexity. However, note now that there can be locally optimal solutions, which are not necessarily globally optimal. A general approach to solve quasiconvex problems is to employ *convex feasibility problems*. That is, instead of minimizing f_0, it can upper bounded by t, i.e., $f_0(\mathbf{x}) \leq t$, and can be added as a constraint to the original problem. Hence, the minimization problem boils down to be a convex feasibility check problem, where we check if the upper bound t is feasible or not:

$$\begin{aligned} \textbf{find } \mathbf{x} \textbf{ subject to } & f_o(\mathbf{x}) \leq t \\ & f_i(\mathbf{x}) \leq 0, \quad i = 1, \ldots, m \\ & \mathbf{a}_i^T\mathbf{x} = \mathbf{b}_i, \quad i = 1, \ldots, p. \end{aligned} \tag{A.8}$$

By trying different upper bounds t, we can solve the quasiconvex problem optimally in an iterative manner by applying a feasibility check at each iteration. The corresponding optimization methods is called *bisection* and summarized in Algorithm 6.

Algorithm 6 Bisection Method

initiate $\epsilon > 0, t_{\text{min}}, t_{\text{max}}$.
repeat
Step 1. Set $t \Leftarrow (t_{\text{min}} + t_{\text{max}})/2$.
Step 2. Solve convex feasibility problem (A.8) with t.
Step 3. Check if feasible, then set $t_{\text{max}} \Leftarrow t$,
else, set $t_{\text{min}} \Leftarrow t$.
Step 4. Quit if $|t_{\text{min}} - t_{\text{max}}| \leq \epsilon$.

Next, we review commonly used convex optimization models.

Linear Program A subclass of convex optimization problems is *linear program*, in which the objective and constraint functions are all affine and have the following form

$$\min_{\mathbf{x}} \ \mathbf{c}^T\mathbf{x} + d \quad \textbf{subject to} \ \mathbf{G}\mathbf{x} \preceq \mathbf{h}, \\ \mathbf{A}\mathbf{x} = \mathbf{b} \tag{A.9}$$

where $d \in \mathbb{R}$, $\mathbf{b} \in \mathbb{R}^p$, $\mathbf{c} \in \mathbb{R}^n$, $\mathbf{h} \in \mathbb{R}^m$, $\mathbf{G} \in \mathbb{R}^{m\times n}$ and $\mathbf{A} \in \mathbb{R}^{p\times n}$.

Linear-fractional Program If we change the objective function of LP with a ratio of two affine functions, i.e., $\frac{\mathbf{c}^T\mathbf{x}+d}{\mathbf{e}^T\mathbf{x}+f}$, then the corresponding problem is called *linear-fractional program.* Moreover, as the objective is quasiconvex, the linear-fractional program is a quasiconvex problem.

Quadratic Program If we let the objective and constraint functions to be quadratic (convex) and affine, respectively, then we have a *quadratic program* (QP) with the from

$$\min_{\mathbf{x}} \ \frac{1}{2}\mathbf{x}^T\mathbf{P}\mathbf{x} + \mathbf{q}^T\mathbf{x} + r \ \textbf{subject to} \ \mathbf{G}\mathbf{x} \preceq \mathbf{h}, \\ \mathbf{A}\mathbf{x} = \mathbf{b} \tag{A.10}$$

where $\mathbf{q} \in \mathbb{R}^n$, $\mathbf{P} \in \mathbf{S}^n_+$. In addition to the objective, if the constraints are also quadratic, then we call this as *quadratically constrained quadratic program*, i.e.,

$$\min_{\mathbf{x}} \ \frac{1}{2}\mathbf{x}^T\mathbf{P}\mathbf{x} + \mathbf{q}^T\mathbf{x} + r \ \textbf{subject to} \ \frac{1}{2}\mathbf{x}^T\mathbf{P}_i\mathbf{x} + \mathbf{q}_i^T\mathbf{x} + r_i \leq 0, \ i = 1, \ldots, m \\ \mathbf{A}\mathbf{x} = \mathbf{b}, \tag{A.11}$$

which simply minimize a quadratic function over an intersection region of ellipsoids.

Summing up, QP includes LP as a special case, by taking $\mathbf{P} = 0$; QCQP includes QP (and therefore also LP) as a special case, by taking $\mathbf{P}_i = 0$, $i = 1, \ldots, m$. Recall that all three LP, QP, QCQP are convex optimization problems.

Second Order Cone Programming The *second order cone programming* (SCOP) is a modified version of QCQP such that we have

$$\min_{\mathbf{x}} \ \mathbf{c}^T\mathbf{x} \quad \textbf{subject to} \ \|\mathbf{C}_i\mathbf{x} + \mathbf{f}_i\|_2 \leq \mathbf{w}_i^T\mathbf{x} + s_i, \ \ i = 1, \ldots, m \\ \mathbf{A}\mathbf{x} = \mathbf{b}, \tag{A.12}$$

where $s_i \in \mathbb{R}$, $\mathbf{w}_i \in \mathbb{R}^n$, $\mathbf{f}_i \in \mathbb{R}^{n_i}$, and $\mathbf{C}_i \in \mathbb{R}^{n_i\times n}$. The inequality constraint in (A.12) is called as the second order constraint since it indicates the affine function

$(\mathbf{C}_i\mathbf{x}+\mathbf{f}_i, \mathbf{w}_i^T\mathbf{x}+s_i)$ lies in the second order cone in $\mathbb{R}^{n_i+1}$. When $\mathbf{c}_i = 0, \forall i$, the SOCP is equivalent to a QCQP; and when $\mathbf{A}_i = 0, \forall i$, then the SOCP reduces to a LP. Hence, the SOCPs are generalized versions of QCQPs and LPs.

Semidefinite Programming A cone $C \subseteq \mathbb{R}^n$ is called *proper* if it is convex, closed, solid (i.e., it has non-empty interior), pointed (i.e., it contains no line). A proper cone C is used to define a *generalized inequality*, which is a partial ordering on $\mathbb{R}^n$. The association between C and the partial ordering on $\mathbb{R}^n$ is defined by

$$x \preceq_C y \Leftrightarrow y - x \in C \quad \text{or} \quad x \prec_C y \Leftrightarrow y - x \in \textbf{int}\, C \tag{A.13}$$

where **int** denotes the interior. The positive semidefinite cone $\mathbf{S}_+^n$ is a proper cone in $\mathbf{S}^n$, and the generalized inequality becomes the usual matrix inequality, i.e., $\mathbf{X} \preceq_C \mathbf{Y}$ translated in that $\mathbf{Y} - \mathbf{X}$ is positive semidefinite.

The condition

$$\mathbf{A}(\mathbf{x}) = x_1\mathbf{A}_1 + \cdots + x_n\mathbf{A}_n \preceq \mathbf{B} \tag{A.14}$$

where $\mathbf{B}$, $\mathbf{A} \in \mathbf{S}^n$, is called *linear matrix inequality.* The solution set of a LMI is convex and it is the inverse image of the positive semidefinite cone under the affine function $f(\mathbf{x}) = \mathbf{B} - \mathbf{A}(\mathbf{x})$.

Next, we generalize the definiton of the convex problem (A.6) using generalized inequalities as

$$\begin{aligned} \min_{\mathbf{x}} \; f_0(\mathbf{x}) \quad \textbf{subject to} \; & f_i(\mathbf{x}) \preceq_{C_i} 0, \;\; i = 1, \ldots, m \\ & \mathbf{a}_i^T\mathbf{x} = \mathbf{b}_i, \;\; i = 1, \ldots, p \end{aligned} \tag{A.15}$$

where $C_i \subseteq \mathbb{R}^{c_i}$ are proper cones, and $f_i : \mathbb{R}^n \to \mathbb{R}^{c_i}$ are C_i-convex. This problem is called as *convex optimization problem with generalized inequality constraints.*

The *conic form problems* (CFP), which will be later associated with SDP, is the simplest convex optimization problem with generalized inequalities, which have a linear objective and one affine inequality constraint function (so C-convex):

$$\begin{aligned} \min_{\mathbf{x}} \; \mathbf{c}^T\mathbf{x} \quad \textbf{subject to} \; & \mathbf{G}\mathbf{x} + \mathbf{h} \preceq_C 0, \\ & \mathbf{A}\mathbf{x} = \mathbf{b}. \end{aligned} \tag{A.16}$$

When C is the cone of k-dimensional positive semidefinite matrices, the corresponding CFP is called SDP with the form

$$\begin{aligned}\min_{\mathbf{x}}\ \mathbf{c}^T\mathbf{x}\quad \textbf{subject to}\ & \mathbf{G}_1 x_1 + \mathbf{G}_2 x_2 + \cdots + \mathbf{G}_n x_n + \mathbf{H} \preceq 0 \\ & \mathbf{A}\mathbf{x} = \mathbf{b}\end{aligned} \tag{A.17}$$

where $\mathbf{H}, \mathbf{G}_i \in \mathbf{S}^k$. If the LMI in (A.17) is composed of diagonal matrices, then SDP boils down to an LP.

For instance, following the previously introduced cone framework, the inequality constraint in (A.12) can be expressed as

$$\begin{bmatrix}\mathbf{w}_i^T\mathbf{x} + s_i \\ \mathbf{C}_i\mathbf{x} + \mathbf{f}_i\end{bmatrix} \preceq_C 0, \tag{A.18}$$

which can be directly converted to a LMI (SDP constraint) as

$$\begin{bmatrix}\mathbf{w}_i^T\mathbf{x} + s_i \\ \mathbf{C}_i\mathbf{x} + \mathbf{f}_i\end{bmatrix} \preceq_C 0 \;\Leftrightarrow\; \begin{bmatrix}\mathbf{w}_i^T\mathbf{x} + s_i & \mathbf{x}^T\mathbf{C}_i^T + \mathbf{f}_i^T \\ \mathbf{C}_i\mathbf{x} + \mathbf{f}_i & (\mathbf{w}_i^T\mathbf{x} + s_i)\mathbf{I}\end{bmatrix} \preceq 0 \tag{A.19}$$

where $\mathbf{I}$ is an identity matrix with corresponding size.

A.1.3. Lagrangian Duality and Optimality Conditions

Lagrangian Duality Consider the problem (A.4) which is not necessarily convex. We define the *Lagrangian* $\mathcal{L} : \mathbb{R}^n \times \mathbb{R}^m \times \mathbb{R}^p \to \mathbb{R}$ as

$$\mathcal{L}(\mathbf{x}, \lambda, \nu) = f_0(\mathbf{x}) + \sum_{i=1}^{m} \lambda_i f_i(\mathbf{x}) + \sum_{i=1}^{p} \nu_i h_i(\mathbf{x}) \tag{A.20}$$

with $\mathbf{dom}\ \mathcal{L} = \mathcal{D} \times \mathbb{R}^m \times \mathbb{R}^p$ and $\mathcal{D} \triangleq \left(\bigcap_{i=0}^{m} \mathbf{dom}\ f_i \cap \bigcap_{i=0}^{p} \mathbf{dom}\ h_i\right)$. The terms λ_is and ν_is are called Lagrangian multiplier vectors or dual variables. The Lagrangian dual function, $g : \mathbb{R}^m \times \mathbb{R}^p \to \mathbb{R}$, is the minimum value of the $\mathcal{L}$ over $\mathbf{x}$,

$$g(\boldsymbol{\lambda}, \boldsymbol{\nu}) = \inf_{\mathbf{x}\in\mathcal{D}} \mathcal{L}(\mathbf{x}, \boldsymbol{\lambda}, \boldsymbol{\nu}) = \inf_{\mathbf{x}\in\mathcal{D}} \left(f_0(\mathbf{x}) + \sum_{i=1}^{m} \lambda_i f_i(\mathbf{x}) + \sum_{i=1}^{p} \nu_i h_i(\mathbf{x})\right) \tag{A.21}$$

where $\mathcal{D} = \bigcap_{i=0}^{m} \mathbf{dom}\ f_i \cap \bigcap_{i=0}^{p} \mathbf{dom}\ h_i$. If $\mathcal{L}$ is unbounded below in $\mathbf{x}$, then g takes on the value $-\infty$. The dual function is always concave not depending on the convexity of primal problem (A.4), because it is pointwise infimum of affine functions of (λ, ν).

The dual function $g(\lambda, \nu)$ is an inherent lower bound on the optimal value of the primal problem, i.e., $g(\lambda, \nu) \leq p^*$ for any ν and $\lambda \succeq 0$. The maximal lower bound can be found through the *Lagrangian dual problem*:

$$\max \quad g(\boldsymbol{\lambda}, \boldsymbol{\nu}) \quad s.t. \quad \lambda \succeq 0. \tag{A.22}$$

The dual variables (λ^*, ν^*) are called optimal dual multipliers. Independent of the primal problem, dual problem is a convex optimization problem. Given that d^* is the optimal value of dual problem, we have the relation of $d^* \leq p^*$, which is called as the *weak duality*. Note that the weak duality inequality holds when $d^*, p^* = \infty$. The difference $p^* - d^*$ is called the *optimal duality gap*, which is always nonnegative. If $d^* = p^*$, the optimal duality gap is zero and *strong duality* holds.

Optimality Conditions

Certificate of suboptimality: Since $g(\lambda, \nu)$ serves as a lower bound on p^*, the dual feasible point (λ, ν) provides a *proof* or *certificate* that $p^* \geq g(\lambda, \nu)$. Good certificates need strong duality. If $\mathbf{x}$ is primal feasible and (λ, ν) is dual feasible, then

$$f_0(\mathbf{x}) - p^* \leq f_0(\mathbf{x}) - g(\lambda, \nu).$$

This establishes that $\mathbf{x}$ is ϵ-suboptimal where $\epsilon = f_0(\mathbf{x}) - g(\lambda, \nu)$. If the duality gap is zero, (λ, ν) acts as a certificate that proves $\mathbf{x}$ is optimal.

Complementary Slackness: The complementary slackness condition is expressed as

$$\lambda^\star > 0 \rightarrow f_i(\mathbf{x}^\star) = 0 \quad \text{or} \quad f_i(\mathbf{x}^\star) < 0 \rightarrow \lambda^\star = 0, \tag{A.23}$$

which means that the ith optimal multiplier is zero unless the ith constraint is active at the optimum.

KKT Optimality Conditions: Assume that objective and all constraint functions are differantiable. Let $x^\star$ and $(\lambda^\star, \nu^\star)$ be any primal and dual optimal points with zero duality gap. We know fundamentally that as $\mathbf{x}^\star$ minimizes $\mathcal{L}$, the corresponding gradient should vanish at $\mathbf{x}^\star$. Hence, we have the following conditions which are called *Karush-Kuhn-Tucker* (KKT) conditions:

$$f_i(\mathbf{x}^\star) \leq 0, \quad i = 1, \ldots, m \tag{A.24}$$

$$h_i(\mathbf{x}^\star) = 0, \quad i = 1, \ldots, p \tag{A.25}$$

$$\lambda_i^\star \geq 0, \quad i = 1, \ldots, m \tag{A.26}$$

$$\lambda_i^\star f_i(\mathbf{x}^\star) = 0, \;\; i = 1, \dots, m \tag{A.27}$$

$$\nabla f_0(\mathbf{x}^\star) + \sum_{i=1}^{m} \lambda_i^\star \nabla f_i(\mathbf{x}^\star) + \sum_{i=1}^{p} \nu_i^\star \nabla h_i(\mathbf{x}^\star) = 0. \tag{A.28}$$

The KKT conditions must be satisfied by any pair of primal and dual optimal solutions with the given properties above.

A.1.4. Algorithms to Solve Convex Optimization Problems

Unconstrained Optimization The unconstrained optimization problem is defined as

$$\min_{\mathbf{x}} \;\; f(\mathbf{x}), \tag{A.29}$$

where $f : \mathbb{R}^n \rightarrow \mathbb{R}$ is convex and twice-continuously differentiable. A necessary and a sufficient condition for a solution $\mathbf{x}^\star$ to be optimal is $\nabla f(\mathbf{x}^\star) = 0$. The unconstrained minimizations are solved with iterative *descent* algorithms, which means that $f(\mathbf{x}^{(k+1)}) < f(\mathbf{x}^{(k)})$ unless $\mathbf{x}^{(k)}$ is optimal, where the subscript denotes the iteration index. The corresponding update equation is given by

$$\mathbf{x}^{(k+1)} = \mathbf{x}^{(k)} + t^{(k)} \Delta\mathbf{x}^{(k)}, \tag{A.30}$$

where the vector $\Delta\mathbf{x}^{(k)}$ is known as the *search direction* for the kth iteration and the scalar $t^{(k)}$ is the step size, which decides how much we should increment along the search direction. The search directions is called as the *descent direction* when it satisfies the condition that $\nabla f(\mathbf{x}^{(k)})^T \mathbf{x}^{(k)} < 0$, which follows from the convexity of the problem.

The step size can be determined through *exact* or *backtracking line search* methods, where in both cases the objective f is minimized over the ray $\{\mathbf{x} + t\Delta\mathbf{x} | t \geq 0\}$. As the exact line search methods turns out to be costly, in general, f is approximately minimized to reduce it only enough.

There are several descent methods such as gradient descent, steepest descent and Newton. For instance, in the gradient descent method the search direction is simply chosen to be the negative gradient, i.e., $\Delta\mathbf{x} = -\nabla f(\mathbf{x})$, whereas the Newton method employs the second derivative to determines the direction, i.e., $\Delta\mathbf{x} = -\nabla^2 f(\mathbf{x})^{-1} \nabla f(\mathbf{x})$. Referring to [26] for further details, we focus on the *gradient descent* which is employed commonly within this thesis, and summarize the corresponding algorithm with Algorithm 7.

Algorithm 7 Gradient Descent Method

initiate a starting vector $\mathbf{x}$.

repeat

Step 1. Set $\Delta\mathbf{x} \Leftarrow -\nabla f(\mathbf{x})$.

Step 2. Line search to find the step size t through either exact or backtracking method.

Step 3. Update $\mathbf{x} \Leftarrow \mathbf{x} + t\Delta\mathbf{x}$.

Step 4. Check the stopping criterion, and quit if necessary.

Constrained Optimization The *interior point methods* are used to solve the convex optimization problem (A.6) or corresponding KKT conditions by applying gradient methods [26] to a sequence of equality constrained problems. There are several interior point metods such as the *barrier*, *penalty* and *primal-dual interior point* methods.

The KKT conditions for linear equality constrained quadratic problems (LECQP) give a set of linear equations, which can be solved analytically. Newton's method is a technique for solving a linear equality constrained optimization (LECP) problem, with twice differentiable objective, by reducing it to a sequence of LECQP. Interior-point methods solve an optimization problem with linear equality and inequality constraints by reducing it to a sequence of LECPs.

Here, we focus on the barrier method and refer the interested reader on *primal-dual interior point method* to [26]. The aim of "logarithmic barrier" method is to reformulate (A.6) such that an equality constrained suitable for gradient method applications is obtained. We rewrite (A.6) by incorporating the inequality constraint with the objective such that we have

$$\min_{\mathbf{x}} \; f_0(\mathbf{x}) + \sum_{i=1}^{m} I_-(f_i(\mathbf{x})) \quad \textbf{subject to} \quad \mathbf{Ax} = \mathbf{b}, \tag{A.31}$$

where $I_- : \mathbb{R} \rightarrow \mathbb{R}$ is the indicator function for the nonpositive real numbers

$$I_-(u) = \begin{cases} 0, & u \leq 0 \\ \infty, & u > 0. \end{cases}$$

Due to its undifferentiability, $I_-(u)$ is approximated with $\hat{I}_-(u) = (-1/z) \cdot \log(-u)$ where z sets the accuracy of the approximation. Substituting $\hat{I}_-(u)$ in (A.31), we obtain

$$\min_{\mathbf{x}} \; f_0(\mathbf{x}) + \frac{-1}{z} \sum_{i=1}^{m} \log(-f_i(\mathbf{x})) \quad \textbf{subject to} \quad \mathbf{Ax} = \mathbf{b}, \tag{A.32}$$

where the cost function is convex since the added term on top of $f_0(\mathbf{x})$ is convex and differentiable. The term $\phi = -\log(-f_i(\mathbf{x}))$ is called the logarithmic barrier of the problem (A.31), and its domain is the set of points that satisfy the inequality constraints of (A.31) strictly. Now, the problem turns out to be an equality constrained problem, ϕ is a smooth approximation of $I_-(u)$, where the approximation improves as $z \to \infty$.

The basic algorithm of barrier method is as follows:

Algorithm 8 Logarithmic Barrier Methods

initiate Strictly feasible **x**, $z := z^{(0)} > 0$, $\mu > 1$, tolerance $\epsilon > 0$.
repeat
Step 1. Centering step: Compute $\mathbf{x}^\star(z)$ by minimizing $zf_0 + \phi$ subject to $\mathbf{Ax} = \mathbf{b}$, starting at **x**.
Step 2. Update $\mathbf{x} \Leftarrow \mathbf{x}^\star(z)$.
Step 3. Check the stopping criterion and quit if $m/z < \epsilon$.
Step 4. Increase $z \Leftarrow \mu z$.

The centering step is performed by using gradient methods, e.g., Newton's method, starting at current $\mathbf{x}$. The choice of the parameter μ involves a trade-off between the inner and outer iterations. When μ is large, it gives rise to inner iterations and decreases the number of outer iterations. It is proposed by [26] that values 10-20 seem to work well. Lastly, there are several heuristic approaches to determine $z^{(0)}$ [26].

A.1.5. Robust Convex Optimization Methodology

We consider a simplified version of the general optimization problem (A.4), in which we further make the instantaneous data explicit in the input of the objective and constraint functions

$$\min_{\mathbf{x}} \; f_0(\mathbf{x}, \mathcal{I}) \quad \textbf{subject to} \; f_i(\mathbf{x}, \mathcal{I}) \leq 0, \;\; i = 1, \ldots, m, \tag{A.33}$$

where $\mathcal{I}$ denotes the set of all input data to the problem. In general, we assume that this data set $\mathcal{I}$ is perfect in the sense that there is no imperfection with it and the optimized solution achieves the optimal result exactly in reality. However, this is not the case necessarily in real-world implementations, where uncertainty may be well introduced to $\mathcal{I}$. As we want our solution to be feasible for any possible realization of the data set $\mathcal{I}$, we should take the complete possible set of realizations for $\mathcal{I}$ into account during the problem formulation. Hence, we defined an uncertainty set $\mathcal{U}$ such that any $\mathcal{I}$ realization is represented within it, i.e., $\forall \mathcal{I} \in \mathcal{U}$.

We define the robust counterpart of the problem (A.33) as

$$\min_{\mathbf{x}} \; f_0(\mathbf{x}, \mathcal{I}) \quad \textbf{subject to} \; f_i(\mathbf{x}, \mathcal{I}) \leq 0, \;\; i = 1, \ldots, m, \;\; \forall \mathcal{I} \in \mathcal{U}. \tag{A.34}$$

At this point, two issues should be addressed properly: the conditions under which we can design *computationally tractable* solutions, i.e., polynomial in time, and the definition of the uncertainty set $\mathcal{U}$. Ben-Tal *et al.* have aimed to clarify these and developed a general understanding on robust counterpart derivation in a series of papers [9–14], where they have considered various types of generic optimization problems, e.g., LP, QP, SDP, and representative definitions for uncertainty sets. Broadly speaking, they have shown that assuming ellipsoidal uncertainty sets, an exact or approximate tractable convex robust counterpart can be designed for almost each of the aforementioned generic convex problems.

As they are frequently use throughout this thesis, in the following we review the robust counterpart derivation for QP problems in details. For further details for other types of uncertainty and optimization problems, we refer to [9–14].

Robust Quadratic Programming We consider an uncertain QCQP problem of

$$\min_{\mathbf{x}} \; \mathbf{c}^T\mathbf{x} \quad \textbf{subject to} \; \mathbf{x}^T\tilde{\mathbf{P}}_i^T\tilde{\mathbf{P}}_i\mathbf{x} \leq \mathbf{q}_i^T\mathbf{x} + r_i, \;\; i = 1, \ldots, m, \;\; \forall (\tilde{\mathbf{P}}_i, \mathbf{q}_i, r_i) \in \mathcal{U}. \tag{A.35}$$

The uncertainty set $\mathcal{U} \triangleq \mathcal{U}_i \times \cdots \times \mathcal{U}_m$ is as if in the case of constraint-wise uncertainty sets and each $\mathcal{U}_i$ is given by a bounded ellipsoid:

$$\mathcal{U}_i \triangleq \left\{ (\tilde{\mathbf{P}}_i, \mathbf{q}_i, r_i) = (\tilde{\mathbf{P}}_{i,0}, \mathbf{q}_{i,0}, r_{i,0}) + \sum_{j=1}^{k} \eta_j (\tilde{\mathbf{P}}_{i,k}, \mathbf{q}_{i,k}, r_{i,k}) \;\middle|\; \boldsymbol{\eta}^T\mathbf{Q}_j\boldsymbol{\eta} \leq 1, j = 1, \ldots, k \right\},$$

where $\mathbf{Q}_j \succeq 0$ and $\sum_{j=1}^{k} \mathbf{Q}_j \succ 0$. It has been shown in [9] that the robust counterpart of (A.35) associated with the uncertainty set $\mathcal{U}$ can be expressed as an SDP problem:

$$\min_{\mathbf{x}, \lambda_i \in \mathbb{R}} \; \mathbf{c}^T\mathbf{x} \quad \textbf{subject to}$$

$$\begin{bmatrix} r_{i,0} + 2\mathbf{x}^T\mathbf{q}_{i,0} - \lambda_i & \frac{r_{i,1}}{2} + \mathbf{x}^T\mathbf{q}_{i,1} & \cdots & \frac{r_{i,k}}{2} + \mathbf{x}^T\mathbf{q}_{i,k} & \mathbf{x}^T\tilde{\mathbf{P}}_{i,0}^T \\ \frac{r_{i,1}}{2} + \mathbf{x}^T\mathbf{q}_{i,1} & \lambda_i & & & \mathbf{x}^T\tilde{\mathbf{P}}_{i,1}^T \\ \vdots & & \lambda_i & & \vdots \\ \frac{r_{i,k}}{2} + \mathbf{x}^T\mathbf{q}_{i,k} & & & \lambda_i & \mathbf{x}^T\tilde{\mathbf{P}}_{i,k}^T \\ \tilde{\mathbf{P}}_{i,0}\mathbf{x} & \tilde{\mathbf{P}}_{i,1}\mathbf{x} & \cdots & \tilde{\mathbf{P}}_{i,k}\mathbf{x} & \mathbf{I} \end{bmatrix} \succeq 0, i = 1, \ldots, m,$$

where $\mathbf{Q}_j'$s are simply chosen to be identity matrix. A very brief sketch of the proof is given as follows. First, the ith constraint is algebraically modeled such that it is expressed as an explicit non-negative quadratic inequality in terms of $\boldsymbol{\eta}$. Then, the non-negativity of this quadratic function is incorporated with an another quadratic function of $\boldsymbol{\eta}$, i.e., $\boldsymbol{\eta}^T\boldsymbol{\eta} \leq 1$, through the S-lemma. Finally, employing the Schur complement, the above given final SDP formulation is obtained. For further details, we refer to [9].

Although QP and SCOP are closely related in certain problems, the extension of the uncertain solution of QP to SOCP is not that trivial. In essence, it is not possible to derive computationally tractable exact robust counterparts for SOCP [12, 14]. The reason is simply that the left and right hand side of the inequality in (A.12) is affected by the same uncertainty, i.e.,

$$\|\mathbf{C}_i\mathbf{x} + \mathbf{f}_i\|_2 \leq \mathbf{w}_i\mathbf{x} + s_i, \quad (\mathbf{C}_i, \mathbf{f}_i, \mathbf{w}_i, s_i) \in \mathcal{U}_i.$$

There, *side-wise uncertainty* is assumed, meaning that the uncertainty affecting each side is assumed to be independent:

$$(\mathbf{C}_i, \mathbf{f}_i) \in \mathcal{U}_i^{\text{left}}, \ (\mathbf{w}_i, s_i) \in \mathcal{U}_i^{\text{right}}, \ \mathcal{U}_i = \mathcal{U}_i^{\text{left}} \times \mathcal{U}_i^{\text{right}}.$$

However, such an assumption disturbs the exactness of the solution and results in a conservative approximate solution.

Finally note that a more generalized ellipsoidal uncertainty set considers the intersection of finitely many ellipsoids. Although such a generalized set leads to a computationally intractable robust counterpart for QP [9], it has been shown in [11] that LP can have a tractable robust counterpart.

A.2. Some Useful Derivative Expressions and Properties

Consider any complex vector $\mathbf{x}$, and real vectors $\mathbf{x}_\mathrm{r}$ and $\mathbf{x}_\mathrm{i}$, such that $\mathbf{x} = \mathbf{x}_\mathrm{r} + j\mathbf{x}_\mathrm{i}$. Then,

$$\frac{\partial f}{\partial \mathbf{x}^T} = \frac{1}{2}\left(\frac{\partial f}{\partial \mathbf{x}_\mathrm{r}^T} - j\frac{\partial f}{\partial \mathbf{x}_\mathrm{i}^T}\right), \tag{A.36}$$

$$\frac{\partial f}{\partial \mathbf{x}^H} = \frac{1}{2}\left(\frac{\partial f}{\partial \mathbf{x}_\mathrm{r}^T} + j\frac{\partial f}{\partial \mathbf{x}_\mathrm{i}^T}\right) \tag{A.37}$$

(See Appendix A in [47]). Through (A.36) and (A.37), we have the relations

$$\frac{\partial f^*}{\partial \mathbf{x}} = \left(\frac{\partial f}{\partial \mathbf{x}^*}\right)^* \text{ and } \frac{\partial f^*}{\partial \mathbf{x}^*} = \left(\frac{\partial f}{\partial \mathbf{x}}\right)^*. \tag{A.38}$$

Some useful examples:

- $$\frac{\partial \mathbf{x}^H \mathbf{Y}\mathbf{x}}{\partial \mathbf{x}} = \frac{\partial(\mathbf{x}_\mathrm{r}^T\mathbf{Y}\mathbf{x}_\mathrm{r} + j\mathbf{x}_\mathrm{r}^T\mathbf{Y}\mathbf{x}_\mathrm{i} - j\mathbf{x}_\mathrm{i}^T\mathbf{Y}\mathbf{x}_\mathrm{r} + \mathbf{x}_\mathrm{i}^T\mathbf{Y}\mathbf{x}_\mathrm{i})}{\partial \mathbf{x}}$$
$$= \frac{1}{2}\Big(\mathbf{Y}^T\mathbf{x}_\mathrm{r} - j\mathbf{Y}^T\mathbf{x}_\mathrm{i}\Big) - \frac{j}{2}\Big(j\mathbf{Y}^T\mathbf{x}_\mathrm{r} + \mathbf{Y}^T\mathbf{x}_\mathrm{i}\Big) = \mathbf{Y}^T\mathbf{x}^*,$$
- Similarly, $\dfrac{\partial \mathbf{x}^H \mathbf{Y}\mathbf{x}}{\partial \mathbf{x}^T} = \mathbf{x}^H\mathbf{Y}$, $\dfrac{\partial \mathbf{x}^H \mathbf{Y}\mathbf{x}}{\partial \mathbf{x}^*} = \mathbf{Y}\mathbf{x}$, $\dfrac{\partial \mathbf{x}^H \mathbf{Y}\mathbf{x}}{\partial \mathbf{x}^H} = \mathbf{x}^T\mathbf{Y}^T$,
- $$\frac{\partial \mathbf{x}^H\mathbf{x}}{\partial \mathbf{x}} = \frac{\partial(\mathbf{x}_\mathrm{r}^T\mathbf{x}_\mathrm{r} + \mathbf{x}_\mathrm{i}^T\mathbf{x}_\mathrm{i})}{\partial \mathbf{x}} = \mathbf{x}_\mathrm{r} - j\mathbf{x}_\mathrm{i} = \mathbf{x}^*,$$
- $$\frac{\partial \mathrm{Re}(\mathbf{y}^T\mathbf{x})}{\partial \mathbf{x}} = \frac{1}{2}\frac{\mathbf{y}^T\mathbf{x}_\mathrm{r} + j\mathbf{y}^T\mathbf{x}_\mathrm{i} + \mathbf{y}^H\mathbf{x}_\mathrm{r} - j\mathbf{y}^H\mathbf{x}}{\partial \mathbf{x}} = \frac{1}{2}\mathbf{x}.$$

A.3. Construction of $\mathbf{N}_\mathsf{r}$ and $\mathbf{R}_{\mathbf{\Lambda},i}$

Using the elements of $\mathbf{n}_\mathsf{r}$ we can construct an $M_\mathsf{r} \times N_\mathsf{c}$ matrix $\mathbf{N}_\mathsf{r}$ such that $\mathbf{N}_\mathsf{r}\mathbf{g} = \mathbf{G}\mathbf{n}_\mathsf{r}$. At first, we index the non-zero entries of $\mathbf{G}$ column by column starting from the first row, e.g.,

$$\text{if } \mathbf{G} = \begin{bmatrix} g_{1,1} & g_{1,2} & 0 \\ g_{2,1} & g_{2,2} & 0 \\ 0 & 0 & g_{3,3}, \end{bmatrix} \text{ then } \mathbf{G}_\mathsf{indexed} = \begin{bmatrix} 1 & 3 & 0 \\ 2 & 4 & 0 \\ 0 & 0 & 5 \end{bmatrix}.$$

Next, we create each row of $\mathbf{N}_\mathsf{r}$ by mapping from the corresponding row of $\mathbf{G}_\mathsf{indexed}$. While creating the kth row, we check the ℓth column of $\mathbf{G}_\mathsf{indexed}$ and place the $n_{\mathsf{r},\ell}$ to the kth row and $(\mathbf{G}_\mathsf{indexed}[k,\ell])$th column of $\mathbf{N}_\mathsf{r}$, where $k = 1, \ldots, M_\mathsf{r}$ and $\ell = 1, \ldots, N_\mathsf{c}$.

For instance,

$$\mathbf{G} = \begin{bmatrix} g_{1,1} & g_{1,2} & 0 \\ g_{2,1} & g_{2,2} & 0 \\ 0 & 0 & g_{3,3} \end{bmatrix} \text{ and } \mathbf{n}_\mathsf{r} = \begin{bmatrix} n_{\mathsf{r},1} \\ n_{\mathsf{r},2} \\ n_{\mathsf{r},3} \end{bmatrix} \Rightarrow \mathbf{N}_\mathsf{r} = \begin{bmatrix} n_{\mathsf{r},1} & 0 & n_{\mathsf{r},2} & 0 & 0 \\ 0 & n_{\mathsf{r},1} & 0 & n_{\mathsf{r},2} & 0 \\ 0 & 0 & 0 & 0 & n_{\mathsf{r},3} \end{bmatrix} \text{ and } \mathbf{g} = \begin{bmatrix} g_{1,1} \\ g_{2,1} \\ g_{1,2} \\ g_{2,2} \\ g_{3,3} \end{bmatrix}.$$

In order to construct $\mathbf{R}_{\mathbf{\Lambda},i}$ for a given i, we benefit from $\mathbf{N}_\mathsf{r}$ above. For each noise term, we build a set of row indices that the noise term is associated within $\mathbf{N}_\mathsf{r}$. For example, for the above $n_{\mathsf{r},1}, n_{\mathsf{r},2}, n_{\mathsf{r},3}$ these sets become, $\mathcal{N}_1 = \{1,2\}, \mathcal{N}_2 = \{1,2\}, \mathcal{N}_3 = \{3\}$, respectively. Next, for each $k \in \{1, \ldots, M_\mathsf{r}\}$, we generate a vector $\mathbf{x}_{\mathsf{n},i,k}$ that is composed of the elements $\tilde{\mathbf{\Lambda}}[i,\ell]$, $\ell \in \mathcal{N}_k$. For example, for $k = 1$, $\mathbf{x}_{\mathsf{n},i,1} = [\tilde{\mathbf{\Lambda}}[i,1]\ \tilde{\mathbf{\Lambda}}[i,2]]^T$, for

$k = 2$, $\mathbf{x}_{\mathsf{n},i,2} = [\tilde{\mathbf{\Lambda}}[i,1]\ \tilde{\mathbf{\Lambda}}[i,2]]^T$, and for $k = 3$, $\mathbf{x}_{\mathsf{n},i,3} = [\tilde{\mathbf{\Lambda}}[i,3]]^T$. Then, $\mathbf{R}_{\mathbf{\Lambda},i}$ is simply written as

$$\mathbf{R}_{\mathbf{\Lambda},i} \triangleq \mathrm{bdiag}\{\mathbf{x}_{\mathsf{n},i,1}\mathbf{x}_{\mathsf{n},i,1}^H, \mathbf{x}_{\mathsf{n},i,2}\mathbf{x}_{\mathsf{n},i,2}^H, \ldots, \mathbf{x}_{\mathsf{n},i,M_\mathsf{r}}\mathbf{x}_{\mathsf{n},i,M_\mathsf{r}}^H\}. \tag{A.39}$$

A.4. An Example of Relay Cooperation

Assume we have $N_\mathsf{r} = 3$ relays, where the first one has $M_{\mathsf{r},1} = 2$ antennas and the rest have a single antenna each. Then, without any relay cooperation, the relay gain matrix $\mathbf{G}$ has a structure of

$$\mathbf{G} = \begin{bmatrix} \clubsuit & \clubsuit & 0 & 0 \\ \clubsuit & \clubsuit & 0 & 0 \\ 0 & 0 & \clubsuit & 0 \\ 0 & 0 & 0 & \clubsuit \end{bmatrix},$$

where $\clubsuit$ denotes a non-zero entry. In the following we show the corresponding changes in $\mathbf{G}$ for two specific examples: 1) The first relay broadcasts its received signal at its first antenna to other relays, 2) The second and the third relays exchange received signals.

$$\mathbf{G}_1 = \begin{bmatrix} \clubsuit & \clubsuit & 0 & 0 \\ \clubsuit & \clubsuit & 0 & 0 \\ \clubsuit & 0 & \clubsuit & 0 \\ \clubsuit & 0 & 0 & \clubsuit \end{bmatrix}, \quad \text{and} \quad \mathbf{G}_2 = \begin{bmatrix} \clubsuit & \clubsuit & 0 & 0 \\ \clubsuit & \clubsuit & 0 & 0 \\ 0 & 0 & \clubsuit & \clubsuit \\ 0 & 0 & \clubsuit & \clubsuit \end{bmatrix}$$

Note that as depicted with $\mathbf{G}_1$, the gain matrix does not need to have a symmetric structure. Moreover, in $\mathbf{G}_2$, after exchange of received signals between two relays, they jointly start to act like a MIMO relay with two antennas. With this, we would like to emphasize the adaptability of the coherent multiuser relaying to heterogeneous nodes within the network.

A.5. Starting Vector Calculation for Single S-D Pair

From (3.36), the optimum vector $\bar{\mathbf{g}}$ is a nonzero eigenvector of $\mathbf{T}_i$; in other words, this eigenvector is a projection of $\mathbf{h}_{ir}$, i.e.,

$$\bar{\mathbf{g}}_{\mathrm{opt}} = c \cdot \mathbf{\Sigma}_{\breve{\mathbf{B}}_i}^{\frac{-1}{2}} \mathbf{U}_{\breve{\mathbf{B}}_i}^H \mathbf{h}_{ir}^*,$$

where c is a positive constant. Assuming that there is no interference, $\breve{\mathbf{B}}_i$ simplifies to

$$\breve{\mathbf{B}}_i = \sigma_{n_\mathsf{r}}^2 \mathbf{I}_{N_\mathsf{r}} + \frac{\sigma_{n_\mathsf{d}}^2}{P_\mathsf{r}} \cdot (\mathbf{\Gamma}_i^{-1})^H \mathbf{M} \mathbf{\Gamma}_i^{-1},$$

and correspondingly, $\mathbf{U}_{\breve{\mathsf{B}}_i} = \mathbf{U}_{\breve{\mathsf{B}}_i}^H = \mathbf{I}_{N_\mathsf{r}}$, and

$$\mathbf{\Sigma}_{\breve{\mathsf{B}}_i} = \sigma_{n_\mathsf{r}}^2 \mathbf{I}_{N_\mathsf{r}} + \sigma_{n_\mathsf{d}}^2 \cdot \mathrm{diag}\left\{ \left[\frac{\mathbf{h}_{\mathsf{s}1}^H \mathbf{h}_{\mathsf{s}1}}{|f_{1\mathsf{d}}^{(i)}|^2 P_\mathsf{r}} \cdots \frac{\mathbf{h}_{\mathsf{s}N_\mathsf{r}}^H \mathbf{h}_{\mathsf{s}N_\mathsf{r}}}{|f_{N_\mathsf{r}\mathsf{d}}^{(i)}|^2 P_\mathsf{r}} \right] \right\}.$$

Hence, we have

$$\bar{\mathbf{g}}_{\mathrm{opt},k} = \frac{h_{\mathsf{s}k}^{(i)*}}{\left(\frac{\sigma_{n_\mathsf{d}}^2 + \sigma_{n_\mathsf{r}}^2 \mathbf{h}_{\mathsf{s}k}^H \mathbf{h}_{\mathsf{s}k}}{P_\mathsf{r} |f_{k\mathsf{d}}^{(i)}|^2} \right)^{\frac{1}{2}}}.$$

Mapping this result to the original physical gain vector $\mathbf{g}_{\mathrm{opt}}$, we obtain

$$\mathbf{g}_{\mathrm{opt}} = c \cdot \mathbf{\Gamma}_i^{-1} \mathbf{U}_{\breve{\mathsf{B}}_i} \mathbf{\Sigma}_{\breve{\mathsf{B}}_i}^{-\frac{1}{2}} \mathbf{\Sigma}_{\breve{\mathsf{B}}_i}^{-\frac{1}{2}} \mathbf{U}_{\breve{\mathsf{B}}_i}^H \mathbf{h}_{ir} = c \cdot (\mathbf{\Gamma}_i^{-1})^H \mathbf{\Sigma}_{\breve{\mathsf{B}}_i}^{-1} \mathbf{h}_{ir}.$$

Finally, extracting the kth element of $\mathbf{g}_{\mathrm{opt}}$, which corresponds to the optimal gain factor for the kth relay in the absence of interference, we conclude our derivation

$$\mathbf{g}_{\mathrm{opt},k} = \frac{h_{\mathsf{s}k}^{(i)*} \cdot f_{k\mathsf{d}}^{(i)*}}{\sigma_{n_\mathsf{r}}^2 |f_{k\mathsf{d}}^{(i)}|^2 + (\sigma_{n_\mathsf{d}}^2 / P_\mathsf{r}) \cdot \mathbf{h}_{\mathsf{s}k}^H \mathbf{h}_{\mathsf{s}k}}.$$

Note that the constant c is decided according to the relay sum transmit power constraint.

A.6. Proof of Lemma 5.2.1

Assume that for given C and N_r, there exists a cluster relay configuration $\tilde{\mathcal{R}} = \{\tilde{N}_{\mathsf{r},1}, \ldots, \tilde{N}_{\mathsf{r},C}\}$ with a total amount of channel use $\sum_{c=1}^{C} f_{\mathsf{cu}}^{\mathsf{u}}(\tilde{N}_{\mathsf{r},c})$, which is smaller than or equal to $\sum_{c=1}^{C} f_{\mathsf{cu}}^{\mathsf{u}}(N_{\mathsf{r},c})$ obtained by the proposed OCC $\mathcal{R} = \{N_{\mathsf{r},1}, \ldots, N_{\mathsf{r},C}\}$. Let $\tilde{N}_{\mathsf{r},c} = N_{\mathsf{r},c} + M_{\mathsf{r},c}$ and $f_{\mathsf{cu}}^{\mathsf{u}}(\tilde{N}_{\mathsf{r},c}) = f_{\mathsf{cu}}^{\mathsf{u}}(N_{\mathsf{r},c}) + \chi_c$ for all $c \in \{1, \ldots, C\}$. Moreover, we define $\gamma_+(N_{\mathsf{r},c}) = 4NN_{\mathsf{r},c}$ and $\gamma_-(N_{\mathsf{r},c}) = 4N(N_{\mathsf{r},c} - 1)$ as functions of $N_{\mathsf{r},c}$ and they are respectively equal to the absolute channel use difference obtained by incrementing or decrementing $N_{\mathsf{r},c}$ by 1. Incorporating these definitions, we can define the following bounds for χ_c:

$$\begin{cases} \chi_c \geq M_{\mathsf{r},c} \gamma_+(n_i), \text{ if } M_{\mathsf{r},c} \geq 0 \\ \chi_c \geq M_{\mathsf{r},c} \gamma_-(n_i), \text{ if } M_{\mathsf{r},c} < 0, \end{cases}$$

for $M_{\mathsf{r},c} \in \mathbb{Z}$. Henceforth, we drop the $(N_{\mathsf{r},c})$ designation from $\gamma_+(N_{\mathsf{r},c})$ and $\gamma_-(N_{\mathsf{r},c})$ for notational simplicity.

Next, we rearrange the total amount of channel use needed for CSI dissemination of the set $\tilde{\mathcal{R}}$ as

$$\begin{aligned}\sum_{c=1}^{C} f^{\mathrm{u}}_{\mathrm{cu}}(\tilde{N}_{\mathsf{r},c}) &= \sum_{c\in\mathcal{M}} f^{\mathrm{u}}_{\mathrm{cu}}(\tilde{N}_{\mathsf{r},c}) + \sum_{c\in\mathcal{M}'} f^{\mathrm{u}}_{\mathrm{cu}}(\tilde{N}_{\mathsf{r},c}),\\ &= \sum_{c\in\mathcal{M}} \Big(f^{\mathrm{u}}_{\mathrm{cu}}(N_{\mathsf{r},c}) + M_{\mathsf{r},c}\gamma_+ + \varepsilon_{+,c}\Big) + \sum_{c\in\mathcal{M}'} \Big(f^{\mathrm{u}}_{\mathrm{cu}}(N_{\mathsf{r},c}) + M_{\mathsf{r},c}\gamma_- + \varepsilon_{-,c}\Big),\end{aligned} \tag{A.40}$$

where $\epsilon_{+,c}, \epsilon_{-,c} \in \mathbb{R}_+\ \forall c$, $\mathcal{M}$ is the index set of clusters with positive $M_{\mathsf{r},c}$, and $\mathcal{M}'$ is the index set of remainder clusters. Further substituting the following trivial equalities

$$\sum_{c=1}^{C} f^{\mathrm{u}}_{\mathrm{cu}}(N_{\mathsf{r},c}) = \sum_{c\in\mathcal{M}} f^{\mathrm{u}}_{\mathrm{cu}}(N_{\mathsf{r},c}) + \sum_{c\in\mathcal{M}'} f^{\mathrm{u}}_{\mathrm{cu}}(N_{\mathsf{r},c}),$$

$$\sum_{c=1}^{C} \varepsilon_c = \sum_{c\in\mathcal{M}} \varepsilon_{+,c} + \sum_{c\in\mathcal{M}'} \varepsilon_{-,c}, \quad \text{for } \epsilon_c \in \mathbb{R}^+\ \forall c,$$

in (A.40), we express the total channel use for $\tilde{\mathcal{R}}$ as the sum of the total channel use for $\mathcal{R}$, additional effects of the bounds on $M_{\mathsf{r},c}$ and the equality satisfying auxiliary ε_c values:

$$\sum_{c=1}^{C} f^{\mathrm{u}}_{\mathrm{cu}}(\tilde{N}_{\mathsf{r},c}) = \sum_{c\in\mathcal{M}} M_{\mathsf{r},c}\gamma_+ + \sum_{c\in\mathcal{M}'} M_{\mathsf{r},c}\gamma_- + \sum_{c=1}^{C} f^{\mathrm{u}}_{\mathrm{cu}}(N_{\mathsf{r},c}) + \sum_{c=1}^{C} \varepsilon_c. \tag{A.41}$$

If we can show that the sum of the first two summands of (A.41) are positive, then we can conclude the proof. Hence, focusing on these two summands in (A.41), we re-write them as

$$\sum_{c\in\mathcal{M}} M_{\mathsf{r},c}\gamma_+ + \sum_{c\in\mathcal{M}'} M_{\mathsf{r},c}\gamma_- = \underbrace{\sum_{c\in\mathcal{M}\cap\mathcal{T}} M_{\mathsf{r},c}\gamma_+ + \sum_{c\in\mathcal{M}\cap\mathcal{T}'} M_{\mathsf{r},c}\gamma_+}_{(1)} + \underbrace{\sum_{c\in\mathcal{M}'\cap\mathcal{T}} M_{\mathsf{r},c}\gamma_- + \sum_{c\in\mathcal{M}'\cap\mathcal{T}'} M_{\mathsf{r},c}\gamma_-}_{(2)} \tag{A.42}$$

where $\mathcal{T}$ and $\mathcal{T}'$ are the index sets of clusters according to the OCC with $\lceil N_{\mathsf{r}}/C \rceil$ and $\lfloor N_{\mathsf{r}}/C \rfloor = \lceil N_{\mathsf{r}}/C \rceil - 1$ relays, respectively. Now, as $\{\mathcal{M}\cap\mathcal{T}\} \cup \{\mathcal{M}\cap\mathcal{T}'\} = \mathcal{M}$, after substituting the definition of γ_+ in (A.42)-(1) and after some algebraic manipulations, we obtain

$$\text{(A.42)} - (1):\ \sum_{c\in\mathcal{M}\cap\mathcal{T}} M_{\mathsf{r},c}\gamma_+ + \sum_{c\in\mathcal{M}\cap\mathcal{T}'} M_{\mathsf{r},c}\gamma_+ = 4N\Bigg(\sum_{c\in\mathcal{M}} M_{\mathsf{r},c}\left\lceil \tfrac{N_{\mathsf{r}}}{C} \right\rceil - \sum_{c\in\mathcal{M}\cap\mathcal{T}'} M_{\mathsf{r},c}\Bigg).$$

With similar reasonings and manipulations, we also have

$$(\text{A.42}) - (2): \quad \sum_{c\in\mathcal{M}'\cap\mathcal{T}} M_{\mathsf{r},c}\gamma_- + \sum_{c\in\mathcal{M}'\cap\mathcal{T}'} M_{\mathsf{r},c}\gamma_- = 4N\Big(\sum_{c\in\mathcal{M}'} M_{\mathsf{r},c}\Big(\Big\lceil\tfrac{N_\mathsf{r}}{C}\Big\rceil - 1\Big) - \sum_{c\in\mathcal{M}'\cap\mathcal{T}'} M_{\mathsf{r},c}\Big).$$

Combining these two equations, and noting that the condition $\sum_{c\in\mathcal{M}} |M_{\mathsf{r},c}| = \sum_{c\in\mathcal{M}'} |M_{\mathsf{r},c}|$ should be satisfied since $\sum_{c=1}^{C} \tilde{N}_{\mathsf{r},c} = \sum_{c=1}^{C} N_{\mathsf{r},c} = N_\mathsf{r}$, we re-write and simplify (A.42) as

$$\begin{aligned}
&\sum_{c\in\mathcal{M}} M_{\mathsf{r},c}\gamma_+ + \sum_{c\in\mathcal{M}'} M_{\mathsf{r},c}\gamma_- = \\
&= 4N\Big(\sum_{c\in\mathcal{M}} M_{\mathsf{r},c}\Big\lceil\tfrac{N_\mathsf{r}}{C}\Big\rceil - \sum_{c\in\mathcal{M}\cap\mathcal{T}'} M_{\mathsf{r},c} + \sum_{c\in\mathcal{M}'} M_{\mathsf{r},c}\Big(\Big\lceil\tfrac{N_\mathsf{r}}{C}\Big\rceil - 1\Big) - \sum_{c\in\mathcal{M}'\cap\mathcal{T}'} M_{\mathsf{r},c}\Big), \\
&= 4N\Big(-\sum_{c\in\mathcal{M}\cap\mathcal{T}'} M_{\mathsf{r},c} - \sum_{c\in\mathcal{M}'} M_{\mathsf{r},c} - \sum_{c\in\mathcal{M}'\cap\mathcal{T}'} M_{\mathsf{r},c}\Big), && \text{(A.43)} \\
&= 4N\Big(\sum_{c\in\mathcal{M}} |M_{\mathsf{r},c}| - \sum_{c\in\mathcal{M}\cap\mathcal{T}'} |M_{\mathsf{r},c}| + \sum_{c\in\mathcal{M}'\cap\mathcal{T}'} |M_{\mathsf{r},c}|\Big), && \text{(A.44)} \\
&= 4N\Big(\sum_{c\in\mathcal{M}\cap\mathcal{T}} |M_{\mathsf{r},c}| + \sum_{c\in\mathcal{M}'\cap\mathcal{T}'} |M_{\mathsf{r},c}|\Big), && \text{(A.45)}
\end{aligned}$$

where (A.43) follows from

$$\sum_{c\in\mathcal{M}} M_{\mathsf{r},c}\Big\lceil\tfrac{N_\mathsf{r}}{C}\Big\rceil + \sum_{c\in\mathcal{M}'} M_{\mathsf{r},c}\Big\lceil\tfrac{N_\mathsf{r}}{C}\Big\rceil = \Big\lceil\tfrac{N_\mathsf{r}}{C}\Big\rceil\Big(\sum_{c\in\mathcal{M}} |M_{\mathsf{r},c}| - \sum_{c\in\mathcal{M}'} |M_{\mathsf{r},c}|\Big) = 0.$$

Moreover, in (A.44) we introduced the absolute values, and finally in (A.45) we used the equality of $\sum_{c\in\mathcal{M}} |M_{\mathsf{r},c}| - \sum_{c\in\mathcal{M}\cap\mathcal{T}'} |M_{\mathsf{r},c}| = \sum_{c\in\mathcal{M}\cap\mathcal{T}} |M_{\mathsf{r},c}|$. As (A.45) results in a positive real value, we obtain that $\sum_{c=1}^{C} f_{\mathrm{cu}}^{\mathrm{u}}(\tilde{N}_{\mathsf{r},c}) > \sum_{i=1}^{C} f_{\mathrm{cu}}^{\mathrm{u}}(N_{\mathsf{r},c})$, which contradicts with our assumption that there exists an $\tilde{\mathcal{R}}$ which has a smaller or equal total number required channel use for CSI dissemination than $\mathcal{R}$. □

A.7. Proof of Lemma 5.2.2

The proof follows from the definition of $f_{\mathrm{cu}}^{\mathrm{u}}$. Let $\mathcal{R} = \{N_{\mathsf{r},1}, \ldots, N_{\mathsf{r},C}\}$ and $\hat{\mathcal{R}} = \{\hat{N}_{\mathsf{r},1}, \ldots, \hat{N}_{\mathsf{r},C+\hat{v}}\}$ be the OCC of networks with respectively C and $C + \hat{v}$ clusters. It is claimed that

$$\sum_{c\in\mathcal{Q}}^{C} f_{\mathrm{cu}}^{\mathrm{u}}(N_{\mathsf{r},c}) + \sum_{i\in\mathcal{Q}'}^{C} f_{\mathrm{cu}}^{\mathrm{u}}(N_{\mathsf{r},c}) > \sum_{i\in\mathcal{V}}^{C+\hat{v}} f_{\mathrm{cu}}^{\mathrm{u}}(\hat{N}_{\mathsf{r},c}) + \sum_{c\in\mathcal{V}'}^{C+\hat{v}} f_{\mathrm{cu}}^{\mathrm{u}}(\hat{N}_{\mathsf{r},c}), \qquad \text{(A.46)}$$

where $\mathcal{Q}, \mathcal{Q}', \mathcal{V}, \mathcal{V}'$ stand for the index set of clusters with $\hat{\eta}_1 = \lceil N_r/C \rceil$, $\hat{\eta}_2 = \lfloor N_r/C \rfloor$, $\hat{\eta}_3 = \lceil N_r/(C+\hat{v}) \rceil$, and $\hat{\eta}_4 = \lfloor N_r/(C+\hat{v}) \rfloor$ relays, respectively. Defining $\hat{\lambda}_1 = \text{mod}(N_r, C)$, $\hat{\lambda}_2 = C - \text{mod}(N_r, C)$, $\hat{\lambda}_3 = \text{mod}(N_r, C+\hat{v})$, $\hat{\lambda}_4 = C+\hat{v} - \text{mod}(N_r, C+\hat{v})$, the inequality (A.46) simplifies down to

$$\hat{\eta}_2(N_r - \hat{\lambda}_2) > \hat{\eta}_4(N_r - \hat{\lambda}_4). \tag{A.47}$$

The inequality (A.47) can be further modified to the following expression after some algebraic manipulations and setting $\hat{v} = 1$ with loss of generality,

$$\begin{aligned}\left(\left\lfloor \tfrac{N_r}{C} \right\rfloor - \left\lfloor \tfrac{N_r}{C+1} \right\rfloor\right)(2N_r - C) > C\left(\left\lfloor \tfrac{N_r}{C} \right\rfloor - \left\lfloor \tfrac{N_r}{C+1} \right\rfloor\right)\left(\left\lfloor \tfrac{N_r}{C} \right\rfloor + \left\lfloor \tfrac{N_r}{C+1} \right\rfloor\right) \\ - \left\lfloor \tfrac{N_r}{C+1} \right\rfloor\left(\left\lfloor \tfrac{N_r}{C+1} \right\rfloor + 1\right).\end{aligned} \tag{A.48}$$

Now, we focus on two different cases separately: $C > N_r/2$ and $C \leq N_r/2$.

i) The case of $C > N_r/2$*:* For any $C > N_r/2$, the difference $\lfloor N_r/(C+1) \rfloor - \lfloor N_r/C \rfloor$ becomes zero. Hence, (A.48) boils down to be $\lfloor N_r/(C+1) \rfloor \left(\lfloor N_r/(C+1) \rfloor + 1\right) > 0$, which always holds since $\lfloor N_r/(C+1) \rfloor \geq 1$.

ii) The case of $C \leq N_r/2$*:* Here, we again consider two sub-cases separately: $C \leq N_r/2$ that leads $\lfloor N/C \rfloor = \lfloor N/(C+1) \rfloor$ and $C \leq N_r/2$ that leads $\lfloor N/C \rfloor \neq \lfloor N/(C+1) \rfloor$. For the former sub-case, it can be trivially shown using the similar reasonings in the case above, that (A.48) holds true.

On other hand, in order to prove the theorem for any $C \leq N_r/2$ resulting in $\lfloor N/C \rfloor \neq \lfloor N/(C+1) \rfloor$, we algebraically manipulate (A.48) and obtain equivalently

$$\frac{2N_r}{C} - 1 > \left(\left\lfloor \tfrac{N_r}{C} \right\rfloor + \left\lfloor \tfrac{N_r}{C+1} \right\rfloor\right) - \frac{\left\lfloor \tfrac{N_r}{C+1} \right\rfloor\left(\left\lfloor \tfrac{N_r}{C+1} \right\rfloor + 1\right)}{C\left(\left\lfloor \tfrac{N_r}{C} \right\rfloor - \left\lfloor \tfrac{N_r}{C+1} \right\rfloor\right)}. \tag{A.49}$$

To prove that (A.49) holds true, it is sufficient to show that

$$\frac{2N_r}{C} - 1 \geq \left(\left\lfloor \tfrac{N_r}{C} \right\rfloor + \left\lfloor \tfrac{N_r}{C+1} \right\rfloor\right), \tag{A.50}$$

as the latter term on the right hand side of (A.50) is always larger than zero. Redefining $\lfloor N_r/C \rfloor$ and $\lfloor N_r/(C+1) \rfloor$ as $N_r/C - \hat{\epsilon}_1$ and $N_r/(C+1) - \hat{\epsilon}_2$, $0 \leq \hat{\epsilon}_1 < 1$, $0 \leq \hat{\epsilon}_2 < 1$, respectively, (A.50) becomes

$$\frac{N_r}{C} - \frac{N_r}{C+1} \geq 1 - \hat{\epsilon}_1 - \hat{\epsilon}_2, \tag{A.51}$$

which can be shown to hold true immediately, and this concludes the proof of the inequality (A.46) for $\hat{v} = 1$. The proof for any $\hat{v} > 1$ can written trivially by induction. □

A.8. Proof for Rate Performance of Optimum Cluster Configuration

As the proof has similarities with the proof of lemma 5.2.1, in the following we present only a short sketch by dropping the details.

Assume that for given C and N_r, there exists a cluster relay configuration $\tilde{\mathcal{R}} = \{\tilde{N}_{\mathsf{r},1}, \ldots, \tilde{N}_{\mathsf{r},C}\}$ with $\sum_{c=1}^{C} f_{\text{rate}}(\tilde{N}_{\mathsf{r},c})$ which is larger than $\sum_{c=1}^{C} f_{\text{rate}}(N_{\mathsf{r},c})$ obtained by the optimum cluster configuration $\mathcal{C} = \{N_{\mathsf{r},1}, \ldots, N_{\mathsf{r},C}\}$. Note that the term $1/(2C)$ is dropped for notational simplicity. Let $\tilde{N}_{\mathsf{r},c} = N_{\mathsf{r},c} + M_{\mathsf{r},c}$ and $f_{\text{rate}}(\tilde{N}_{\mathsf{r},c}) = f_{\text{rate}}(N_{\mathsf{r},c}) + M_{\mathsf{r},c} \cdot \chi_c$ for all $c \in \{1, \ldots, C\}$. The value χ_c is bounded as

$$\begin{cases} \chi_c \leq M_{\mathsf{r},c}\tilde{\lambda}_+, \text{ if } M_{\mathsf{r},c} \geq 0 \\ \chi_c \leq M_{\mathsf{r},c}\tilde{\lambda}_-, \text{ if } M_{\mathsf{r},c} < 0, \end{cases}$$

for $M_{\mathsf{r},c} \in \mathbb{Z}$, where $\tilde{\lambda}_+ = f_{\text{rate}}(N_{\mathsf{r},c} + 1) - f_{\text{rate}}(N_{\mathsf{r},c})$, $\tilde{\lambda}_- = f_{\text{rate}}(N_{\mathsf{r},c}) - f_{\text{rate}}(N_{\mathsf{r},c} - 1)$, and $\lambda_1 \leq \lambda_2$. The last condition for λ_1, λ_2 follows from the claim of sublinearity.

Next, following similar steps as for lemma 5.2.1, we obtain

$$\sum_{c=1}^{C} f_{\text{rate}}(\tilde{N}_{\mathsf{r},c}) = \sum_{c \in \mathcal{M}} M_{\mathsf{r},c}\tilde{\lambda}_+ + \sum_{c \in \mathcal{M}'} M_{\mathsf{r},c}\tilde{\lambda}_- + \sum_{c=1}^{C} f_{\text{rate}}(N_{\mathsf{r},c}) - \sum_{c=1}^{C} \varepsilon_c, \quad \text{(A.52)}$$

where $\varepsilon_c \in \mathbb{R}_+ \; \forall c$. The sum of the first two summands at the right hand side of (A.52) are equal to

$$\sum_{c \in \mathcal{P} \cap \mathcal{T}} |M_{\mathsf{r},c}|\tilde{\lambda}_+ + \sum_{c \in \mathcal{M} \cap \mathcal{T}'} |M_{\mathsf{r},c}|\breve{\lambda}_+ - \sum_{c \in \mathcal{M}' \cap \mathcal{T}} |M_{\mathsf{r},c}|\tilde{\lambda}_- - \sum_{c \in \mathcal{M}' \cap \mathcal{T}'} |M_{\mathsf{r},c}|\breve{\lambda}_- \quad \text{(A.53)}$$

where $\breve{\lambda}_+ = f_{\text{rate}}(N_{\mathsf{r},c}) - f_{\text{rate}}(N_{\mathsf{r},c} - 1)$ and $\breve{\lambda}_- = f_{\text{rate}}(N_{\mathsf{r},c} - 1) - f_{\text{rate}}(N_{\mathsf{r},c} - 2)$. Since $\tilde{\lambda}_+ \leq \breve{\lambda}_+ = \tilde{\lambda}_- \leq \breve{\lambda}_-$, (A.53) can be re-written as

$$\sum_{c \in \mathcal{P}} M_{\mathsf{r},c}\tilde{\lambda}_+ + \sum_{c \in \mathcal{P}'} M_{\mathsf{r},c}\tilde{\lambda}_- = \tilde{\kappa}\Big(- \sum_{c \in \mathcal{M} \cap \mathcal{T}} |M_{\mathsf{r},c}| - \sum_{c \in \mathcal{M}' \cap \mathcal{T}'} |M_{\mathsf{r},c}|\Big) \leq 0 \quad \text{(A.54)}$$

for $\tilde{\kappa} \in \mathbb{R}_+$. Combining (A.52) and (A.54), we have $\sum_{c=1}^{C} f_{\text{rate}}(\tilde{N}_{\mathsf{r},c}) \leq \sum_{c=1}^{C} f_{\text{rate}}(N_{\mathsf{r},c})$, which contradicts with our assumption on the existence of $\tilde{\mathcal{R}}$. This concludes the proof of our claim that the OCC maximizes the upper bound of average information rate performance. Our claim will be further validated through computer simulations in the following performance evaluation section.

Notation and Symbols

Notation

x	scalar
$\mathbf{x}$	vector
$\mathbf{X}$	matrix
$\mathbf{X}^T$	matrix transpose of $\mathbf{X}$
$\mathbf{X}^*$	complex conjugate of $\mathbf{X}$
$\mathbf{X}^H$	complex conjugate transpose of $\mathbf{X}$
$\mathbf{X}^\dagger$	pseudo inverse of $\mathbf{X}$
$\mathbb{R}$	set of real numbers
$\mathbb{C}$	set of complex numbers
$\mathbb{Z}$	set of integers
$\mathbb{R}_+$	set of non-negative real numbers
$\mathbb{Z}_+$	set of non-negative integers
$\mathbb{R}^{n\times n}$	$n \times n$ matrix of real numbers
$\mathbf{S}_+^n$	set of $n \times n$ positive semidefinite matrices
$\mathbb{C}^{n\times n}$	$n \times n$ matrix of complex numbers
$\mathbb{E}\{x\}$	expectation of x
$\text{diag}\{\mathbf{x}\}$	a diagonal matrix with $\mathbf{x}$ on its diagonal
$\text{bdiag}\{\mathbf{X}_1, \ldots, \mathbf{X}_n\}$	a block-diagonal matrix with $\mathbf{X}_1, \ldots, \mathbf{X}_n$ on its diagonal
$\text{Re}\{x\}$	real part of x
$\text{Im}\{x\}$	imaginary part of x
$\mathbf{x}[i]$	the ith element of $\mathbf{x}$
$\mathbf{X}[i,j]$	the (i,j)th element of $\mathbf{X}$

$\mathbf{X}[i,:]$	the ith row of $\mathbf{X}$
$\mathbf{X}[:,i]$	the ith column of $\mathbf{X}$
$\|x\|$	absolute value of x
$\|\mathbf{X}\|$	determinant of X
$\|\mathcal{S}\|$	cardinality of set $\mathcal{S}$
$\\| \mathbf{x} \\|$	Euclidean norm of vector $\mathbf{x}$
$\mathbf{e}_i$	all-zero vector with a single one in the ith position
$\mathbf{E}_{i,j}$	all-zero matrix with a single one at position $[j,i]$
$\mathbf{I}_n$	$n-$dimensional identity matrix
$(\cdot)^\star$	the argument is the optimal value of an optimization problem
$\mathrm{Tr}(\mathbf{X})$	trace of $\mathbf{X}$
$\mathrm{rank}(\mathbf{X})$	rank of $\mathbf{X}$
$\mathrm{null}(\mathbf{X})$	nullspace of $\mathbf{X}$
$\mathrm{vec}(\mathbf{X})$	vector generated by the concatenation of the columns of $\mathbf{X}$
$\mathbf{max}\{x_1,\ldots,x_n\}$	maximum of $x_1,\ldots,x_n$
$\mathbf{min}\{x_1,\ldots,x_n\}$	minimum of $x_1,\ldots,x_n$
$\tan(\phi)$	tangent of the angle ϕ
$\oplus$	exclusive OR
$\odot$	element-wise multiplication
$\otimes$	Kronecker product
$\succeq$	generalized inequality ($\mathbf{X} \succeq 0$ indicates positive semi-definiteness)
$\equiv$	notation of equivalence
$\mathcal{N}(\mu,\sigma^2)$	Gaussian distribution with mean μ and variance σ^2
$\mathcal{CN}(\mu,\sigma^2)$	complex Gaussian distribution with mean μ and variance σ^2
$\nabla_{\mathbf{x}} f$	gradient of function f with respect to $\mathbf{x}$
$f:\mathbb{R}^n \rightarrow \mathbb{R}^m$	function f takes as a real $n-$vector as argument and returns a real $m-$vector
$f:\mathbb{C}^n \rightarrow \mathbb{R}^m$	function f takes as a complex $n-$vector as argument and returns a real $m-$vector

Symbols - Part 1

$\mathbf{D}$	compound source-to-destination direct-link channel matrix
$\mathbf{F}$	compound relay-to-destination channel matrix

$\mathbf{F}_k$	channel matrix between the kth relay and all destinations
$\mathbf{F}^{(c)}$	channel matrix between the cth relay cluster and all destinations
$\mathbf{G}$	compound relay gain matrix
$\mathbf{G}_k$	relay gain matrix of the kth relay
$\mathbf{H}$	compound source-to-relay channel matrix
$\mathbf{H}_k$	channel matrix between all sources and the kth relay
$\mathbf{H}_{\mathsf{srd}}$	equivalent two-hop channel matrix
$\tilde{\mathbf{H}}_{\mathsf{srd}}$	effective equivalent two-hop channel matrix
$\mathbf{H}^{(c)}$	channel matrix between all sources and the cth relay cluster
$\mathbf{\Phi}_{\mathsf{x}}$	LO phase offset matrix of node x
$\mathbf{\Psi}_c$	phase rotation matrix for the cth cluster
$\mathbf{\Psi}$	compound relay phase rotation matrix
$\mathbf{Q}_{\mathsf{ir}}$	channel matrix between interferers and relays
$\mathbf{Q}_{\mathsf{id}}$	channel matrix between interferers and destinations
$\mathbf{\Theta}_{\mathsf{r}}$	matrix of relays' LO phase offset variation
$\mathbf{\Theta}_{\mathsf{s}}$	matrix of sources' LO phase offset variation
$\mathbf{V}$	nullspace of compound interference matrix
$\mathbf{Z}$	compound channel matrix
$\mathbf{Z}_{\mathcal{I}}$	compound interference matrix
$\mathbf{f}_{k\mathsf{d}}$	vector of channel coefficients from the kth relay to destinations
$\mathbf{f}_{\mathsf{r}i}$	vector of channel coefficients from all relays to the ith destination
$\mathbf{g}$	relay gain vector
$\mathbf{g}_{\mathsf{v}}$	reduced relay gain vector in the nullspace
$\mathbf{g}_{\mathsf{oci}}$	relay gain vector with out-of-cluster interference
$\mathbf{g}_{\mathsf{zf}}$	zero-forcing relay gain vector
$\mathbf{g}_{\mathsf{zf-dl}}$	zero-forcing relay gain vector with direct S-D links
$\mathbf{h}_{i\mathsf{r}}$	vector of channel coefficients from the ith source to all relays
$\mathbf{h}_{\mathsf{s}k}$	vector of channel coefficients from all source to the kth relay
$\mathbf{h}_{\mathsf{srd}}$	vectorized representation of $\mathbf{H}_{\mathsf{srd}}$
$\mathbf{h}_{\mathsf{srd},i}$	equivalent channel vector for the ith destination
$\mathbf{n}_{\mathsf{d}}$	destination noise vector vector
$\mathbf{n}_{\mathsf{e}}$	equivalent two-hop noise vector
$\mathbf{n}_{\mathsf{r}}$	relay noise vector vector
$\mathbf{p}$	relay transmit power vector vector
$\mathbf{r}$	compound relays transmit signal vector

$\mathbf{s}$	compound source transmit signal vector
$\boldsymbol{\theta}$	random phase uncertainty vector at the relays
$\mathbf{u}$	interference signal vector
$\mathbf{w}$	normalized relay gain vector
$\mathbf{y}_{\mathsf{d}}$	compound received signal vector at the destinations
$\mathbf{y}_{\mathsf{r}}$	compound received signal vector at the relays
$\mathbf{y}_{\mathsf{r},c}$	compound received signal vector at the cth relay cluster
C	number of relay clusters
$\mathcal{C}_{\text{QoS}}^{(i)}$	QoS constraint for the ith S-D link
$\mathcal{C}_{\text{SINR}}^{(i)}$	SINR constraint for the ith S-D link
$\mathcal{C}_{\text{power}}$	relay transmit power constraint
$\mathscr{C}_c$	the cth relay cluster
$\mathscr{C}_{\text{global}}$	global relay cluster
$\mathscr{C}_{\text{local}}$	local relay cluster
d_i	the ith destination terminal
$\mathcal{F}_{\text{mm}}$	feasibility check for max-min problem
$\mathcal{F}_{\text{mm-zf}}$	feasibility check for max-min problem with zero-forcing
f_{x}	carrier frequency of node x
$f_{i,k}^{(m)}$	the mth channel coefficient of relay k–destination i link
g_k	relay gain coefficient of the kth relay
$g_{\mathbf{azf},k}$	asymptotic zero-forcing complex relay gain factor of relay k
$\mathcal{H}$	channel coefficients set over the whole network
$h_{\mathsf{srd},i}^{(j)}$	the jth element of $\mathbf{h}_{\mathsf{srd},i}$
$h_{k,i}^{(m)}$	the mth channel coefficient of source i–relay k link
M_{r_k}	number of antennas at the kth relay terminal
M_{r}	total number of relay antennas
N	number of source-destination pairs
N_{c}	number of degrees of freedom
N_{it}	number of outer-iterations for gradient algorithm
N_{r}	number of relays
$N_{\mathsf{r},c}$	number of relays in the cth cluster
$N_{\mathsf{r,ex}}$	excess number of relays
$n_{\mathsf{d},i}$	noise at the ith destination
$\phi_{\mathsf{x}}^{(t)}$	LO phase offset of node x with respect to a global phase reference at time t

$\mathcal{P}_{\text{csi}}$	robust optimization problem against noisy CSI
$\mathcal{P}_{\text{csi}}^{\text{down}}$	robust optimization problem against noisy CSI in the downlink
$\mathcal{P}_{\text{csi}}^{\text{up}}$	robust optimization problem against noisy CSI in the uplink
$\mathcal{P}_{\text{mm}}$	minimum link rate maximization problem
$\mathcal{P}_{\text{mm}}^{\text{cluster}}$	minimum link rate maximization problem for clustered network
$\mathcal{P}_{\text{mm-zf}}$	minimum link rate maximization problem with zero-forcing
$\mathcal{P}_{\text{mm-zf}}^{\text{cluster}}$	minimum link rate maximization problem with zero-forcing for clustered network
$\mathcal{P}_{\text{mp}}$	relay sum transmit power minimization problem
$\mathcal{P}_{\text{pr}}$	robust optimization problem against phase uncertainty
$\mathcal{P}_{\text{robust}}^{\text{fb}}$	robust optimization problem against uncertain cluster feedback
$\mathcal{P}_{\text{sum}}$	sum rate maximization problem
$\mathcal{P}_{\text{sum}}^{\text{cluster}}$	sum rate maximization problem for clustered network
$\mathcal{P}_{\text{sum-zf}}$	sum rate maximization problem with zero-forcing
$\mathcal{P}_{\text{sum-zf}}^{\text{cluster}}$	sum rate maximization problem with zero-forcing for clustered network
P_{outage}	outage probability
$P_{\text{outage}}^{\text{mm}}$	outage probability with max-min link-rate problem
$P_{\text{outage}}^{\text{rs}}$	outage probability with relay selection
P_{r}	relay sum transmit power
$P_{\mathsf{r},c}$	transmit power of the cth relay cluster
$P_{\mathsf{r},k}$	transmit power of relay k
P_{s}	source transmit power
$\mathcal{R}$	relay set over the whole network
P_{global}	sum transmit power of global relay cluster
R_i	achievable rate for the ith S-D link
P_{local}	sum transmit power of local relay cluster
R_{min}	minimum link-rate over all S-D links
$P_{\text{local},k}$	transmit power of the kth relay in local cluster
R_{outage}	outage rate
r_k	the kth relay terminal
R_{sum}	achievable sum rate of all S-D links
s_i	the ith source terminal
s_i	the ith source terminal's symbol
σ_p^2	variance of phase noise
$\sigma_{n_\mathsf{r}}^2$	variance of relay noise
$\sigma_{n_\mathsf{d}}^2$	variance of destination noise

$\tilde{\theta}_{\mathsf{x}}^{(t)}$	random phase offset induced by phase noise at node x and time t
$\sigma_{\mathbf{D}}^2$	variance of elements of $\mathbf{D}$
$\sigma_{\mathbf{F}}^2$	variance of elements of $\mathbf{F}$
$\sigma_{\mathbf{H}}^2$	variance of elements of $\mathbf{H}$
SINR_i	SINR observed at the ith destination
$\mathrm{SINR}_i^{\mathtt{wi-dl}}$	SINR observed at the ith destination with direct S-D links
$\mathcal{U}_{\mathbf{X}}$	spherical uncertainty set for vector $\mathbf{x}$
$y_{\mathsf{d},i}$	received signal at the ith destination
w_k	normalized relay gain coefficient of the kth relay
$\mathcal{Z}_i$	interference index set related with the ith destination

Symbols - Part 2

$\vec{\mathbf{G}}_i$	the ith pair's interference matrix for one-way relaying
$\overleftrightarrow{\mathbf{G}}_i$	the ith pair's interference matrix for two-way relaying
$\mathbf{H}_{\mathsf{n}}$	channel matrix from terminal n to relay
$\overleftrightarrow{\mathbf{V}}_i$	the ith pair's nullspace matrix for two-way relaying
$\vec{\mathbf{W}}_i$	precoding matrix for the ith pair in multiuser one-way relaying
$\overleftrightarrow{\mathbf{W}}_i$	precoding matrix for the ith pair in multiuser two-way relaying
$\mathbf{W}_{\mathsf{n}}$	precoding matrix for terminal n with SPC at relay
$\mathbf{\Lambda}_{\mathsf{n}}$	signal covariance matrix of terminal n
$\mathbf{\Lambda}_{\mathsf{R}}$	signal covariance matrix of relay with XoR precoding
$\mathbf{\Lambda}_{\mathsf{R}_i}$	relay's signal covariance matrix for the ith pair
$\mathbf{\Sigma}_i$	singular value matrix for the ith pair
$\boldsymbol{\eta}_{\mathsf{n}}$	channel uncertainty vector for terminal n
$\mathbf{h}_{\mathsf{n}}$	channel vector from terminal n to relay
$\mathbf{n}_{\mathsf{R}}$	relay noise vector
$\mathbf{n}_{\mathsf{n}}$	noise vector at terminal n
$\boldsymbol{\phi}$	compound angle vector between rate dimensions
$\mathbf{r}_{\mathsf{R}}$	received signal vector at relay
$\mathbf{r}_{\mathsf{n}}$	received signal vector at terminal n
$\mathbf{s}_{\mathsf{n}}$	transmit symbol vector of terminal n
$\mathbf{s}_{\mathsf{Rn}}$	relay's SPC transmit symbol vector to terminal n
$\mathbf{s}_{\mathsf{R}_i}$	relay's transmit symbol vector to the ith pair n
$\mathbf{x}_{\mathsf{n}}$	transmit bit sequence of terminal n

$\mathbf{x}_{\mathsf{R}}$	transmit bit sequence of relay
A	terminal A
$\mathcal{A}$	set of terminal As
α_i	the ith time-share parameter for one-way relaying
α_{BC}	time-share parameter for BC phase in two-way relaying
B	terminal B
α_{MAC}	time-share parameter for MAC phase in two-way relaying
$\mathcal{B}$	set of terminal Bs
$\mathcal{C}_{\mathrm{BC}}^{\mathrm{mu-one}}$	capacity region of BC phase in multiuser one-way relaying
$\mathcal{C}_{\mathrm{BC}}^{\mathrm{mu-two}}$	capacity region of BC phase in multiuser two-way relaying
$\mathcal{C}_{\mathrm{BC}}^{(a,b)}$	capacity region of BC phase for dimensions a and b
$\mathcal{C}_{\mathrm{BC}}$	capacity region of BC phase for single pair
$\mathcal{C}_{\mathrm{BC}}^{\mathrm{SPC}}$	capacity region of BC phase for single pair with SPC
$\mathcal{C}_{\mathrm{BC}}^{\mathrm{wc}}$	worst-case capacity region of BC phase
$\mathcal{C}_{\mathrm{MAC}}^{\mathrm{mu-one}}$	capacity region of MAC phase in multiuser one-way relaying
$\mathcal{C}_{\mathrm{MAC}}^{\mathrm{mu-two}}$	capacity region of MAC phase in multiuser two-way relaying
$\mathcal{C}_{\mathrm{MAC}}^{(a,b)}$	capacity region of MAC phase for dimensions a and b
$\mathcal{C}_{\mathrm{MAC}}$	capacity region of MAC phase for single pair
$\mathcal{C}_{\mathrm{mu}}^{\mathrm{one-way}}$	overall capacity region of multiuser one-way relaying
$\mathcal{C}_{\mathrm{mu}}^{\mathrm{two-way}}$	overall capacity region of multiuser two-way relaying
$\mathcal{C}_{\mathrm{one-way}}$	overall capacity region of one-way relaying
$\mathcal{C}_{\mathrm{two-way}}$	overall capacity region of two-way relaying
$\mathcal{C}_{\mathrm{two-way}}^{\mathrm{SPC}}$	overall capacity region of two-way relaying with SPC
$\overrightarrow{\Gamma}_{\mathsf{Rn}}$	mutual information from relay to terminal n (one-way relaying)
$\overrightarrow{\Gamma}_{\mathsf{nR}}$	mutual information from terminal n to relay (one-way relaying)
$\overleftrightarrow{\Gamma}_{\mathsf{Rn}}$	mutual information from relay to terminal n (two-way relaying)
$\overleftrightarrow{\Gamma}_{\mathsf{nR}}$	mutual information from terminal n to relay (two-way relaying)
$\overrightarrow{\Gamma}_{\mathsf{AB},\mathrm{MAC}}$	MAC sum upper bound at relay (one-way relaying)
$\overleftrightarrow{\Gamma}_{\mathsf{AB},\mathrm{MAC}}$	MAC sum upper bound at relay (two-way relaying)
K	number of terminal pairs
N_{n}	number of antennas at terminal n
N_{R}	number of antennas at relay
P_i	the ith pair
P_{R}	transmit power of relay
P_{n}	transmit power of terminal n

$\phi_{a,b}$	angle between dimensions a and b
$\mathcal{P}^{\text{sum}}_{\text{one-way}}$	sum rate maximization problem for one-way relaying
$\mathcal{P}^{\text{sum}}_{\text{two-way}}$	sum rate maximization problem for two-way relaying
$\mathcal{P}^{\text{sum},i}_{\text{two-way}}$	sum rate maximization sub-problem for two-way relaying
$\mathcal{P}^{\text{mm}}_{\text{two-way}}$	max-min problem for two-way relaying
$\mathcal{P}_{\text{MAC},\mathsf{n}}$	rate maximization problem for terminal n within MAC
$\mathcal{P}_{\text{MAC},\mathsf{AB}}$	link-rate constrained MAC sum rate maximization problem
$\mathcal{P}_{\text{BC},\mathsf{n}}$	BC rate maximization problem for terminal n
$\text{R}^{\leftrightarrow}_{\mathsf{Rn}}$	achievable rate from relay to terminal n (two-way relaying)
$\text{R}^{\rightarrow}_{\mathsf{Rn}}$	achievable rate from relay to terminal n (one-way relaying)
$\text{R}^{\rightarrow}_{\mathsf{nR}}$	achievable rate from terminal n to relay (one-way relaying)
$\text{R}^{\leftrightarrow\text{wc}}_{\mathsf{Rn}}$	worst-case achievable rate from relay to terminal n
$\text{R}^{\leftrightarrow}_{\mathsf{nR}}$	achievable rate from terminal n to relay (two-way relaying)
$\text{R}^{\rightarrow}_{\mathsf{AB},\text{MAC}}$	achievable MAC sum rate at relay (one-way relaying)
$\text{R}^{\rightarrow}_{\mathsf{mn}}$	achievable rate from terminal m to n (one-way relaying)
$\text{R}^{\leftrightarrow}_{\mathsf{mn}}$	achievable rate from terminal m to n (two-way relaying)
$\text{R}^{\leftrightarrow}_{\mathsf{AB},\text{MAC}}$	achievable MAC sum rate at relay (two-way relaying)
$\text{R}^{\text{sum}}_{\text{two-way}}$	achievable sum rate for single pair two-way relaying
$\text{R}^{\text{mu-one}}_{\text{sum}}$	achievable sum rate of multiuser one-way relaying
$\text{R}^{\text{mu-two}}_{\text{sum}}$	achievable sum rate of multiuser two-way relaying
$\vec{\mathcal{R}}$	achievable rate tuple for multiuser one-way relaying
$\overleftrightarrow{\mathcal{R}}$	achievable rate tuple for multiuser two-way relaying
ρ_{n}	size of channel uncertainty for terminal n
σ^2_{n}	noise variance at teminal n
σ^2_{R}	variance of relay noise
$\sigma^2_{H_{\mathsf{n}}}$	variance of terminal n's channel coefficients
SNR_{n}	signal-to-noise ratio at terminal n
$\mathcal{U}_{\mathsf{n}}$	channel uncertainty set for terminal n
ζ	path loss coefficient

Acronyms

AF	Amplify-and-Forward
AWGN	Additive White Gaussian Noise
AZF	Asymptotic Zero Forcing
BC	Broadcast Channel
BF	Beamforming
CDF	Cumulative Distribution Function
CDMA	Code Division Multiple Access
CFP	Conic Form Problems
CMUR	Clustered Multi-user Relaying
CQ	Complex Conjugate
CSIT	Channel State Information at the Transmitter
CSI	Channel State Information
DF	Decode-and-Forward
DMT	Diversity Multiplexing Trade-off
DPC	Dirty Paper Coding
DTP	Date Transmission Phase
EFP	Estimation and Feedback Phase
FDD	Frequency Division Duplex
FDMA	Frequency Division Multiple Access
GRQ	Generalized Rayleigh Quotient
HRP	Hierarchical Relaying Protocol
KKT	Karush Kuhn Tucker
LMI	Linear Matrix Inequality
LO	Local Oscillator

LP	Linear Programming
MAC	Multiple Access Channel
MD	Multiuser Diversity
MIMO	Multiple-Input Multiple-Output
MISO	Multiple-Input Single-Output
MMSE	Minimum Mean Square Error
MRC	Maximum Ratio Combining
MUR	Multi-user Relaying
NP	Non-polynomial
OCC	Optimum Cluster Configuration
OFDMA	Orthogonal Frequency Division Multiple Access
P-CSMA	Prioritized Carrier Sense Multiple Access
QCQP	Quadratically Constrained Quadratic Program
QP	Quadratic Program
QoS	Quality of Service
S-D	Source-Destination
SDMA	Space Division Multiple Access
SDP	Semidefinite Programming
SDR	Semidefinite Relaxation
SINR	Signal-to-Interference plus Noise Ratio
SM	Spatial Multiplexing
SNR	Signal-to-Noise Ratio
SOCP	Second Order Cone Programming
SPC	Superposition Coding
SVD	Singular Value Decomposition
TDD	Time Division Duplex
TDMA	Time Division Multiple Access
WLAN	Wireless Local Area Network
XOR	Exclusive OR
ZF	Zero Forcing
i.i.d.	Independent and Identically distributed

Bibliography

[1] T. Abe, H. Shi, T. Asai, and H. Yoshino, "Relay techniques for MIMO wireless networks with multiple source and destination pairs," *EURASIP Journal on Wireless Communications and Networking*, vol. 2006, pp. 19, April 2006.

[2] C. Akçaba, P. Kuppinger, and H. Bölcskei, "Distributed transmit diversity in relay networks", to be submitted to *IEEE Transactions on Information Theory*, 2009.

[3] I. F. Akyildiz, W. Su, Y. Sankarasubramaniam, and E. Cayirci, "A survey on sensor networks," *IEEE Communications Magazine*, vol. 40, pp. 102 - 114, August 2002.

[4] K. Azarian, H. El Gamal, and P. Schniter, "On the achievable diversity-multiplexing tradeoff in half-duplex cooperative channels," *IEEE Transactions on Information Theory*, vol. 51, no. 12, pp. 4152-4172, December 2005.

[5] B. C. Banister and J. R. Zeidler, "A simple gradient sign algorithm for transmit antenna weight adaptation with feedback," *IEEE Transactions on Signal Processing*, vol. 51, no. 5, pp. 1156-1171, May 2003.

[6] S. Barbarossa, M. Pompili, and G. B. Giannakis, "Channel-independent synchronization of orthogonal frequency division multiple access systems," *IEEE Journal Selected Areas Communications*, vol. 20, no. 2, pp. 474- 486, February 2002.

[7] A. Beck, "Convexity properties associated with nonconvex quadratic matrix functions and applications to quadratic programming," *Journal of Optimization Theory and Applications*, vol. 142, no. 1, pp. 1–29, July 2009.

[8] J. J. Beek *et al.*, "A time and frequency synchronization scheme for multiuser OFDM," *IEEE Journal Selected Areas Communications*, vol. 17, no. 11, pp. 1900-1914, November 1999.

[9] A. Ben-Tal, A. Nemirovski,"Robust convex optimization," *Mathematics of Operations Research*, vol. 23, no. 4, 1998.

[10] —, "Robust solutions to uncertain programs," in *Operations Research Letter*, vol. 25, pp. 1-13, February 1999.

[11] —, "Robust solutions to linear programming problems contaminated with uncertain data," *Mathematical Programming*, vol. 88, pp. 411-424, 2000.

[12] —, and C. Roos, "Robust solutions of uncertain quadratic and conic-quadratic problems," *SIAM Journal on Optimization*, vol. 13, no. 2, pp. 535–560, 2002.

[13] —, "Selected topics in robust convex optimization," *Mathematical Programming*, vol. 1, no. 1, pp. 125-158, July 2007.

[14] —, "Robust optimization - methodology and applications," *Mathematical Programming*, vol. 92, no. 3, pp. 453–480, May 2002.

[15] M. Bengtsson and B. Ottersten, "Optimal downlink beamforming using semidefinite optimization," in *Proc. 37th Annual Allerton Conference on Communications, Control, and Computing*, September, 1999, pp. 987–996.

[16] S. Berger, "Coherent cooperative relaying in low mobility wireless multiuser networks," Ph.D. Dissertation, ETH Zürich, 2009.

[17] S. Berger and A. Wittneben, "A coherent amplify-and-forward relaying demonstrator without global phase reference", in *Proc. IEEE International Symposium on Personal, Indoor and Mobile Radio Communications*, Cannes, France, September 2008, pp. 1–5.

[18] —, "Impact of noisy carrier phase synchronization on linear amplify-and-forward relaying", in *Proc. IEEE Global Communications Conference*, November 2007, pp. 795–800.

[19] —, "Carrier phase synchronization of multiple distributed nodes in a wireless network", in *Proc. 8th IEEE Workshop on Signal Processing Advances in Wireless Communications*, June 2007, pp. 1–5.

[20] —, "When do non-regenerative two-hop relaying networks require a global phase reference?", accepted to *IEEE Global Communications Conference 2009.*

[21] —, "Cooperative distributed multiuser MMSE relaying in wireless ad-hoc networks," in *Proc. the 39th Asilomar Conference on Signals, Systems and Computers, 2005*, October 28 - November 1, 2005, pp. 1072–1076.

[22] D. Bertsekas and R. Gallager, "Data Networks", Prentice-Hall, International Editions, 1987.

[23] H. Bölcskei, R. U. Nabar, Ö. Oyman, and A. J. Paulraj, "Capacity scaling laws in MIMO relay networks," *IEEE Transactions Wireless Communications*, vol. 5, no. 6, pp. 1433–1444, June 2006.

[24] H. Bölcskei, "MIMO-OFDM wireless systems: basics, perspectives, and challenges," *IEEE Transactions on Wireless Communications*, vol. 13, no. 4, pp. 31–37, August 2006.

[25] M. Botros and T. N. Davidson, Convex conic formulations of robust downlink precoder designs with quality of service constraints", *IEEE Journal on Selected Topics in Signal Processing*, vol. 1, no. 2, pp. 714–724, December 2007.

[26] S. Boyd and L. Vandenberghe, "Convex Optimization", Cambridge University Press, 2004.

[27] D. R. Brown III, G. B. Prince, and J. A. McNeill, "A method for carrier frequency and phase synchronization of two autonomous cooperative transmitters, in *IEEE Workshop on Signal Processing Advances in Wireless Communications*, June 2005, pp. 260 - 264.

[28] V. R. Cadambe and S. A. Jafar, "Interference alignment and the degrees of freedom for the K user interference channel," ArXiv e-prints, vol. 707, July 2007.

[29] G. Caire and S. Shamai, "On the achievable throughput of a multiantenna Gaussian broadcast channel", *IEEE Transactions on Information Theory*, vol. 49, no. 7, pp. 1691-1706, July 2003.

[30] C. Chae, T. Tang, R. W. Heath, and S. Cho, "MIMO relaying with linear processing for multiuser transmission in fixed relay networks," *IEEE Transactions on Signal Processing*, vol. 56, no. 2, pp. 727–738, February 2008.

[31] M. Chen and A. Yener, "Multiuser two-way relaying: detection and interference management strategies", *IEEE Transactions on Wireless Communications*, under review, March 2009.

[32] M. Conti and S. Giordano, "Multihop ad hoc networking: the theory," *IEEE Communications Magazine*, vol. 45, no. 4, pp. 78–86, April 2007.

[33] A. del Coso, O. Simeone, Y. Barness and C. Ibars, "Space-time coded cooperative multicasting with maximal ratio combining and incremental redundancy," in *Proc. IEEE International Communications Conference*, Glasgow, UK, June 2007, pp. 6079-6084.

[34] M. Costa, "Writing on dirty paper," *IEEE Transactions on Information Theory*, vol. 29, no. 3, pp. 439–441, May 1983.

[35] T. M. Cover and A. A. El Gamal, "Capacity theorems for the relay channel," *IEEE Transactions on Information Theory*, vol. 25, pp. 572-584, September 1979.

[36] A. D. Dabbagh and D. J. Love, "Precoding for multiple antenna Gaussian broadcast channels with successive zero-forcing, " *IEEE Transactions on Signal Processing*, vol. 55, no. 7, pp. 3837–3850, July 2007.

[37] A. F. Dana and B. Hassibi, "On the power efficiency of sensory and ad-hoc wireless networks", *IEEE Transactions on Information Theory*, vol. 52, no. 7, pp. 2890-2914, July 2006.

[38] T. N. Davidson, Z. Q. Luo, and J. F. Sturm, "Linear matrix inequality formulation of spectral mask constraints with applications to FIR filter design", *IEEE Transactions on Signal Processing*, vol. 50, no. 11, pp. 2702–2715, November 2002.

[39] K. K. Delgado, "The complex gradient operator and the $\mathbb{CR}$-calculus," Jacobs School of Engineering, University of California, San Diego, Lecture Supplement, 2006.

[40] A. Demir, A. Mehrotra, and J. Roychowdhry "Phase noise in oscillators: A unifying theory and numerical methods for characterization", *IEEE Transactions on Circuits and System - I: Fundamental Theory and Applications*, vol. 47, no. 5, pp. 655–674, May 2000.

[41] Y. Ding, "On efficient semidefinite relaxations for quadratically constrained quadratic programming", Master of Mathematics Thesis, University of Waterloo,, 2007.

[42] L. El-Ghaoui and H. Lebret, "Robust solutions to least-square problems to uncertain data matrices," *SIAM Journal Matrix Anal. Applications*, vol. 18, pp. 1035-1064, 1997.

[43] L. El-Ghaoui, F. Oustry, and H. Lebret, "Robust solutions to uncertain semidefinite programs," *SIAM Journal on Optimization*, vol. 9, pp. 33-52, 1998.

[44] C. Eşli and A. Wittneben, "One- and two-way decode-and-forward relaying for wireless multiuser MIMO networks," in *Proc. IEEE Global Communications Conference*, New Orleans, USA, December 2008, pp. 1–6.

[45] Y. T. Feng and D. R. J. Owen, "Conjugate gradient methods for solving the smallest eigenpair of large symmetric eigenvalue problems,", *International Journal for Numerical Methods in Engineering*, vol. 39, pp. 2209–2229, 1996.

[46] P. Fertl, A. Hottinen, and G. Matz, "Perturbation-based distributed beamforming for wireless relay networks," ArXiv, 0809.5182v1, September 2007.

[47] R. F. H. Fischer, "Precoding and Signal Shaping for Digital Transmission," John Wiley and Sons Inc., 2002.

[48] C. Fragouli, J. Widmer, and J.-Y. Le Boudec, "Network coding: an instant primer", *ACM SIGCOMM Computer Communication Review*, vol. 36, no. 1, pp 63–68, January 2006.

[49] A. Goldsmith, S. Jafar, N. Jindal, and S. Vishwanath, "Capacity limits of MIMO channels", *IEEE Journal on Selected Areas in Communications*, vol. 21, no. 5, pp. 684–702, June 2003.

[50] I. Hammerström , M. Kuhn, and A. Wittneben, "Cooperative diversity by relay phase rotations in block fading environments", in *Proc. IEEE Workshop on Signal Processing Advences in Wir. Communications*, July 2004, pp. 293–297.

[51] —, "Impact of relay gain allocation on the performance of cooperative diversity networks," in *Proc. IEEE Vehicular Technology Conference*, Los Angeles, USA, September 2004, pp. 1815–1819.

[52] —, "Distributed MIMO for cellular networks with multihop transmission protocols," in *Proc. Asilomar Conference on Signals, Systems and Computers*, CA, October 2006, pp. 671–675.

[53] —, "Impact of relay gain allocation on the performance of cooperative diversity networks", in *Proc. IEEE Vehicular Technology Conference*, vol. 3, September 2004, pp. 1815–1819.

[54] I. Hammerstrom and A. Wittneben, "On the optimal power allocation of nonregenerative OFDM relay links," in *Proc. IEEE Conference on Communications, Istanbul*, Turkey, June 2006, pp. 4463–4468.

[55] I. Hammerstrom, M. Kuhn, C. Eşli, J. Zhao, A. Wittneben, and G. Bauch, "MIMO two-way relaying with transmit CSI at the relay," in *Proc. IEEE 8th Workshop on Signal Processing Advances in Wireless Communications*, June 2007, pp. 1–5.

[56] I. Hammerström, "Cooperative relaying and adaptive scheduling for low mobility wireless access networks," Ph.D. thesis, ETH Zürich, No. 16775, Logos Verlag Berlin, ISBN 978-3-8325-1466-2, pp. 243, 2006.

[57] V. Havary-Nassab, S. Shahbazpanahi, A. Grami,and L. Zhi-Quan, "Distributed beamforming for relay networks based on second-order statistics of the channel state information", *IEEE Transactions Signal Processing*, vol. 56, no. 9, pp. 4306–4316, September 2008.

[58] Y. Huang and S. Zhang, "Complex matrix decomposition and quadratic programming", *Mathematical Operations Research*, vol. 32, pp. 758-768, 2007.

[59] A. P. Iserte, D. P. Palomar, A. P. Neira, and M. A. Lagunas, "A robust maximin approach for MIMO communications with imperfect channel state information based

on convex optimization", *IEEE Transactions on Signal Processing*, vol. 54, no. 1, pp. 346–360, January 2006.

[60] N. Jindal and Z. Q. Luo, "Capacity Limits of Multiple Antenna Multicast," in *Proc. IEEE International Symposium on Information Theory*, July 2006, pp. 1841–1845.

[61] U. T. Jönsson, "A Lecture on the S-Procedure," Lecture notes at the Royal Institute of Technology, Sweden, May 2001 (revised 2006).

[62] E. Karipidis, N. D. Sidiropoulos, Z. Q. Luo, "Far-field multicast beamforming for uniform linear antenna arrays", *IEEE Transactions on Signal Processing*, vol. 55, no. 10, pp. 4916–4927, October 2007.

[63] —, "Quality of service and Max-min fair transmit beamforming to multiple cochannel multicast groups", *IEEE Transactions on Signal Processing*, vol. 56, no. 3, pp. 1268–1279, March 2008.

[64] N. Khajehnouri and A. H. Sayed, "Multicast relay strategies with local and global power constraints for wireless networks", in *Proc. IEEE Signal Processing Advances in Wireless Communications*, July. 2006, pp. 1–5.

[65] —, "Distributed MMSE relay strategies for wireless sensor networks," *IEEE Transactions on Signal Processing*, vol. 55, no. 7, pp. 3336–3348, 2007.

[66] S. J. Kim, P. Mitran, and V. Tarokh, "Performance bounds for bidirectional coded cooperation protocols," *IEEE Transactions on Information Theory*, vol. 54, no. 11, pp. 5235–5241, November 2008.

[67] R. Knopp and P. Humblet, "Information capacity and power control in single-cell multiuser communications," in *IEEE International Conference on Communications*, vol. 1, Seattle, WA, June 1995, pp. 331-335.

[68] T. Koike-Akino, P. Popovski, and V. Tarokh, "Denoising maps and constellations for wireless network coding in twoway relaying systems," in *Proc. IEEE Global Communications Conference,* New Orleans, USA, December 2008.

[69] M. Kuhn, S. Berger, I. Hammerstrom, and A. Wittneben, "Power line enhanced cooperative wireless communications," *IEEE Journal Selected Areas in Communications*, vol. 24, no.7, pp. 1401–1410, July 2006.

[70] J. N. Laneman and G. W. Wornell, "Exploiting distributed spatial diversity in wireless networks," in *Proc. Allerton Conference on Communication, Control and Computing*, Monticello, Illinois, October 2000.

[71] —, "Distributed Space-Time-Coded Protocols for Exploiting Cooperative Diversity in Wireless Networks," *IEEE Transactions on Information Theory*, vol. 49, no. 10, pp. 2415-2425, October 2003.

[72] N. J. Laneman, D. N. Tse, and G. W. Wornell, "Cooperative diversity in wireless networks: Efficient protocols and outage behavior", *IEEE Transactions on Information Theory*, vol. 50, pp. 3062–3080, December 2004.

[73] P. Larsson, "Large-scale cooperative relaying network with optimal coherent combining under aggregate relay power constraints", in *Proc. Future Telecommunications Conference, Beijing, China*, December 2003.

[74] P. Larsson, N. Johansson, and K.-E. Sunell, "Coded bidirectional relaying", in *Proc. IEEE Vehicular Technology Conference*, vol. 2, pp. 851-855, 2006.

[75] C. K. Lo, S. Vishwanath, and R. W. Heath, Jr., "Rate bounds for MIMO relay channels using precoding, " in *Proc. Global Communications Conference*, December 2005.

[76] M. Lobo, L. Vandenberghe, S. Boyd, and H. Lebret, "Applications of second-order cone programming," *Linear Algebra and its Applications*, vol. 284, no. 1-3, pp. 193–228, November 1998.

[77] J. Lofberg, "YALMIP : A Toolbox for Modeling and Optimization in MATLAB," in *Proc. the CACSD Conference, Taipei, Taiwan*, 2004.

[78] Z.-Q. Luo and W. Yu, "An introduction to convex optimization for communications and signal processing", *IEEE Journal on Selected Areas in Communications*, vol. 24, no. 8, pp. 1426–1438, August 2006.

[79] I. Maric, A. Goldsmith, and M. Medard, "Information-theoretic relaying for multicast in wireless networks," in *Proc. IEEE Military Communications Conference*, October 2007, pp. 1–7.

[80] V. I. Morgenshtern and H. Bölcskei, "Crystallization in Large Wireless Networks", *IEEE Transactions on Information Theory*, vol. 53, no. 10, pp. 3319–3349, October 2007.

[81] R. Mudumbai, G. Barriac, and U. Madhow, "On the feasibility of distributed beamforming in wireless networks, *IEEE Transactions on Wireless Communications*, vol. 6, no. 5, pp. 1754-1763, May 2007.

[82] F. Munier, T. Eriksson, and A. Svensson, " Receiver algorithms for OFDM systems in phase noise and AWGN", in *Proc. IEEE International Symposium on Personal, Indoor and Mobile Radio Communications*, vol. 3, September 2004, pp. 1998–2002.

[83] O. Munoz, A. Agustn, and J. Vidal, "Cellular capacity gains of cooperative MIMO transmission in the downlink", in *Proc. International Zurich Seminar on Communication*, February 2004, pp. 22–26.

[84] O. Munoz, J. Vidal, and A. Agustin, "Linear transceiver design in nonregenerative relays with channel state information", *IEEE Transactions on Signal Processing*, vol. 55, no. 6, pp. 2593–2604, June 2007.

[85] R. U. Nabar, Ö. Oyman, H. Bölcskei, and A. J. Paulraj, "Capacity scaling laws in MIMO wireless networks," in *Proc. Allerton Conference on Communications, Control, and Computing*, Monticello, IL, October 2003, pp. 378–389.

[86] R. Navid, T. H. Lee, and R. W. Dutton, "An analytical formulation of phase noise of signals with Gaussian-distributed jitter", *IEEE Transactions on Circuits and Systems - II: Express Briefs*, vol. 52, no. 3, pp. 149–155, March 2005.

[87] Y. Nesterov and A. Nemirovskii, "Interior Point Polynomial Algorithms in Convex Programming," *Studies in Applied and Numerical Mathematics Series*, 13, 1987.

[88] T. J. Oechtering and H. Boche, "Bidirectional regenerative half-duplex relaying using relay selection," *IEEE Transactions on Wireless Communications*, vol. 7, no. 5, pp. 1879–1888, May 2008.

[89] ——, "Optimal time-division for bidirectional relaying using superposition encoding," *IEEE Communications Letters*, vol. 12, no. 4, pp. 265–267, April, 2008.

[90] T. J. Oechtering, C. Schnurr, I. Bjelakovic, and H. Boche, "Broadcast capacity region of two-phase bidirectional relaying," *IEEE Communications Letters*, vol. 54, no. 1, pp. 454–458, January 2008.

[91] T. J. Oechtering, "Spectrally efficient bidirectional decode-and-forward relaying for wireless networks," Ph.D. Dissertation, Technical University of Berlin, 2007.

[92] F. Ono and K. Sakaguchi, "Space time coded MIMO network coding," in *Proc. IEEE International Symposium on Personal, Indoor and Mobile Radio Communications,* Cannes, France, September 2008.

[93] R. Pabst, *et. al.* , "Relay-based deployment concepts for wireless and mobile broadband radio," *IEEE Communications Magazine*, vol. 42, no. 9, pp. 80–89, September 2004.

[94] M. Payaro, A. Pascual-Iserte, and M. A. Lagunas, "Mutual Information Optimization and Capacity Evaluation in MIMO Systems with Magnitude Knowledge and Phase Uncertainty, in *Circuits, Systems, and Signal Processing (special issue: Signal Processing for Uncertain Systems)*, vol. 26, no. 4, pp. 527–549, ISSN 0278-081X, August 2007.

[95] —, Robust power allocation designs for multiuser and multiantenna downlink communication systems through convex optimization", *IEEE Journal on Selected Areas in Communications*, vol. 25, no. 7, pp. 1390–1401, September 2007.

[96] P. Popovski and H. Yomo, "The anti-packets can increase the achievable throughput of a wireless multi-hop network," in *Proc. IEEE International Conference on Communications*, Istanbul, Turkey, June 2006.

[97] T. Q . S. Quek, H. Shin., M. Z. Win, and M. Chiani, "Optimal power allocation for amplify-and-forward relay networks via conic programming," in *Proc. IEEE Conference on Communications*, Glasgow, UK, 2007, pp. 5058–5063.

[98] —, "Robust wireless relay networks: slow power allocation with guaranteed QoS," *IEEE Journal of Selected Topics in Signal Processing,* , vol. 1, no. 4, pp. 700–713, December 2007.

[99] B. Rankov and A. Wittneben, "Spectral efficient protocols for non-regenerative half-duplex relaying, in *Proc. 43rd Allerton Conference on Communications, Control, and Computing*, September 2005.

[100] ——, "Spectral efficient signaling for half-duplex relay channels, in *Proc. IEEE Asilomar Conference on Signals, Systems, and Computers*, November 2005.

[101] ——, "Achievable rate regions for the two-way relay channel,in *Proc. IEEE International Symposium on Information Theory*, July 2006.

[102] ——, "Spectral efficient protocols for half-duplex fading relay channels," *IEEE Journal on Selected Areas in Communications*, vol. 25, no. 2, pp. 379–389, February 2007.

[103] B. Rankov, "Spectral efficient cooperative relaying strategies for wireless networks," PhD thesis, ETH Zürich, 2006.

[104] Y. E. Sagduyu, D. Guo, and R. Berry, "Throughput optimal control for relay-assisted wireless broadcast with network coding," in *Proc. IEEE International Workshop on Wireless Network Coding*, San Francisco, June 2008.

[105] O. Sahin and E. Erkip, "On achievable rates for interference relay channel with interference cancellation," in *Proc. IEEE Asilomar Conference on Signals, Systems and Computers*, November 2007, pp. 805–809.

[106] A. A. Samad, T. N. Davidson and A. B. Gershman, "Robust transmit eigen beamforming based on imperfect channel state information", *IEEE Transactions on Signal Processing*, vol. 54, no. 5, pp. 1596–1609, May 2006.

[107] H. Sato, "Information transmission through a channel with relay," The Aloha System, University of Hawai, Honolulu, Tech. Rep. B76-7, pp. 549-552, March 1976.

[108] A. Sendonaris, E. Erkip, and B. Aazhang, "Increasing uplink capacity via user cooperation diversity," in *Proc. IEEE International Symposium on Information Theory*, Cambridge, USA, August 1998, p. 156.

[109] ——, "User cooperation diversity-Part I: System description," *IEEE Transactions on Communications*, vol. 51, no. 11, pp. 1927-1938, November 2003.

[110] ——, "User cooperation diversity-Part II: Implementation aspects and performance analysis," *IEEE Transactions on Communications*, vol. 51, no. 11, pp. 1939-1948, November 2003.

[111] C. E. Shannon, "Two-way communication channels," in *Proc. 4th Berkeley Symposium Math. Statistics and Probability*, vol. 1, 1961, pp. 611-644.

[112] J. R. Shewchuk, "An introduction to the conjugate gradient method without the agonizing pain," School of Computer Science, Carnegie Mellon University, Pittsburgh, Pennsylvania, USA, $1\frac{1}{4}$. edition, 1994.

[113] N. D. Sidiropolous, T. N. Davidson and Z. H. T. Luo, "Transmit beamforming for physical-layer multicasting", *IEEE Transactions on Signal Processing*, vol. 54, no. 6, pp. 2239–2251, June 2006.

[114] S. Simoens, O. Munoz, J. Vidal, and A. Del Coso, "Capacity bounds for Gaussian MIMO relay channel with channel state information," in *Proc. IEEE 9th Workshop on Signal Processing Advances in Wireless Communications*, July 2008, pp. 441–445.

[115] Q. H. Spencer, A. L. Swindlehurst, and M. Haardt, "Zero-forcing methods for downlink spatial multiplexing in multiuser MIMO channels," *IEEE Transactions on Signal Processing*, vol. 52, no. 2, pp. 461–471, February 2004.

[116] J. F. Sturm, "Using SeDuMi 1.02, a Matlab toolbox for optimization over symmetric cones," *Optimization Methods and Software*, vol. 11–12, 1999, pp. 625–653.

[117] E. Telatar, "Capacity of multi-antenna Gaussian channels," *European Transactions on Telecommunications*, 10:585–596, 1999.

[118] J. Thukral and H. Bölcskei, "Distributed spatial multiplexing with 1-bit feedback," in *Proc. Allerton Conference on Communications, Control, and Computing*, September 2007.

[119] D. Tse and P. Viswanath, "Fundamentals of wireless communications", Cambridge University Press, 2005.

[120] Y.-S. Tu and G. Pottie, "Coherent cooperative transmission from multiple adjacent antennas to a distant stationary antenna through AWGN channels, in *Proc. 55th IEEE Vehicular Technology Conference Spring*, vol. 1, 2002, pp. 130-134.

[121] R. H. Tutuncu, K.C. Toh, and M.J. Todd, "Solving semidefinite-quadratic-linear programs using SDPT3," *Mathematical Programming Series B*, 95, 2003,, pp. 189–217.

[122] T. Unger and A. Klein, "Duplex Schemes in Multiple Antenna Two-Hop Relaying," *EURASIP Journal on Advances in Signal Processing*, vol. 2008, Article ID 128592.

[123] L. Vandenberghe and S. Boyd, "Semidefinite programming," *SIAM Review*, vol. 38, no. 1, pp. 49–95, 1996.

[124] R. Vaze and R. W. Heath, "Capacity Scaling for MIMO Two-Way Relaying," in *Proc. IEEE International Symposium on Information Theory*, pp. 1451–1455, June 2007.

[125] S. Vishwanath, N. Jindal, and A. Goldsmith, "Duality, achievable rates, and sum-rate capacity of Gaussian MIMO broadcast channels", *IEEE Transactions on Information Theory*, vol. 49, no. 10, pp. 2658–2668, October 2003.

[126] S. Vorobyov, A. Gershman, and Z.-Q. Luo, "Robust adaptive beamforming using worst-case performance optimization: a solution to the signal mismatch problem", *IEEE Transactions Signal Processing*, vol. 51, pp. 313-323, 2003.

[127] S. Vorobyov, Y. Rong, and A. Gershman, "Robust adaptive beamforming using probability-constrained optimization", in *Proc. 13th IEEE Statistical Signal Processing Workshop*, Bordeaux, France, July 17-20, 2005, pp. 934–939.

[128] N. Vucic and H. Boche, "Probabilistically constrained robust power allocation in downlink multiuser MISO systems", in *Proc. Asilomar Conference on Signals, Systems and Computers*, October 2008.

[129] ——, "Robust QoS-constrained optimization of downlink multiuser MISO systems", *IEEE Transactions on Signal Processing*, vol. 57, no. 2, pp. 714–725, February 2009.

[130] N. Vucic, H. Boche, and S. Shi, "Robust transceiver optimization in downlink multiuser MIMO systems", *IEEE Transactions on Signal Processing*, vol. 57, no. 9, pp. 3576–3587, September 2009.

[131] B. Wang, J. Zhang, and A. Hst-Madsen, "On the capacity of MIMO relay channels", *IEEE Transactions on Information Theory*, vol. 51, pp. 29-43, January 2005.

[132] *Wireless LAN Medium Access Control (MAC) and Physical Layer (PHY) Specifications, Amendment 5: Enhancements for Higher Throughput*, IEEE Std. 802.11n/D5.0, August 2008.

[133] A. Wittneben and B. Rankov, "Distributed antenna systems and linear relaying for gigabit MIMO wireless," in *Proc. IEEE 60th Vehicular Technology Conference, Fall 2004*, vol. 5, September 2004, pp. 3624–3630.

[134] A. Wittneben, "Coherent multiuser relaying with partial relay cooperation," in *Proc. IEEE Wireless Communications and Networking Conference, 2006*, vol. 2, 2006, pp. 1027–1033.

[135] A. Wolfgang, N. Seifi, and T. Ottosson, "Resource allocation and linear precoding for relay assisted multiuser MIMO systems", in *Proc. International ITG Workshop on Smart Antennas*, February 2008, pp. 162–168.

[136] Y. Wu, P. Chou, and S.-Y. Kung, "Information exchange in wireless networks with network coding and physical-layer broadcast", in *Proc. 39th Annual Conference on Information Sciences and Systems*, March 2005.

[137] R. Wyrembelski, T. J. Oechtering, I. Bjelakovic, C. Schnurr, and H. Boche, "Capacity of Gaussian MIMO Bidirectional Broadcast Channels," in *Proc. IEEE International Symposium on Information Theory*, Toronto, Canada, July, 2008, pp. 584–588.

[138] R. Wyrembelski, T. J. Oechtering, and H. Boche, "Decode-and-forward strategies for Bidirectional Relaying," in *Proc. IEEE International Symposium on Personal, Indoor and Mobile Radio Communications,* Cannes, France, September 2008, pp. 1–6.

[139] V. A. Yakubovich, "The S-procedure in nonlinear control theory," *Vestnik Leningrad University Math*, vol. 4, pp. 73-93, 1977.

[140] H. Yang, "Conjugate gradient methods for the Rayleigh quotient minimization of generalized eigenvalue problems," Computing, 51, 1993.

[141] E. Yilmaz, M. O. Sunay, "Effects of transmit beamforming on the capacity of multi-hop MIMO relay channels", in *Proc. IEEE International Symposium on Personal, Indoor and Mobile Radio Communications*, September 2007.

[142] T. Yoo and A. Goldsmith, "On the optimality of multiantenna broadcast scheduling using zero-forcing beamforming", *IEEE Journal Selected Areas in Communications*, vol. 24, no. 3, pp. 528-541, March 2006.

[143] S. Zhang, S.-C. Liew, and P. P. Lam, "Hot topic: Physical-layer network coding", in *Proc. 12th International Conference on Mobile Computing and Networking*, Los Angeles, USA, 2006, pp 358–365,

[144] Y. Zhao, R. Adve, and T. J. Lim, "Improving amplify-and-forward relaying networks: Optimal power allocation versus selection," in *Proc. IEEE Int. Conference on Information Theo.*, Seattle, USA, July 2006, pp. 1234–1238.

[145] J. Zhao, M. Kuhn, A. Wittneben, and G. Bauch, "Optimum time-division in MIMO two-way decode-and-forward relaying systems," in *Proc. 42nd Annual Asilomar Conference on Signals, Systems, and Computers,* California, USA, October 2008.

[146] —, "Self-interference aided channel estimation in two-way relaying systems," in *Proc. IEEE Global Communications Conference*, New Orleans, November 2008.

[147] J. Zhao, "MIMO Two-Way Relaying Systems," Ph.D. Dissertation, ETH Zürich, 2009.

[148] G. Zheng, K-K. Wong, and T-S. Ng, "Robust linear MIMO in the downlink: A worst-case optimization with ellipsoidal uncertainty regions", *EURASIP Journal on Advances in Signal Processing*, Article ID 609028, 2008.

[149] E. Zimmermann, P. Herhold, and G. Fettweis,"On the performance of cooperative relaying protocols in wireless networks," *European Transactions on Telecommunications*, vol. 16, no. 1, January 2005.

Curriculum Vitae

Name:	Celal Eşli
Date of Birth:	April 21, 1981
Birthplace:	Istanbul, Turkey

Education

09/2005 – 03/2010	**ETH Zürich, Zürich, Switzerland** Ph.D. studies in the *Department of Information Technology and Electrical Engineering, Communications Technology Laboratory.*
09/2003 – 07/2005	**Boğaziçi University, Istanbul, Turkey** M.Sc. in the *Department of Electrical and Electronics Engineering, Wireless Communications Laboratory.*
09/1999 – 07/2003	**Boğaziçi University, Istanbul, Turkey** B.Sc. in the *Department of Electrical and Electronics Engineering.*

Experience

09/2005 – 03/2010	**ETH Zürich, Zurich, Switzerland** Research Assistant in the Communications Technology Laboratory with Prof. Dr. Armin Wittneben.
09/2003 – 02/2004	**EKOM Telecommunications Systems Inc., Istanbul, Turkey** Software Engineer (digital signal processing).
09/2003 – 07/2005	**Boğaziçi University, Istanbul, Turkey** Research Assistant in the Wireless Communications Laboratory (BUSIM) with Prof. Dr. Hakan Deliç.

Publications

Patents

- **C. Eşli**, G. Psaltopoulos, J. Wagner, and A. Wittneben, User cooperation aided multicasting for cellular communications. Disclosure with NOKIA, Radio Implementation Patent Board, FI, 2008.
- J. Wagner, G. Psaltopoulos, **C. Eşli**, and A. Wittneben, Intra-virtual antenna array communication through secondary MIMO systems beyond 10 GHz. Disclosure with NOKIA, Radio Implementation Patent Board, FI, 2008.

Journals

- **C. Eşli** and A. Wittneben, A hierarchical AF protocol for distributed orthogonalization in multiuser relay networks, *IEEE Tran. on Vehicular Tech.*, under review, submitted in Sept. 2009.
- **C. Eşli**, J. Wagner, and A. Wittneben, Distributed gradient methods for relay gain allocation in coherent AF relaying networks, *IEEE Tran. on Wireless Commun.*, under review, submitted in Apr. 2009.
- **C. Eşli**, Mutlu Koca, and H. Deliç, Iterative joint tone-interference cancellation and decoding for MIMO-OFDM, *IEEE Tran. on Vehicular Tech.*, vol. 57, no. 5, pp. 2843-2855, Sept. 2008.
- **C. Eşli** and H. Deliç, Performance analysis for OFDMA in the presence of tone interference, *IEEE Tran. on Commun.*, vol. 55, no. 5, pp. 845-849, May 2007.
- **C. Eşli** and H. Deliç, Coded OFDM with transmit diversity for digital terrestrial TV broadcasting, *IEEE Tran. on Broadcasting*, vol. 52, pp. 586-596, Dec. 2006.
- **C. Eşli** and H. Deliç, Anti-jamming performance of space-frequency coding in partial-band noise, *IEEE Tran. on Vehicular Techn.*, vol. 55, no. 2, pp. 466-476, Mar. 2004.
- **C. Eşli**, B. Özgül and H. Deliç, Space-frequency coded HIPERLAN/2, *IEEE Tran. on Consumer Electronics*, vol. 50, no. 4, pp. 1162-1168, Nov. 2004.

Conferences

- Azadeh Ettefagh, Marc Kuhn, **C. Eşli**, and Armin Wittneben, Performance of a cluster-based MAC protocol in multiuser MIMO WLANs, in *Proc. ITG WSA*, Bremen, Germany, Feb. 2010.
- R. Rolny, J. Wagner, **C. Eşli**, and A. Wittneben, Distributed gain matrix optimization in non-regenerative MIMO relay channels, in *Proc. ASILOMAR Conf.*, Pacific Grove, USA, Oct. 2009.

- **C. Eşli**, J. Wagner, and A. Wittneben, Distributed gradient based gain allocation for coherent multiuser AF relaying networks, in *Proc. IEEE ICC'09*, Dresden, Germany, June 2009.
- **C. Eşli** and A. Wittneben, One- and two-way decode-and-forward relaying for wireless multiuser MIMO networks, in *Proc. IEEE GLOBECOM'08*, New Orleans, USA, Nov. 2008.
- **C. Eşli** and A. Wittneben, Robust gain allocation against phase uncertainty at the relays for multiuser cooperative networks, in *Proc. ASILOMAR Conf.*, Pacific Grove, USA, Oct. 2008.
- **C. Eşli** and A. Wittneben, Multiuser MIMO two-way relaying for cellular communications, in *Proc. IEEE PIMRC'08*, Cannes, France, Sept. 2008.
- **C. Eşli** and A. Wittneben, A cluster-based multiuser cooperative network, in *Proc. IEEE GLOBECOM'07*, Washington DC, USA, Nov. 2007.
- **C. Eşli** and A. Wittneben, Distributed multiuser cooperative network with heterogenous relay clusters, in *Proc. ASILOMAR Conf.*, Pacific Grove, USA, Oct. 2007.
- **C. Eşli**, H. Deliç, and M. Koca, Bit error probability performance of MIMO-OFDM in multi-tone interference, in *Proc. IEEE AFRICON'07*, Windhoek, Namibia, Oct. 2007.
- **C. Eşli**, M. Koca, and H. Deliç, Iterative joint tone-interference cancellation and decoding for MIMO-OFDM, in *Proc. EUSIPCO'07*, Poznan, Poland, Sept. 2007.
- I. Hammerström, M. Kuhn, **C. Eşli**, J. Zhao, A. Wittneben, and G. Bauch, MIMO two-way relaying with transmit CSI at the relay, in *Proc. IEEE SPAWC'07*, Helsinki, Finland, June 2007.
- **C. Eşli**, S. Berger, and A. Wittneben, Optimizing zero-forcing based gain allocation for wireless multiuser networks, in *Proc. IEEE ICC'07*, Glasgow, Scotland, June 2007.
- **C. Eşli** and H. Deliç, Space-frequency coding in the presence of partial-band noise jamming, in *Proc. IEEE MILCOM'05*, Atlantic City, USA, Oct. 2005.
- **C. Eşli** and H. Deliç, OFDMA under multi-tone interference, in *Proc. IEEE SPAWC'05*, New York, USA, June 2005.
- **C. Eşli** and H. Deliç, Performance of space-frequency coded COFDM in narrowband interference, in *Proc. IEEE CONTEL'05*, Zagreb, Croatia, June 2005.
- **C. Eşli**, G. Ay, B. Özgül, and H. Deliç, Space-frequency coding for the HIPERLAN/2 standard, in *Proc. IEEE MELECON'04*, Dubrovnik, Croatia, May 2004.